Praise for W/

'Highly persuasive ... a well-organised and solid dossier that alerts us to legalised chemical trickery'

Joanna Blythman, *The Spectator*

'A bombshell book'

Daily Mail

'The industry attitudes and tactics [Guillaume Coudray] reveals strongly resemble those of the tobacco industry in the 1970s and 1980s.'

Financial Times

'[An] eye-opening and important exposé ... I couldn't put it down. It read[s] like a detective story unravelling a conspiracy against the eaters of the world ... a book full of righteous anger.'

BEE WILSON, from her foreword

Praise for the French edition

'This investigation is a service to public health: its strength and originality lie in the author's deconstruction of the extraordinary web of manipulations, factual distortions, propaganda and doubt manufacturing, which have enabled industrial meat-processors to keep using chemical additives that should have been banned if regulations on carcinogenic substances were simply enforced.'

Le Monde

'An indictment, a very good investigation into industrial processed meat, which is now declared "a proven carcinogen".'
Le Figaro

'[A book] that reveals the dark side of ham.'
Le Canard Enchaîné

'For several years, Guillaume Coudray has been investigating the nitro-meat industry. His book reveals the underpinnings of a scandal, of which people on modest means are the first victims. ... Digging into government archives, Coudray uncovers shameful practices in a processed-meat sector unhinged by voracious industrial appetites.'
L'Humanité

WHO POISONED YOUR BACON?

THE DANGEROUS HISTORY OF MEAT ADDITIVES

GUILLAUME COUDRAY

Translated by David Watson

ICON

ABOUT THE AUTHOR

Guillaume Coudray is a French director and producer of documentary films. A political scientist by training, he is a graduate of Sciences Po and a former research grantee at the National Foundation for Political Science. His decade-long investigation into meat-processing has led to widespread press coverage and a hard-hitting documentary film broadcast in more than fifteen countries, which has had a major impact on the meat-processing industry. He lives in Paris.

Follow him on twitter at @g_coudray

ABOUT THE TRANSLATOR

David Watson is a freelance translator and editor. He has translated over 30 fiction and non-fiction books from French, including works by Alina Reyes, Agota Kristof, André Gide and a number of titles in the recent reissue of the Inspector Maigret novels by Georges Simenon. He lives in London.

This expanded edition published in the UK and USA in 2022
by Icon Books Ltd, Omnibus Business Centre,
39–41 North Road, London N7 9DP
email: info@iconbooks.com
www.iconbooks.com

Previously published in the UK in 2021 by Icon Books Ltd

Previously published in French under the title *Cochonneries*

© Editions La Découverte, Paris, 2017

Sold in the UK, Europe and Asia
by Faber & Faber Ltd, Bloomsbury House,
74–77 Great Russell Street,
London WC1B 3DA or their agents

Distributed in the UK, Europe and Asia
by Grantham Book Services, Trent Road,
Grantham NG31 7XQ

Distributed in the USA
by Publishers Group West,
1700 Fourth Street, Berkeley, CA 94710

Distributed in Australia and New Zealand
by Allen & Unwin Pty Ltd, PO Box 8500,
83 Alexander Street, Crows Nest, NSW 2065

Distributed in South Africa
by Jonathan Ball, Office B4, The District,
41 Sir Lowry Road, Woodstock 7925

ISBN: 978-178578-786-7

Typeset in Sabon by Marie Doherty

Printed and bound in Great Britain
by Clays Ltd, Elcograf S.p.A.

CONTENTS

PART 1
In the Pink: How Your Bacon Ended Up
Full of Nitrate and Nitrite

PART 2
How to Make Cancer Acceptable:
Bring On Botulism

FOREWORD

In Victorian England, pickled cucumbers were dyed green with copper-based dyes. Buyers preferred pickles with a lovely fresh green colour and sellers knew that by using plenty of dye, they could make a decent profit. What few people realised was that the green copper dyes were extremely poisonous.

Anyone today can see that it would be crazy to add green copper dye to pickles, no matter how appealing they may look. So why are we so uncritical of the countless bright pink meats in our supermarkets that have been processed using chemicals that increase their carcinogenic properties? As Guillaume Coudray reveals in this eye-opening and important exposé, our willingness to forgive bacon, ham and hot dogs for their cancer-causing properties has been carefully promoted by meat industry stakeholders who have spun us decades of lies.

I first read this book in French nearly three years ago. Even though my French is shaky, I couldn't put it down. It read like a detective story unravelling a conspiracy against the eaters of the world.

Like anyone who reads the newspapers, I had long been aware of dramatic headlines linking the consumption of processed meats and colon cancer. Like anyone who enjoys an occasional bacon sandwich, I often pretended those headlines did not exist. It was only when I read Coudray's devastatingly clear and far-reaching reportage that I understood how systematically the harm caused by 'nitro-meats' has been covered up by industry.

This is a book full of righteous anger. Coudray makes you feel how deeply wrong it is that the processed meat industry chooses to cure its meats with nitrate and nitrite even though it is well established that these chemicals give rise to cancer-causing compounds. What we are talking about here is preventable human suffering and death – all for the sake of achieving a rosy-red colour and speeding up production.

But this is also a book that offers reassurance to lovers of cured meats. The chink of light is Coudray's revelation that thanks to alternative techniques, 'the meat companies could produce perfectly safe processed meats without any need for harmful additives'. As of 1993, Prosciutto di Parma has been produced without any nitrate or nitrite, since when there has not been a single case of botulism associated with consuming the ham – contrary to all the warnings issued by the nitro-meat industry. The question is what it will take for other producers to follow suit.

As Coudray writes, when saltpetre (aka potassium nitrate) was first used in the curing of hams, there was no scandal, because the ham-makers did not know any better, just like those poor Victorians who poisoned themselves with green pickles. But now that we know better, to quote Maya Angelou, we should do better.

Bee Wilson

A NOTE ON TERMINOLOGY: SALT, NITRATE, NITRITE AND 'NITRITED CURING SALT'

Salt

Historically, meat was cured using *common salt*. This is the white substance we all know and call simply 'salt', 'sea salt' or 'cooking salt'. Its chemical name is sodium chloride. Even today, many types of processed meat are made using simple salting and maturation.

Nitrate

To speed up production and obtain a more uniform red colour, the salt can be supplemented by a derivative of nitrogen called *nitrate*. Nitrate comes in a variety of forms (ammonium nitrate, strontium nitrate, magnesium nitrate …), all of which give similar results in meat. The most commonly used is nitrate of potash or *potassium nitrate*. Historically, this was known as *saltpetre*.* In Europe, its E-number is E 252. Sometimes sodium nitrate (E 251) is also used.

Nitrite

Another derivative of nitrogen gives the same results as nitrate, but even more rapidly: *nitrite*. Many variants work well (ammonium nitrite, strontium nitrite, magnesium nitrite …). Nitrite can also be derived synthetically from nitrogen contained in the air or in vegetables. The cheapest is the most widely used: nitrite of soda or *sodium nitrite*. Its European E-number is E 250.

'Nitrited curing salt'

Sodium nitrite is so powerful it only requires a gram of powder to colour several kilograms of meat. In meat-processing it is never used in a pure state but only ever mixed with common salt. This mixture is known as *nitrited salt*, *nitrite curing salt* or *nitrited curing salt*. In Europe, nitrited curing salt contains between 99.1% and 99.5% common salt and between 0.5% and 0.9% nitrite. In the USA, nitrited curing salt is dosed at 6% and must be remixed with salt before use.

---

* *Saltpeter* in US spelling. This and other American spellings are retained when quoting US sources in this book. Similarly, the plural use of 'nitrates'/'nitrites' is retained where appropriate in quoted material. Otherwise these terms are given in the singular.

PROLOGUE

It is 12 October 2015. One by one, the delegates walk into the small auditorium. Twenty-two researchers from all over the world. A number of observers file in around them and take up positions at the tables by the windows. We are in Lyon, at the home of the International Agency for Research on Cancer (IARC). Founded by Charles de Gaulle, this agency is one of the jewels in the crown of the World Health Organization (WHO). Unlike other bodies in the cancer field, the IARC has virtually no involvement in drug development. De Gaulle wanted this organization to focus on one particular aim: to uncover the causes of cancer. Back in 1963, this was described as a 'new approach':[1] it was no longer enough to seek a cure for cancer, its root causes also had to be identified.

Since its creation, the IARC has held large scientific conferences two to four times a year, and each one examines a specific group of substances. The sessions always follow a tried-and-tested procedure. In the preceding six months, the experts review all the known scientific results, eliminating those that are least relevant, in order to distil the essential results from thousands of documents. The meeting in Lyon is the final sprint. In the course of one week, the experts have to thrash out the issues and reach a conclusion. At the end of the deliberations a voluminous report – a monograph, in the jargon of the IARC – is produced, but everything boils down to those few lines of text that deliver the final decision, rather like a judge giving a verdict at the end of a long trial.

On this day in October 2015, the IARC experts decided to classify all processed meats in 'group 1: carcinogenic to humans'.[2] 'Processed meats' includes all products made of 'meat preserved by smoking, curing or salting, or addition of chemical preservatives',[3] in particular, bacon and bologna, salami, ham, smoked sausages, and corned beef. This was the first time the WHO had officially designated a whole type of foodstuff as 'carcinogenic'.[4] Two weeks later, when the new classification was released to the public, it made headlines everywhere. A large 'X' made of bacon appeared on the cover of *Time* magazine, which informed its readers that the global figure for deaths attributable to the consumption of processed meat had been estimated at 34,000.[5] On its front page, the *Financial Times* criticized this 'ham hysteria' and advised its readers not to believe this 'scaremongering'[6] but rather to continue to 'savour the bacon'.[7] In Germany, *Die Welt* announced: 'WHO places sausages on the cancer list',[8] and *Die Tageszeitung* suggested a 'Salami tactic: to enjoy better health, use sparingly'.[9] In France, *Le Figaro* and *Le Parisien* proclaimed: 'Charcuterie is carcinogenic'.[10] *The Times* announced on its front page: 'Processed meats blamed for thousands of cancer deaths a year'.[11]

The main form of cancer at stake here is colorectal cancer. It is one of the most frequent and deadliest of cancers: in the US, it is the second most common cause of cancer death. Between 408 and 411 people receive a diagnosis every day – 150,000 new cases a year. Close to two thirds of patients survive, often after surgical removal of the tumours. But 35% of patients die. The figures for Europe are 500,000 new cases per year, with 238,000 deaths.[12] On a global level, there are 1.8 million new cases each year and 861,000 deaths.[13] Every five years, colorectal cancer affects 9 million people and kills approximately 4.3 million.[14]

Even though it is the second most deadly cancer in the world,[15] it is one of the least well known among the general public. Is that because, on top of the taboo about cancer, people have an instinctive repulsion when it comes to faecal matter? For colorectal cancer is quite simply cancer of the large bowel: the colon is the tube that leads out of the small intestine. It is where the final acts of digestion take place (recuperation of water and vitamins, compacting of waste, storage prior to excretion). The rectum is the bottom end of the digestive tract: its last fifteen centimetres.

THE INDUSTRY MOBILIZES

The 'group 1 carcinogen' classification marks a turning point in the history of meat products. By using data assembled in the 1990s, the IARC indicated that a 50g portion of processed meat consumed every day increases the risk of colorectal cancer by 18%.[16] Studies based specifically on recent British data have produced even more alarming figures: in spring 2019, researchers at the IARC and the Cancer Epidemiology Unit at Oxford University showed that a daily 50g portion of processed meat leads to a 42% increase in the risk of colorectal cancer.[17] A daily 25g portion leads to an increase of 19%.[18] A 25g helping is not a lot of processed meat: a rasher of cooked bacon weighs between 8g and 15g. A slice of ham weighs 10g for the smaller ones, 40g for the larger ones. A frankfurter-style sausage generally weighs around 40g, a slice of salami 10g, a single slice of bologna sausage around 30g. As well as these there are the 'hidden' processed meats: slices of pepperoni on pizzas, the fillings in pork pies, pieces of ham in salads or ready meals ... they all need to be factored in.[19]

The IARC's assessment immediately had an impact on sales. One month after the release of the results, the *Guardian* headline read: 'UK shoppers give pork the chop after processed meats linked to cancer'.[20] But the global processed meat industry fought back, mobilizing its PR companies to conduct a damage-limitation exercise. In an interview with the Spanish daily *El Pais* entitled 'The public has a choice: believe us or believe the industry',[21] Doctor Kurt Straif, who led the programme at the IARC, was scathing about these PR campaigns run by powerful industrial interests exercised purely by the impact on their bottom line.

The meat companies churned out press statements full of mock surprise or incredulity; the scientific data were skilfully reviewed and 'reframed':[22] the aim was to reassure consumers, convince them that there was really nothing to worry about, that the population was not exposed to any serious risk. Some industry voices portrayed the IARC as 'just a lab' and diminished its findings as just one study among many. They made out that the mechanisms that cause cancer are virtually unknown, or that the results indicated that processed meats represented only a 'theoretical danger', unrelated to any 'real risk'. In each and every country, they explained that the IARC conclusions did not apply to local habits, that the consumers studied by the IARC were 'statistical entities', theoretical figments.

In fact, behind the figures there are real victims. In 2017, in the *British Medical Bulletin*, a researcher at the Institute of Food Research in Norwich calculated that the levels of risk published by the IARC meant that, 'for every 100 male regular processed meat consumers we might expect approximately one additional case of colorectal cancer'[23] (the levels were slightly lower in women). One case in every 100 consumers: the defenders of processed meat might argue that that is not such a big deal. But at the scale of a city, one male consumer in a hundred amounts to a very big deal. At the national scale, it is huge. As the *British Medical Bulletin* explains: 'At the population level, differences in risk of this magnitude are of considerable significance for public health.'[24] The epidemiologists estimate that of the 110 to 115 new cases of colorectal cancer that appear on average each day in the UK, about ten are directly related to the consumption of processed meats. Similarly, in the USA, it is estimated that 10.3% of cases of cancer of the colon are the direct result of the consumption of processed meat.[25] (By way of comparison: in the same population, 13.5% of cancers of the colon are caused by cigarette smoking and 17% by alcohol.[26]) The carcinogenic impact of processed meats is understood in such precise detail that public health specialists have been able to show that in the United States a simple 6g reduction in the daily intake of processed meats would lead to a saving over ten years of a billion dollars on the cost of healthcare.[27]

OLD NEWS

In each of the producing countries, the processed meat industries have one powerful ally: the agricultural administrations are terrified by the economic consequences that a long-term collapse in sales would entail. The day after the IARC published its results, the French minister of agriculture declared: 'I don't want a report like this creating even more panic among people.'[28] His concern was justifiable: in France, more than 70% of pork is consumed in the form of charcuterie. The implications of a drastic fall in sales are scary: 58% of French pork production takes place in a single region, Brittany, where the industry employs 30,000 people. The picture is no rosier in Germany, and that is the reason why the minister of agriculture in Berlin took the IARC to task over its results.[29] Similarly, the Italian minister of agriculture attacked the cancer specialists and their 'unjustified scaremongering',[30] and the Australian minister of agriculture used the word 'farce'. In a radio interview he proclaimed: 'If you got everything that the World Health Organization said was carcinogenic and took it out of your daily requirements, well you are kind of heading back to a cave.'[31]

Though it seemed to take some by surprise, the IARC's classification was in fact the culmination of a long series of results proving the carcinogenicity of processed meats. At the end of the 1960s, the line of specialists working on bowel cancer was still: 'the presence of carcinogenic factors swallowed and present in high concentration in the stool has been postulated but no such carcinogen has been identified'.[32] By the 1970s they had gathered enough evidence to be able to accuse processed meats of being responsible for a considerable number of cancers – even if they couldn't yet produce actual figures. Later we will describe how the health authorities had to take on the processed meat industry during that decade. Since then, epidemiological research has never stopped. In the 1990s, a large collection of biological samples was established together with data on individual behaviours (smoking and drinking, physical activity, eating habits): known as the European Prospective Investigation into Cancer and Nutrition (EPIC), the project concentrated on cancer and food-related factors analysed over a long period.

EPIC by name, epic by nature: the study covered half a million people and was based on 8 million samples taken at 23 collection centres across Europe. At the beginning of the study, there wasn't a single case of cancer. As the years went by, cancers began to appear. A retrospective analysis of the accumulated data allowed precise links to be established between certain behaviours and the occurrence of certain cancers. The epidemiologists who headed up EPIC obtained definitive proof of the role of processed meat in cancer induction. From 2002, they provided preliminary figures suggesting that eating 30g of processed meat every day was likely to increase risk of cancer by 36%.[33] In 2003, the WHO published a preliminary recommendation aimed at limiting the quantities of processed meat consumed.[34] But the real turning point was the publication of two reports, in 1997 and then in 2007, by the World Cancer Research Fund (WCRF): having conducted an analysis involving a team of the best cancer specialists and epidemiologists, the WCRF concluded its evaluation with the recommendation: 'avoid processed meat'.[35] In 2012, the American Cancer Society recommended that people should 'minimize consumption of processed meats such as bacon, sausage, luncheon meats, and hot dogs'.[36] Meanwhile in Europe, summing up the state of scientific knowledge on the prevention of colorectal cancer, the Belgian Superior Health Council recommended 'avoiding as much as possible processed red meat'[37] and only consuming such products 'rarely, if at all'.[38]

NITRO-MEATS

In fact, it is not necessary to give up eating processed meats to reduce the risk of cancer. What needs to be reduced – eliminated altogether if possible – is *carcinogenic processed meat*. This truism contains a secret that the processed meat industry tries hard to obscure in its many communications on the subject: when it comes to cancer, not all processed meats are equal. Some are very dangerous, others less so, while others still have been shown to have no harmful effects in laboratory tests.[39]

Processed meats are made especially dangerous by the use of two food additives: nitrite and nitrate, which accelerate the curing process and give the meat an appetizing colour. Nitrate and nitrite are chemical substances composed of a nitrogen atom bound to three (for nitrate) or two (for nitrite) oxygen atoms. Nitrate is abundant in the natural world: some vegetables (such as spinach, beets, lettuce) contain a high concentration of nitrate, which may be transformed into nitrite by the action of micro-organisms. This transformation (the chemical term is reduction) occurs in the oral cavity; that is why saliva contains low concentrations of nitrite that is continually swallowed in a highly diluted solution.[40]

Nitrate and nitrite are not carcinogenic in themselves. But under certain conditions they can be transformed. Then they give rise to free radicals, in particular nitric oxide (NO). Highly reactive, and known for its role as an electron-thief or oxidizer, this oxide of nitrogen reacts with a wide range of biological molecules (lipids, proteins, DNA). When the nitric oxide reacts with the components of meat – especially with iron and with amides or amines derived from meat proteins – it leads to the production of certain carcinogenic compounds that the chemists call nitroso compounds or N-nitroso compounds (scientists use the terms interchangeably, some preferring the abbreviation 'NOCs' for 'nitric-oxide-releasing compounds').[41] These molecules result from the interaction between nitrite and elements in the meat itself; that is why these nitroso compounds are said to be 'neo-formed'. They damage the cells lining the bowel.[42] It is a long process, which typically takes ten to fifteen years: over time, the slow accumulation of damage leads to the proliferation of cancerous cells.[43]

In 2015, the French cancer specialist Denis Corpet was the chairman of the group of experts commissioned by the IARC to evaluate the mechanisms that lead to the appearance of tumours. Four years later, in early 2019, Professor Corpet expressed his indignation: 'The failure of governments globally to engage on this public health scandal is nothing less than a dereliction of duty – both in regards to the number of cases that could be avoided by ridding nitrites from processed meats – and in the potential to reduce the strain on increasingly stretched and under-funded public health services.'[44] In a letter addressed to the European commissioner for health, he reiterated that nitro-additives, by reacting with meat, give birth to carcinogenic nitroso compounds. 'The research team I led at the university of Toulouse demonstrated that while an experimental nitrite-cured ham promotes colorectal carcinogenesis in rats, the same ham cured without nitrite does not',[45] he explained. When the lab rats were fed with two separate batches of ham (one made by using nitrite, the other without it), only the nitro version led to the development of tumours.[46] Corpet concluded his letter to the European authorities by saying: 'The addition of nitrites to foods such as ham and bacon is thus central to cancer risk – and marks out those meats that contain these chemicals as significantly more dangerous than other processed meats.'[47]

ALL IN A GOOD CAUSE: THE BOTULISM ARGUMENT

On the occasions when the industrial meat-processors implicitly recognize that their nitrited meat is carcinogenic, they claim it is *all in a good cause.* According to them, nitro-additives are there to protect consumers from an *even greater danger* than cancer: they insist that nitrate and nitrite are needed to prevent botulism, a dangerous illness caused by a bacterium. They allege that the use of nitrate and nitrite is the only safe way to destroy germs that might be lurking in sausages and ham, waiting to kill those unwise enough to eat meat that hasn't been 'treated'. The CEO of a large French industrial meat-processor gave this explanation of why he refused to stop using sodium nitrite in ham and frankfurters: 'We don't stick it in out of habit or just for the fun of it. We put it in to combat a very grave evil: botulism, which is fatal ... This isn't a bout of indigestion we're talking about. With botulism we're talking death! ... With botulism, it's game over!'[48]

So it's a cost–benefit calculation, a case of the lesser of two evils: like a responsible parent, the meat-processing industry would rather risk giving cancer to a minority of consumers in order to protect the others. It's hard to find an equivalent elsewhere: would it be acceptable if orange juice killed thousands of people each year to protect everyone else from some mysterious 'orange poison'? Would parents feed their children chips if the flesh of the potatoes had been rendered carcinogenic by the introduction of some additive? Would the risk of bird flu be enough to induce consumers to eat chicken injected with a carcinogenic solution?

But the worst thing about this is that the botulism argument is factitious. In fact, manufacturers have a whole arsenal of techniques to prevent bacterial infections. 'We conclude that any effect of nitrite on product safety and stability may be compensated for by modification of formulations and processes',[49] explained the German biologist Friedrich-Karl Lücke, a specialist in processed meats, in 2007. For decades, in article after article, the experts in meat microbiology have been echoing this: simply by using adapted production techniques, the meat companies could produce perfectly safe processed meats without any need for harmful additives.[50] And those who like their ham pink can rest assured: there are alternative colouring methods. Back in 2008, in response to the report of the World Cancer Research Fund, three of the most respected European meat scientists wrote: 'It is now known that acceptable alternatives for the use of nitrate and nitrite exist in relation to color development, flavor and microbiological safety.'[51] Ten years later, chemists and biochemists were still making the same point: 'the use of nitrate and nitrite may be substituted by modifications of product composition, and processes'.[52]

From artisan pork butchers to small family enterprises to sizable factories producing hundreds of thousands of tons of meat a year, manufacturers large and small throughout Europe are working perfectly well without using nitrate and nitrite and yet have never recorded a single case of the 'inevitable' botulism that the nitro companies wield as their weapon of fear. To do without carcinogenic additives, they use different methods: they buy raw meat of better quality, they apply stricter rules of hygiene, they use longer periods of refrigeration and maturation, they adapt their equipment to meet these requirements. In the UK and Denmark a handful of industrial manufacturers make bacon and ham without using nitro-additives. In Germany and Holland organic producers likewise eschew all additives. In Italy the best products (Parma and San Daniele dried hams) are made without nitrate and nitrite. Similarly in Spain the top-notch meat products (authentic *chorizo* and *lomo*, most authentic *bellota* hams) are not treated with nitro-additives. In France and Belgium, artisan and industrial meat-processors are vying to bring nitrate- and nitrite-free products to the market. Some of them cater for purely regional markets, others already distribute on a national scale, such as the Biocoop group, which, in autumn 2017, launched an excellent ham without nitro-additives that is made in Brittany. It is a ham that doesn't cheat: it has its own true colour. The distributor is quite happy to make the case for it on its packaging: 'Pale ham, is that normal? Yes, if you don't add nitrited curing salt the ham retains its natural colour, that is, grey!'[53]

This proves that the sector can change, that 'virtuous' processed meats can be successful. But most of the market leaders baulk at the prospect: processing meat without using nitro-additives takes more time and requires more care. Adapting equipment does not come cheap: machines need to be changed, refrigeration units revamped, production processes revised – a complete overhaul. Why undertake such expense only to end up with a product that is less pink and thus likely to sell less?

IT'S ENDEMIC

Listening to the industry spokespeople, you would think that they have been straining every sinew over the last few decades to minimize the use of additives that make processed meats carcinogenic.[54] In fact, the number of nitro-treated products is growing relentlessly. It is the strange paradox of carcinogenic meats: the more it has been understood how dangerous they are, the more they have grown in numbers. Thus nowadays, in the UK, the 'farmhouse pâté', 'Brussels pâté' and 'Ardennes pâté' sold in supermarkets is often treated with sodium nitrite. The chemical produces an appetizing pink colour which makes for an appealing aspect when sold sliced and wrapped in transparent packaging. And yet in France and Belgium (for example in the Ardennes) the pâté is often not treated: it isn't pink, but grey. Another example: rillettes (potted meat), a traditional product of rural western France. The technical guides make clear that it is not necessary to use preservatives in rillettes, as the cooking pasteurizes them.[55] That is why the first *Code of Practice for Charcuterie*, published in 1969, explicitly forbade the use of nitro-additives in rillettes: only salt and spices were allowed.[56] Even though aware of the cancer risk, the professional organizations gave the nod to nitrite curing in rillettes. A reference manual says: 'This technique is not of particular interest *unless you are seeking to bring pinkness to meat pieces in the final product*.'[57] For many years now, the number of carcinogenic rillettes being sold in shops has increased, especially in cut-price and discount stores. So the injustice deepens: as with ham and sausages, it is always the cheapest products – that is, those consumed by households on modest means – which are treated the most with nitro-additives.

By way of an excuse for not banning these dangerous additives, the European Commission explains that it has not taken any measures because of 'the need for certain traditional foods to be maintained on the market'.[58] But in fact this liberal attitude essentially benefits products that are far from 'traditional': since nitrite curing is allowed, European industrial meat-processors have consistently come up with new products which rely on the use of this miracle additive. Instead of favouring healthier options, the manufacturers are permanently competing to develop new formulae and new nitro-meats. You simply have to peruse the supermarket shelves to clock the appearance of new items with labels that list sodium nitrite, sodium nitrate and potassium nitrate. All tastes and all ages are catered for: nitro treatment is used even in products aimed specifically at children and teenagers. And in this incessant escalation, there is one constant theme: nitro treatment lowers costs, accelerates production, simplifies the work of factories, prolongs shelf life and is the quickest way to achieve that lovely colour that customers like so much.

As for cancer, the organizations that represent the industry point out that they fund research. And the result of this work? Researchers confirm that the simplest solution would be to ban nitro-additives altogether. But as far as the industry is concerned, that is out of the question: there's too much to lose. So the biochemists suggest adding supplementary chemicals to counteract the carcinogenic action (especially tocopherol, a compound which has powerful antioxidant properties).[59] For years, nitro-meat manufacturers have been floating this idea: they claim that if we wait just a bit longer, a new revolutionary method will be developed that will allow the risk to be nullified, and this will suit everyone: for the public, fewer deaths; for the industry, no change to the look of the product and no costly adaptation of processing technology.

But if you delve into the history of the link between cancer and processed meats, you discover that studies into such inhibiting techniques were being announced as long ago as the 1970s, when the industry was first confronted with evidence of the carcinogenicity of nitro-meats.[60] Early formulae of anti-cancer tocopherol were already developed and patented in the late 1970s.[61] So how do we explain that the processed meat industry is still at this preliminary stage of development? Whom does this inertia benefit the most? How can we tolerate the fact that hundreds or thousands of deaths have been caused by this procrastination? Why trust the industry lobby when we learn that their current statements are almost verbatim repetitions of what was said back in 1975: 'To date, no substitute for nitrite has been discovered';[62] or else 'Researchers are still trying to find a replacement'?[63] Already in the 1970s, cancer specialists were critical of these dilatory tactics.[64]

As the manufacturers buy themselves more time, consumers are being poisoned. Systematic nitrite curing hurts everyone: consumers, who are made ill; health services, which have to expend valuable resources in expensive treatments; pig farmers, who are impoverished by allowing the processors to use meat of mediocre quality. And the meat curers themselves: when they make sincere efforts to produce healthier food, they are discouraged by having to compete against nitro products, with their perfect pink colour and their unbeatable prices – because they are produced quickly and with less care. The only winners are a few giant industrial companies who, thanks to nitrite, can rapidly produce meats that look as tasty as sweets and stay that way for a long time.

DON'T BRING HOME THE BACON

Rather than force the meat industry to give up nitrite curing, most countries prefer what might be termed 'the diagnostic and therapeutic option'. People over 45 are encouraged to provide stool samples and, if necessary, undergo a colonoscopy. When pre-cancerous cells are detected, the patients are operated on. Promotional campaigns riff on the theme of: '90% of colorectal cancers are treatable when detected early.' This is all very comforting, but contrarians might point instead to a crueller statistic: even in countries with an advanced hospital system, four out of ten people diagnosed with bowel cancer don't survive five years after diagnosis.[65] And beyond the statistics, each of these 'cases' represents enormous hardship for individuals and their families. Sometimes, the surgeon removes a piece of intestine and diverts one end of the colon through an opening in the belly. Who can hope to lead a normal life with a plastic colostomy bag attached to their abdomen to collect their bodily waste?

Rather than a genuine strategy based on tackling the *causes* of cancer, this combination of screening and treatment of patients *already affected* by cancer is passed off as 'prevention'. And too bad if the less well-off in society, who happen to be the largest consumers of processed meat, bear the brunt (in the UK, recent studies have shown that those in the most deprived social categories are disproportionately more prone to cancer than the rest of the population).[66]

And what does it matter if, following the USA and Europe, the number of cases of colorectal cancer is increasing in the Global South as they begin to adopt Western eating habits? As early as 1971, the British surgeon and epidemiologist Denis Burkitt, a pioneer in the identification of the role of food in colorectal cancer, was pointing out: 'The rise in bronchial carcinoma accompanied an increase in cigarette smoking, and likewise the rise in colon carcinoma accompanied a progressive adaptation to a North American type of diet.'[67] He noted, for example: 'Rural Africans rarely develop cancer of the large bowel, but when these same people move into a city and start eating Western-style food, their susceptibility to this type of malignancy increases dramatically and eventually matches the high rates found in Europeans and Americans.'[68] The same goes for Asian populations: 'The Japanese, especially in rural areas, also have very few malignancies of the colon. But when they migrate to Hawaii or California, their children have almost as many bowel cancers as the general population of these areas.'[69] According to the historian Robert Proctor, Denis Burkitt was one of the first to denounce the lack of any nutritional prevention policy for cancer: Burkitt 'suggests that we have a leaky faucet, an overflowing sink, and many experts busily mopping the floor. But why, he says, are there so few trying to turn off the tap?'[70]

The artisan producers who use nitro-additives often do so with regret: they would prefer to sell food that isn't detrimental to health. I hope that this book, by introducing them to the secret history of nitrate and nitrite, will offer them encouragement: they will discover how traditional meat curing was taken over by industrial groups obsessed with speed and volume, often indifferent to the health of their customers and willing to employ any ruse so as not to have to give up their recipes for 'accelerated meat curing'. These nitro-meats look lovely, but they are dangerous. It is time for real meat curers to take back control of their salt.

PART 1

IN THE PINK: HOW YOUR BACON ENDED UP FULL OF NITRATE AND NITRITE

CHAPTER 1

MIRACLE ADDITIVES

Even before we engage our sense of taste or smell, we use our eyes to select what we are going to eat. We react positively to the green colour of vegetables. The sight of a ripe apple or raspberry makes the mouth water, but if these fruits are painted blue we find them repulsive. As for meat, millions of years of evolution as carnivores have taught us to judge the freshness of flesh by its colour: our instincts interpret certain colours as offering guarantees against pathogens. 'Red' signifies quality; 'pink' expresses safety.

Unfortunately, the natural colour of ham and sausages is not pink: it is grey or brown, the same as pork after it has been cooked. Which is why meat curers have constantly been tempted to use artificial means to recreate the colour of fresh meat. The pink hue of hams/sausages/pâtés is the result of two products: potassium nitrate (chemical symbol KNO_3) and 'nitrited curing salt', a mixture of cooking salt and sodium nitrite ($NaNO_2$).

NITRATE, NITRITE AND IRON

As we saw earlier, nitrate and nitrite are not *directly* carcinogenic: even with repeated ingestion, nitrate and nitrite do not cause tumours in either animals or humans. But under certain conditions these substances can give rise to several carcinogenic agents. The best known are *nitroso compounds*, which can form during the processing, cooking or digestion of meat. Firstly, there are the *nitrosamines*, molecules formed by the combination of nitrosating agents (nitrite, nitrous acid, nitrogen oxides) and amines (amines form from the breakdown of amino acids, peptides and proteins). The other group of nitroso compounds contains *nitrosamides*, which form when a nitro agent reacts with an amide (an organic compound similar to amines). Nitrosamines and nitrosamides work even in low doses: they target the DNA of cells and cause lesions which can lead to tumours.

Industrial meat-processors claim that by adding vitamin C to their products they reduce the risk posed by nitroso compounds. This technical solution was invented in the 1950s and is in wide-spread use today: a great many nitro-processed meat products are supplemented with vitamin C (in the form of ascorbate) to speed up their production and to try to reduce the frequency of nitrosa-mines. Nevertheless, nitro-processed meat remains carcinogenic, as it contains other agents that generate tumours: those that result when nitro elements encounter iron contained in meat (special-ists call it 'haem iron'). When it is consumed in excess, the iron trace element has a pro-oxidant effect that stimulates cancerous cells. That is why the IARC, when it examined processed meat in October 2015, classed *untreated* red meat in category 2 ('probably carcinogenic'). The carcinogenic effect of haem iron is activated when meat is treated with nitrate or nitrite, as the element iron reacts with nitric oxide to create nitrosyl haem, an agent that is key to carcinogenesis.[1]

For producers of nitro-processed meat, acknowledging this mechanism means, in a sense, accepting their own death warrant: whereas the nitrosamine risk manifests itself only under certain conditions in cooking and digestion, the nitrosyl haem risk is potentially latent in every particle of nitro-meat. The only solution is to stop using nitro-additives and return to traditional methods of curing meat, using only meat and common salt.

THE NATURAL PIGMENT OF RAW HAM

Archaeologists have shown that humans have been salting pork in Europe at least since the Bronze Age (tenth century BCE), especially the Celts. In the area of modern-day France, excavations have uncovered several sites of salting workshops.[2] For example, in his depiction of Gaul in 18 CE, the geographer Strabo writes: 'These people produce magnificent cuts of salted pork that are exported as far as Rome itself.'[3]

In many regions of Europe, ham is still made following ancestral methods, that is, using only salt and no additives. This is the case for certain Spanish hams (most traditional *bellota* and *pata negra*), but it is mainly in Italy that the most famous examples of nitrate- and nitrite-free hams are to be found today. These meat products are created using the ancient procedures: after having been rubbed down with salt, the meat first of all takes on a brown tinge. Then, after a few weeks, without any other external intervention, a red colour starts to emerge, which becomes more and more intense. Humans have been exploiting this phenomenon for millennia, but it is only in the last twenty years that Italian and Japanese scientists have managed to work out the biochemical process involved: when an artisan makes a raw ham without nitrate or nitrite, he is unwittingly bringing a new pigment into being. Through the action of an enzyme present in the flesh, a part of the element iron contained in the meat is replaced with the element zinc. The scientists call this natural pigment of cured meat 'zinc protoporphyrin' (Zn-pp or ZPP).[4]

This pigment does not appear only in ham: in the past it was zinc protoporphyrin that gave dried beef that red colour that our ancestors called 'brési' or 'brazi'; today it is zinc protoporphyrin that gives an intense colour to the traditional (nitrate- and nitrite-free) sausages that are still found in Auvergne, Corsica, Spain and especially in Italy and Hungary. But Zn-pp entails certain constraints. Choice of ingredients, precision of method, control of temperature, acidity, humidity: contemporary Italian *salumerie* demonstrate the care and attention required to produce an authentic salami. If not done well, the maturing process can end up with a poor-tasting product that doesn't keep very long. Furthermore, this traditional method of production has another major disadvantage: it is *slow*. The zinc protoporphyrin pigment forms gradually throughout the whole period of production but grows most rapidly during the maturing process.[5] It achieves a satisfactory colour only after several months and continues to improve with time, because the longer the maturing process is, the more the quantity of Zn-pp pigment increases. The taste improves at the same time as the colour: in Spain, true *pata negra* without nitrate or nitrite generally takes 24 months to reach the market. As with wine, traditional hams improve with age: in Auvergne, up until just a few decades ago, dried hams would be strung from the ceilings of houses of well-off peasants, maturing for years in anticipation of a wedding or some other major occasion.

SPEEDING THINGS UP

Almost everywhere, traditional methods of production have been replaced by an accelerated process. In France, the most striking example involves the famous 'Bayonne ham'. Today, most Bayonne hams are treated with potassium nitrate (saltpetre), but traditionally they were produced using only salt, with neither nitrate nor nitrite.[6] Until the end of the 1960s, the Fraud Prevention Office prohibited the use of the expression '*real* Bayonne ham' if the producers made use of nitro-additives.[7] This is how the situation was presented in 1965 by the French meat-curing expert and chemical engineer René Pallu: 'We know that true Bayonne hams are cured purely with salt; these hams have a pleasing colour, a good texture for slicing and they keep perfectly.'[8] He explains that several conditions need to be met for quality production to take place: hams should be sourced from heavy, fattened animals; the pigs should be rested and not worked too hard before slaughter. He emphasizes that Bayonne hams are traditionally obtained by 'a slow and progressive drying and maturing process of between six and twelve months carried out in the open air in winter and spring or in air-conditioned drying kilns during the hot season'. For Pallu, these are 'So many conditions that seem to us rarely fulfilled in this day and age, where *quick and easy solutions* are at a premium.'[9] He concludes: 'In the majority of factories they not only do not select the hams to begin with, but they add saltpetre and sugary substances to the salt in order to facilitate a rapid development of the colour; furthermore, they subject the hams to a speeded-up "drying/maturation" process in air-conditioned environments.'[10]

In the 1960s, specialists referred to this accelerated method using an expression that industrial meat-processing firms have since tried hard to obliterate: 'chemical salting'.[11] This is how it works: once inside the meat, the nitrate and nitrite break down and bring about a release of nitric oxide. Spreading easily through the muscles, this gas reacts with the natural pigment of the meat (myoglobin) and binds to the atoms of iron. A new pigment appears, which chemists call NO-myoglobin (for 'nitric oxide myoglobin'). To put it in simpler terms, biochemists sometimes refer to 'nitroso-pigment'. This pigment has a deep red colour which – visually – closely resembles the zinc protoporphyrin pigment. The two pigments are indistinguishable without subjecting them to a chemical test, and the NO-myoglobin pigment has the added advantage of giving an even more intense colour than the natural colour it is mimicking.[12] As well as this intensity, the nitroso-pigment has a number of technological advantages. When curing is not performed competently, the appearance of the naturally cured ham can be marked with chromatic imperfections (patches of dark grey-brown coloration shading into black).[13] The nitro-additive, on the other hand, *consistently* gives a homogeneous and uniform coloration. Its second essential advantage is that in order to produce hams and sausages using the traditional method the meat has to be taken from older animals, as their muscles contain a lot of myoglobin. Nitro treatment, on the other hand, can produce visually satisfying hams from flesh that contains very little myoglobin, so younger pigs which haven't been exercised as much can be used.

The main advantage lies in the speed of processing: when the 'chemical method' is used, coloration is acquired much more quickly. The NO-myoglobin pigment forms so fast that a nitrate-treated ham can be on the market in less than 100 days. Spanish hams offer a typical example: whereas a real *bellota* ham without nitrate or nitrite requires 24 months to mature (and is often not ready for sale before 30 months), *serrano* hams (produced with the aid of nitrate or nitrite) can – if the producer so desires – be on the market after only three months.[14] This difference can have a significant impact on margins, given savings that can be made on floorspace devoted to drying and maturing, and capital that would otherwise be tied up can be released.

Nitro-additives offer other, even more controversial advantages. Nitrate and nitrite enable a global increase in productivity because they have a disinfectant function. Processing methods can be less rigorous without affecting the appearance of the final product. That is not to say that production is botched, rather that certain rules of traditional curing can be relaxed: the introduction of antiseptic right inside the meat itself prevents the development of bacteria which could lead to a product of inferior quality, or one unfit for sale at all.[15] As the bacteriologists and biochemists working for a French industrial meat-processing company wrote in 1954, nitro-additives 'ensure more or less complete bacteriostasis for polluting bacteria'.[16] Other industry technicians are more euphemistic: nitro-additives help to 'prevent spoilage'.[17] It makes production easier, avoids the need to keep factories completely refrigerated, reduces the constraints on the supply of fresh meat, simplifies handling, facilitates transport and storage, extends the shelf life: the presence of nitro-additives reduces losses, increases volume, lowers prices. This is why this accelerated process has gained traction in every country. In Spain, as recently as a few decades ago *chorizo*, *longanizas* and *sobrasada* were still made without the use of nitrate and nitrite; their coloration was obtained naturally through the addition of red pepper and through the natural occurrence of the zinc protoporphyrin pigment. Today, they are mostly nitro products, that is, treated with potassium nitrate or sodium nitrite – sometimes both. Recently, even producers of *bellota* ham – the high end of the Spanish dried ham market – have adopted nitro-additives, either because they allow them to significantly increase productivity or because their use opens up export markets, particularly in the UK and North America. And European authorities regularly publish new authorizations that allow a more extensive use of nitro treatments.[18] This is the story of processed meat and cancer: 'chemical salting' is so advantageous that it has ubiquitously become the norm.

MEAT CURING DONE THE PROPER WAY

One example from Italy is more encouraging. For several decades, the producers of Parma ham had been using nitro-additives. But in the middle of the 1990s, they collectively decided to revert to traditional (*parmigiana*) production methods, using no other ingredient than salt. This return to the old ways necessitated a long maturing period to allow the zinc protoporphyrin pigment to appear naturally. There has been a plethora of articles in specialist journals musing over the mystery – the 'riddle', some called it – of Parma ham.[19] Even now, certain specialists of nitrite curing find it hard to believe: the colour and stability of Parma ham strike them as 'intriguing, as they seem to occur without the intervention of nitrate and nitrite'.[20] There were any number of different theories: some suspected the ham producers of Parma of cheating, others conjectured that there were nitrifying bacteria in the hams (i.e. bacteria capable of causing nitrate to appear). Could a staphylococcus be responsible for the appearance of the red colour?[21] Some chemists came up with the hypothesis that the long maturation of the hams gave rise to sulphurous components which could interact with the meat.[22] Others convinced themselves that the salt used in Parma was contaminated by nitrate or nitrite.[23] Tests were conducted to confirm that it contained no such chemicals.[24]

Now the 'riddle' has been solved, and the meat scientists confirm that this is not a phenomenon peculiar to Parma ham: the role of zinc protoporphyrin has become clear, and producers everywhere are rediscovering that, although the traditional methods of salting are slower and require greater effort, they give an excellent colour and an aroma that is second to none.

Specialists in meat science had noted previously that Parma hams were often much less salty than hams treated with nitrate or nitrite.[25] From a public health point of view, that is already a plus, but the merits of zinc protoporphyrin don't end there. The absence of nitro-additives also implies the absence of carcinogenic components specific to nitro-processed meat: not only does non-nitro ham not generate nitrosamines and nitrosamides, it is also free of nitrosyl haem. Moreover, biochemists have discovered that natural pigment is able to inhibit the deleterious mechanisms connected to the iron content, to the extent that there are now attempts to enrich modern processed meats with the pigment that they have been deprived of.[26] This encapsulates the absurdity of nitro-additives: not only do they give rise to carcinogenic derivatives, but they also suppress the emergence of protective mechanisms.

COOKED PROCESSED MEAT: THE PINK MIRACLE

So far, we have been examining the so-called 'raw' forms of processed meat: that is, raw (and dried) hams, raw sausages, salamis, etc., all of which are consumed without having been previously cooked. Let us now turn to the other category of processed meats: the products that are sold cooked (such as cooked ham, corned beef, cured ox tongue) and those that are cooked before being eaten (for example bacon, bologna, hot dogs). In the raw products we discussed earlier, the nitro-additives accelerate and copy (rapidly giving an appearance resembling that which would occur naturally). But in the cooked products, the nitro-additives produce an even more interesting marvel: they give colour to products that don't have any.

In its natural state, cooked ham is off-white or brown, like roast pork. That is the hue that the customer would obtain if they made their own cooked ham, sausages or corned beef. But if you treat the meat with nitrate or nitrite before cooking it, everything changes. In a brochure written for meat producers at the beginning of the twentieth century, one of the inventors of nitrited curing salt explained: 'If you make bologna sausage out of fresh meat, it, of course, will be gray. If you roast a piece of beef, it will be gray. If you cook a piece of beef, it will be gray. It is the same with bologna. When bologna is made with fresh meat, it will be gray, just as though you take a piece of fresh meat and boil it. It is impossible to make bologna with a pink color and make it out of fresh meat.'[27] A longtime technical director of meat-processing factories, the American Fred Wilder, explained in 1905 that, without nitro treatment, the meat has 'a dead, slatish appearance, which is very unattractive'.[28] With the use of nitro-additives, this handicap disappears: once nitrited flesh is exposed to heat, the nitrosomyoglobin gives rise to another pigment, called nitrosylhemochrome, which is pink in colour. This new hue doesn't exactly resemble the colour of fresh meat (it is a sort of raspberry pink which has no real natural equivalent), but that doesn't matter: what is important is that the resulting meat is *pink* – any shade of pink will do.

In sales terms, this is a godsend. As a book published in the USA in 1942 by the Swift company put it: 'The retail dealer soon realizes that few meats that he can display are more tempting to the customer than these appealing cured products.'[29] They 'take the eye', according to another American specialist.[30] One manufacturer of nitrite curing salt promised meat-processors in 1951 that this treatment would give the flesh a precious 'eye-appeal'[31] – something almost akin to sex appeal, perhaps!

Nitro coloration offers a host of technical advantages. It is firstly very cheap compared with other existing colouring agents (capsicum, saffron, cochineal/carmine, etc.). But above all it is selective: since the colour only appears because of a reaction between myoglobin and nitric oxide, the nitro-additive colours only the flesh, not the fat.[32] In salami or mortadella, the flecks of fat can remain a nicely contrasting white colour, creating a chequered effect that is very easy on the eye. Nitro coloration is neat and tidy; it doesn't stain the rind or the subcutaneous fat. The effect is particularly distinct when the meat is sold pre-sliced and shrink-wrapped. As two specialists of nitrite curing techniques wrote in 2000, 'the bright pink color of nitrite-cured bacon and ham has long been used as a selling point, particularly since the development of transparent film vacuum packaging'.[33]

TABOO COLOURING

These days, industrial meat-processors try to conceal or minimize as much as possible the colouring function of nitro-additives (who would be happy to know that their food had been made carcinogenic for cosmetic purposes?). But that wasn't always the case. Before the dangers of nitro-additives were identified, producers had no reason to feel timid, and the colouring function was highlighted in the technical manuals. Until the 1960s, the industry made no bones about it: the essential function of nitro-additives was to rapidly colour the meat, to quickly give it a 'cured' taste, to simplify its production while dealing with issues of hygiene, to extend its life by preventing it from oxidizing so that neither its colour nor its taste was altered. For example, just before the Second World War, in written exchanges with the British Department of Health, the Food Manufacturers Federation emphasized that nitrite was used 'only as a colouring agent'.[34] Another industry organization indicated that 'the part played by the saltpetre in curing meat was that of a colouring agent'[35] and proposed a list of scientific articles by way of proof.[36] In his definition of 'the object of adding potassium nitrate', the head of the Food Research Laboratory at the Ministry of Health confirmed: 'The addition of potassium nitrate (saltpetre) to the salt used in curing meat, bacon, etc. preserves the colour of the meat, which would otherwise be a dull brownish grey.'[37]

This is even more explicit if we look at patent applications. In order to ensure legal protection of their inventions, chemists and manufacturers of additives in the USA and Europe had to provide precise descriptions of the effect that their formulae could achieve. The patentees make no bones about the principal role of nitro-additives: the alteration of colour. One of the industry's most respected and prolific inventors – one of his innovations was 'liquid smoke', which has almost universally replaced traditional smoking – explained in 1956 that: 'Originally, the main purpose of curing meat was to preserve the meat without refrigeration; the so-called curing process consisting essentially of the addition of salt. However, it was found later that meat cured with sodium nitrate and/or sodium nitrite produced a product with a desirable heat stable red or pink color.'[38] Likewise, a US patent of 1934 demonstrates the respective functions of salt and nitro-additives: 'It will be understood that the sodium chloride preserves the meat while the nitrite and nitrate mixture serves to color the same properly.'[39]

In a 1952 brochure vaunting the merits of one of the more popular nitrite mixes, Prague Powder®, the manufacturer explained that cooking salt (sodium chloride) is the only ingredient necessary from a sanitary point of view, but that the addition of nitrite can transform the colour: 'If a pork belly is properly treated with salt it keeps perfectly but when it is sliced the lean strips are gray. The same belly treated with salt plus Prague Powder also keeps perfectly but the lean strips are red.'[40] Similarly, the US Department of Agriculture (USDA) explained in 1953 that curing can be done by using only salt, or salt plus sugar, or salt plus saltpetre. It said: 'Remember: Salt preserves the meat. Sugar improves the flavor. Saltpeter (in the small amounts commonly used) merely sets the red color in the lean.'[41]

In a 1965 French technical work, an advert trumpeted: 'Pink ham? It is so easy to obtain using Nitral [which] gives an appetizing and stable deep pink colour.'[42] It was also an 'excellent germicide', and the same firm offered: *Églantine*, 'pinking salt'; *Tourose*, 'instant reddening'; *Radieux*, 'curing and preservative salt'; *Selrose*, 'long-lasting reddening'.[43] The advert explained that *Selrose* 'is both a preservative and a reddening agent … It transforms the colouring molecules in the blood, haemoglobin, into nitroso-haemoglobin, which is more richly coloured.'[44] This firm also vaunted the merits of its 'active reddening agent' *Colorado*, and *Salaisonia rose*, its 'reddening salt': 'Salaisonia acts on the colour of blood in the meat. It fixes this colour like a photographic print is fixed using hyposulphate.'[45] Another company, Berty, brought out the powders *Magie Rose* and *Magique Rose*, and later boasted of the exceptional performance of its 'new miracle reddener'.[46] Another chemist, Colorants Klotz, brought out *Roseline 66* ('the inimitable reddening preservative') and *Vitorose*, whose name evoked its rapid-acting quality ('vite-au-rose', or 'quick-to-pink').[47] There were also *Zulu*

Red, *Parisian Red*, *Indian Red*, *Radio*, *Roujax*, *Rougesec*, *Derosin*, *Cuitrose*, *Yrosy*, *RoseFix*.[48]

Nowadays, all these colour-explicit names have disappeared: additive manufacturers and industrial meat-processors claim that nitro products are not meant for purposes of colouring. Herta, for example, declares on its website that the use of nitrite in its hams is not intended for coloration: according to this manufacturer, nitrite curing is necessary to prevent bacteria from developing, and colour is merely a 'side-effect'.[49]

EXTENDING SHELF LIFE

Industrial meat-processors now assert that the quantities of nitro-additives they use are larger than what is required for colouring. They claim that this is proof that they are not using them for colouring purposes. In this matter too the technical manuals and patents are full of useful information that enables us to examine this argument on its own terms. For the technical texts show that, even if the colouring effect is initiated by relatively low levels of nitrite, the evenness, intensity and stability of the pink colouring are significantly improved when larger quantities of additive are used.[50] At the lower threshold, it is difficult to obtain a harmonious and durable colour that is able to last for weeks under the lights of supermarket shelves. Back in 1936, the chief chemist of the American Meat Packers Association, a lobby organization for industrial meat-processors, suggested that it is not enough to produce 'the desirable pink color which characterizes cured meat'[51] – that colour had to stick. In his patent entitled 'Producing stable color in meats' he insists: 'It is unnecessary to stress here the importance which the purchasing public attaches to bloom or color. The fact is that meat, especially cured meat in which the color has faded or changed, while otherwise entirely wholesome, can be sold only at substantial decrease in price or at a loss.'[52]

By acting upon the haem in the meat, nitric oxide establishes an extremely robust chemical bond. Because of this, the artificial colour of the 'nitroso-pigment' is virtually indestructible. The stabilizing power of the nitro-additives is not limited to visual appeal: they also have an antioxidant effect on the fats that prevents rancidity. Nitrite curing also prevents the occurrence of a certain bitterness which, although it doesn't affect the nutritional quality of the food and poses no health threat, obviously is detrimental to the flavour. Nitrite curing, then, can preserve a stable appearance and taste for several weeks – even several months – and thus extend shelf life.[53] Thanks to this chemical stabilization, in the USA you can find vacuum-packed pre-sliced meats with sell-by dates of longer than three months. Aside from the commercial advantages, this extension of preservation enables meat-processing plants to become more and more integrated and be sited a long way from where the products are consumed.[54] Globally, as the *Wall Street Journal* summarized it in an analysis of pork-belly prices on the Chicago commodity exchange in 1978, chemical treatments have a general stabilizing effect on flesh, living matter that otherwise has too great a tendency to break down: nitro-additives facilitate its commercialization; they 'make processed meat easy to store and transport'.[55]

There are other products that can protect processed meats against oxidation. As early as 1954, biochemists working for the French Olida company stressed that there were alternative antioxidant substances which were able to prevent rancidity.[56] But the advantage of nitro-additives is that they do everything in one go. Thanks to their high chemical reactivity, they are 'polyvalent and multifunctional', to use the technical term. This can be summed up in an equation: coloration + lightning-fast curing + simplified production + prolonged preservation = irresistible industrial advantage. In this sense, nitro-additives really are miraculous.

CHAPTER 2

THE TRUTH BEHIND
THE GOLDEN LEGEND

Nowadays, the vast majority of processed meat products are treated with nitrate or nitrite. According to the industrial companies, this is a practice as old as the invention of hams and sausages themselves. Ancient recipes, however, show that, on the contrary, they were produced without recourse to nitro substances and that this technique was the exception rather than the norm, indeed something of a curiosity, until the eighteenth century. Whereas today nitro-additives are used systematically – in effect, automatically – historical texts indicate that nitro treatment was formerly limited to *certain* products for which a simple, quick and efficient form of coloration was required.

As we have already seen, in 1965 the chemical engineer René Pallu, the administrator of the French technical centre for cured meats, the Centre technique des salaisons, stated categorically that the traditional recipe for Bayonne ham, for example, used only pure salt, without any additives. Likewise, for dried sausages, 'only salt and pepper are indispensable, and saltpetre (or nitrite) and sugared substances should be considered as adjuncts, whose purpose is to improve the appearance and to facilitate or shorten production'.[1]

An example: in 1476, a regulation specified that Parisian char-cutiers could only make their sausages using 'finely minced pork flesh [...] well salted with fine salt and [with nothing else but] good, sharp, well-chosen fennel or other good spices'.[2] There is no nitrate either in the recipe for the famous 'provençal salami' which can be found in the town hall of Arles: salt, pepper, cloves, nutmeg, ginger, 'good white wine' ... the recipe carefully lists fifteen or so ingredients, but for centuries saltpetre did not appear among them.[3] Industrialization transformed production techniques: instead of selected meats, which were dried and matured slowly, the new 'Arles salami' started being produced using meat of inferior quality, just the rough bits trimmed from the bone.[4] Thanks to saltpetre, this modern-day salami – an imitation of the authentic product – could be made in a fraction of the time at a fraction of the cost.

Another example, this time from Italy: the old recipes show that the pinkish colour of mortadella was obtained using a plant-based colorant, usually saffron.[5] But the modern version is nearly always treated with nitrite: the miracle additive ensures a markedly cheaper coloration and simplifies the whole production process. We could go on and list many other types of processed meat which we now know only in their modern chemical version. Is that why those in the industry constantly proclaim and swear that the origins of nitro-meats are to be found in the mists of time, at the dawn of civilization?

THE 5,000-YEAR MYTH

The words 'nitrate' and 'nitrite' derive from 'natron', a mineral abundant in the Nile valley. Egyptians used it to make mummies.[6] This etymology has been exploited by some authors who are keen to anchor the use of modern additives in supposedly ancestral practices. One article in favour of sodium nitrite, for example, claimed that a product called nitre was 'used in cave communities near the Dead Sea'.[7] So the implication is that the application of saltpetre to meat is nothing new – thus supporting the claim that we can't do without it. At a Senate commission called in 1978 to deal with the problem of nitro-meats and cancer, one American industrialist defended nitrite against those who 'are asking for removal of a product we have known for 20 centuries'. He offered up this supposed ancient lineage to argue for a delay: 'our industry needs time and research to replace processes which were begun several thousand years ago.'[8] Some authors justified the use of sodium nitrite by claiming that meats with nitro-additives dated back to 'Homer's time (850 B.C.)'[9] or 'as far back as 1600 B.C.',[10] 'approximately 4000 B.C.',[11] or else that saltpetre has been used 'since Antiquity'[12] or simply 'for ever'.[13] Pitching in, the review *TechniPorc* claimed that 'the use of potassium nitrate or saltpetre (KNO_3) is very ancient. It goes back 5,000 years.'[14]

Blogs, scientific papers, producers' websites: we have lost count of the number of authors who have adopted this symbolic date of '5,000 years' in order to give nitro-processed meat a place in the immemorial culinary history of humanity. The website of a French pro-nitrite lobby organization echoes the claim: 'The use [of nitrate and nitrite] goes back 5,000 years. Back then, men discovered that meat kept better with the addition of saltpetre (potassium nitrate).'[15] Another page on the website claims: 'The use of saltpetre or nitrite in meat products is an ancestral practice, since it goes back to 3000 B.C. So it is not a recent invention.'[16] In the same vein, the former scientific adviser to the American Meat Institute (AMI), today employed by a major Canadian producer of nitrite-cured bacon, regretted that 'nitrite and nitrate in cured and processed meats continues to be perceived as harmful' even though 'nitrite and nitrate have been in use for as long as 5,000 years in the preservation of food'.[17]

This golden legend that meat has been treated with saltpetre 'for ever' appears to be ubiquitous. In one recent article we are informed that Lucius Columella, the famous first-century Roman farmer, apparently stated that 'Hams are rubbed with salt and a little saltpeter for 9–12 days and are then rinsed and hung to dry'. The article goes on to say that 'this is virtually the identical method used today'.[18] In fact, if we look at what Columella actually wrote, he says nothing of the sort: the preparations he describes refer only to salt; there is no mention of saltpetre to be found anywhere in his descriptions of ham curing.[19] The same goes for ham recipes left by Strabo and by Cato the Elder.[20] As for meat curing in Gaul, on which historians have access to numerous primary sources, we know for a fact that saltpetre was not used.[21]

Some authors have suggested that salt used in the past might have been naturally contaminated by nitrate.[22] For example, in the technical treatise *Pigs and Bacon Curing* (first published in 1923), the British specialist Frank Gerrard wrote: 'There is little evidence as to the introduction of saltpetre as a colouring agent and though it may have been introduced accidentally as an impurity in the salt, it was not until comparatively recent times that it was used in its pure state.'[23] The French microbiologist Jacques Rivière, author of a series of studies on nitro-additives published in 1948, made some similar observations: he considered the application of nitro-additives to be recent and judged the ancient custom was to use solely salt and spices.[24] He noted that saltpetre is never mentioned in the oldest texts.[25] But he stated that in certain desert regions, in Asia, nitrate curing might have occurred accidentally: 'It is believed that salting was first practised in the salt deserts of Asia' and 'the desert salts used by the Ancients often contained nitrates and borax in the form of impurities'. Likewise, some in the industry continue to hammer home the idea that nitro-additives have been used 'for ever' on the basis that some deserts in Latin America contain deposits of nitrate.

However, salt is one of the substances most studied by historians of science and technology; researchers have a very precise knowledge of its composition over the ages.[26] In France, at Salins les Bains (Jura) or at Salins de Béarn (Pyrénées) – two sites whose salt has been used to make hams since Neolithic times – the composition of the salt was the object of exact descriptions and reports from the Middle Ages onwards. In Germany, the saltworks of Schwäbisch Hall have produced reports for hundreds of years. As the historian Anthony Bridbury emphasized in his book *England and the Salt Trade in the Later Middle Ages*,[27] the composition of salt is stable, as the underground banks of salt that feed the mines have not changed for millions of years. As for sea salt, even fervent promoters of nitrite curing have been forced to recognize that there is no scientific evidence to support the claim that such salt in the past has been contaminated by nitrate or nitrite.[28]

Even if there are no historical traces, the possibility can't altogether be excluded that some remote populations might at some point have rubbed nitro-minerals onto pieces of meat. So it wouldn't be safe to assert that meat was *never* treated with nitrate during prehistory or antiquity (how could you prove it?). But on the other hand, there is nothing to indicate that this practice did exist. 'The claim […] that saltpetre was regularly added to sausages/salamis more than 2,000 years ago cannot be corroborated by Greek or Roman sources', wrote the German scientist Raphael Koller in 1941 in a book on the history of salting techniques.[29] Nor is there any trace of saltpetre in the 2001 book *Ancient Food Technology*[30] by the American historian Robert Curtis, a specialist in food preservation techniques. And the archaeologist Salima Ikram corroborates this: according to this Cairo professor (she wrote her doctoral thesis on the preservation of meat in the time of the pharaohs), there was no saltpetre in the meats of ancient Egypt![31] Except in mummies …

AN ANCIENT TRADITION … OR A MEDIEVAL CURIOSITY?

In 1929, a report by the British Food Manufacturers' Research Association noted that nitro-additives had never been an indispensable part of the curing of meat.[32] In the mid-twentieth century,

French industry experts were saying quite unambiguously that curing originally employed only salt (common salt or kitchen salt), and that the use of saltpetre came in only much later, with the aim of colouring the meat.[33] Likewise, experts at the German Federal Centre for Meat Research were of the opinion that humans had preserved meat for centuries using salt (sodium chloride), and that the supplementary use of saltpetre arrived only later.[34] Saltpetre curing thus replaced former maturing processes and techniques for colouring meat, of which there were a variety: smoking, strongly pigmented vegetable extracts (juniper, madder, colouring spices such as paprika, saffron, extract of annatto or turmeric), red wine extract, or simply red pepper. The colouring of hams using saffron was widespread until the nineteenth century. In Auvergne, red wine is still the preferred colorant for some traditional charcuterie. Sweet pepper is still often used to give colour to Spanish *chorizo* and Calabrian sausages.

The first vague hints of saltpetre being used regularly for certain types of meat appear at the end of the Middle Ages. A fourteenth-century parchment indicates that game can be treated by rubbing it with 'salt of poite', probably saltpetre.[35] Potassium nitrate was reputed to have multiple applications: one author recommended it as a salve for dog-bites; another suggested it promised a cure for 'almost uncurable diseases', including colic, gout, scabies, fistulas, and 'all tumours and inflammations';[36] one doctor recommended using it against jaundice and asthma, to treat wounds and as a disinfectant.[37]

Historians have shown that pharmaceutical concoctions have often made the leap from apothecaries' counters to domestic kitchens, as remedies and drugs can offer an inexhaustible supply of useful expedients in the production of foodstuffs.[38] Is this the route by which saltpetre found its way into meat?

Other indications suggest that the first uses of nitrate in meat curing were related to the increasing use of firearms: nitrate is the main ingredient of gunpowder. How this explosive mixture originated is shrouded in uncertainty. Were Arab alchemists the first to master the combination of saltpetre and fire? Or did Chinese sorcerers discover how to produce fireworks? Whatever the case, gunpowder found its way to Europe. In the course of time, pyrotechnicians experimented with all sorts of combinations to create a stronger powder, until the final formula of six parts saltpetre to one part sulphur and one or two parts charcoal was arrived at.

Without saltpetre it became impossible to wage war efficiently. And so throughout Europe a whole new industry emerged, dedicated to the collection and refinement of this magical and strategic substance, which was sometimes known as *sal bombardicum*. Did hunters then notice that pieces of game that had been blasted with gunpowder kept longer better than others, and also took on a more attractive hue? This is the hypothesis of historians who link the propagation of nitrate in food with the invention of muskets and rifles. In 1948, the microbiologist Jacques Rivière reviewed different studies on the military history of saltpetre and its use in meat 'to obtain a red colour'.[39] In a report to the French Ministry of Health, Henri Cheftel and Louis Truffert also traced the use of nitro-additives to 'fix the colour red' back to the Middle Ages. Passionate supporters of nitro-additives, these authors wrote: 'This peaceful application of saltpetre, which would have followed shortly after its use in the making of gunpowder, has been shown over the course of centuries to be harmless to humans, which certainly can't be said for the other main use to which it was put.'[40]

According to the British historian Jennifer Stead, the use of saltpetre in meat increased greatly in the seventeenth century, 'when it was found that gunpowder rubbed into hanging game enhanced its keeping qualities'.[41] The American author Mark Kurlansky situates the appearance of saltpetre a century earlier: according to him, Polish hunters worked out how to preserve the game they had killed 'simply by gutting the animal and rubbing the cavity with a blend of salt and gunpowder' in order to 'make them a reddish color, that was thought to be more in keeping with the natural color of meat'.[42] The Polish historian Maria Dembińska even cites a medieval text which describes the case of an entire aurochs sent from the forests of Lithuania to Lake Geneva as a diplomatic gift. After having been killed, the animal was preserved in gunpowder.[43]

A German epidemiologist, Klaus Lauer, has studied recipe books kept in the historical collection of the Culinary Academy of Frankfurt. According to his research, it was at the end of the seventeenth century that saltpetre came to be widely used in meat curing in Germany.[44] Elsewhere in Europe, numerous archive sources confirm this dating. The Irish chemist Robert Boyle wrote in 1664 that 'several curious persons have practised, of salting neats tongues [ox tongues] with saltpeter'.[45] According to Boyle, the treatment was employed exclusively 'to make them look red'. Around the same time, the British doctor William Clarke brought out a book in which he listed the diverse applications of saltpetre. The panacea he described was so powerful and reactive that he believed he had found the central ingredient of the 'great elixir' that generations of alchemists had been searching for in vain. Mixed with clay, saltpetre produces nitric acid (*aqua fortis*) and nitro-hydrochloric acid (*aqua regis*), which dissolve all metals. Prospectors used it to melt and purify gold dust, engravers used it to attack alloys. Glassblowers included saltpetre in their secret preparations, dyers applied it to fix colours, while founders used it to transform base metals: none of the mechanical arts was unaware of the extraordinary transformative power of nitrate and its derivatives. At the very end of his book, William Clarke mentioned a subsidiary use, an innovative trick that he thought would delight cooks of the 1670s: 'It is also a pretty mechanism in cookery, which I shall set down to pleasure our English ladies delighted in such experiments, which is this, nitre giveth a red colour to neats-tongues, collard-beef, bacon or what other meat you will have look red, for which purpose the salt is most used [...] you may please to use the nitre itself, by mixing a small quantity with the other salt, with which you salt your meat, and it will tinge it with this desired red colour, and add to it a more savoury taste, and so both inviteth and pleaseth the appetite.'[46]

These accounts are backed up by ancient guides to curing techniques. For example, a work by John Collins published in London in 1682:[47] over 150 pages, Collins, a book-keeper at the Royal Fishery Company, describes salt, salt production sites, and curing techniques used on different types of fish and meat. The reader is told how hams are cured according to whether the salt originated from salt marshes or salt mines, how artisans make bacon, what type of salt curers prefer for making *sawsedges*, etc. After describing normal curing (with cooking salt), Collins devotes several paragraphs to a particular type of salt ('clod salt' or 'red salt', then sold at saltworks warehouses) which could make bacon very red and colour ox tongues. Other contemporary sources corroborate this description.[48] John Collins explains that this specific 'salt' is 'used for salting bacon and neats-tongues; it makes the bacon redder than other salt, and causes the fat to eat firm'.[49] He goes on to say that, in the absence of this 'red salt', one can just as well use pure saltpetre. Under the heading 'To salt beef and neat tongues red', he writes: 'Lay the tongues and beef in a tray, and almost cover them with salt till there be a brine, then dissolve a small quantity of refined petre-salt in it; to fix tongues allow half an ounce or more, let the tongues or beef lie 12 hours on either side, and it gives it a redness.'[50] He then devotes several pages to saltpetre, outlining where it is extracted and its military function. In particular, he describes a saltpetre refinery on Portsea Island. 'That which falls to the bottom of the pan, is called (as I am informed) petre-salt, or the salt of salt-petre, it resembles common salt, has little or no taste of saltness, but is efficacious in turning what is salted therewith red, as neats tongues, hogs-tongues, martinmas and collard-beef, yea and out of it a spirit may be drawn as red as blood ...'[51]

A SPILLOVER EFFECT

If we look at cookery books we see how the process became gradually more widespread: evidence of saltpetre usage is more frequent from the start of the eighteenth century.[52] In 1710, the apothecary William Salmon described a number of saltpetre recipes, for example the one for 'bacon to dry': 'Cut the Leg with a piece of the Loin (of a young Hog) then with Salt-peter, in fine Pouder and brown Sugar mix'd together, rub it well daily for 2 or 3 days, after which salt it well; so will it look red; let it lye for 6 or 8 Weeks, then hang it up (in a drying-place) to dry.'[53] The technique was primarily reserved for cuts where a particularly intense colouring was required. This is the case with ox tongues, which always take on an unsightly appearance when cooked. In the manuals, it is one of the first cuts that was subjected to systematic nitrate curing (in his novel *The Belly of Paris*, Émile Zola would later evoke 'plump tongues from Strasbourg, red and glazed, blood-red next to the pallor of the sausages and pig's trotters').[54] Before widespread use of saltpetre, charcuterie manuals described the steps to be taken to make 'scarlet tongue' using carmine, obtained by cooking the cochineal fly (*Dactylopius coccus*). It required six to ten kilos of dried flies to make one kilo of carmine, enough to produce twenty litres of colouring solution. Nitrate colouring was less expensive, and more efficient: in 1790, a recipe for scarlet beef states that the saltpetre helps to create a meat that is 'as red as vermillion'.[55]

It should be noted that the use of saltpetre in food was not an isolated phenomenon. With the development of chemical science, a whole host of new substances found their way into foodstuffs. One such was alum, a mixture of aluminium sulphate and potassium sulphate. Long used in early industrial workshops (for dyeing in particular), alum became a very common additive in flour in the eighteenth century, as it gave bread a white colour that the public liked.[56] Then alum was used in sausage making, where it served to fix surface colorants on the skins in order to give them a pleasing 'smoked' appearance.[57]

Throughout Europe, at the end of the eighteenth century, there was a marked rise in the volume of saltpetre imported for military use and a rapid improvement in refining methods. The meat colouring trade became more organized: in France, the first specialist sellers sprang up.[58] Historians of the trade have identified one of the pioneers: in 1792, one Charles Robert, a Carmelite monk prior to the French Revolution and later a pharmacist in Nîmes, was the first Frenchman to sell a saltpetre composition to charcutiers.[59] From the 1820s onwards, saltpetre curing is described being done by small European producers ('a little nitre is useful in the preparation of tongues: this substance gives them a fine colour')[60] as well as in publications aimed at American farmers.[61] In France, the *Manuel Roret* (1827) explains that 'pork, like beef, acquires a greenish colour during curing; if you mix one ounce of saltpetre with five pounds of salt, the muscle fibres acquire an attractive red colour'.[62] Some of these accounts were already advocating universal saltpetre curing. The *Encyclopédie catholique*, published in Paris in 1848, offered six formulae for brine – each of them containing saltpetre. One recipe got a special mention: more concentrated than the others, it gave meat a 'blazing red colour'.[63]

It is not unusual to find texts recommending saltpetre for extending the preservation of poultry,[64] fish[65] or dairy products. Saltpetre made butter keep longer without refrigeration and 'refreshed' it when it had started to go off, and in Britain there was even a saltpetre-based mixture sold under the name of 'cheese powder' which inhibited the fermentation of defective cheeses.[66] The disinfectant powers of saltpetre facilitated storage and prolonged preservation, even in tropical climates. Hence, nitro treatment was recommended for meat products aimed at sailors (known as 'ship meat'),[67] since the military demanded provisions that could keep for four to five years.[68] A Canadian regulation of 1820 specified the saltpetre-curing of barrel-packed beef intended for export.[69] An 1823 enquiry into the curing plants of Ireland stated that they used saltpetre 'only when expressly requested', essentially in orders for the navy: 'The contract for meats destined for the fleet stipulates that they use around two ounces of nitre for every quintal of meat; they sprinkle the cured cuts when transporting them from one barrel to another. In the same contract they guarantee, under caution, that their meat will be preserved six months after its arrival in India.'[70] The chief medical officer of the French navy also recommended nitrate in barrelled meats for soldiers: 'It is the interaction of the saltpetre with the blood that causes the meat to retain this vermillion colour that flatters the eye and banishes all thoughts of rottenness.'[71]

A TECHNIQUE USED INTERMITTENTLY

The Cook's and Confectioner's Dictionary by John Nott (1723) and *The Art of Cookery Made Plain and Easy* by Hannah Glasse (1747) contain a number of recipes where saltpetre is used,[72] but pork products were not systematically treated in this way. In a work published in Vienna in 1777, saffron is still specifically cited as the colorant for sausages, whereas rubbing with saltpetre is recommended for hams.[73] In 1805, the Danish vet Erik Viborg wrote a long essay on the breeding of pigs and the processing of pork. He didn't mention saltpetre in recipes for black puddings, saveloys, mortadellas or sausages.[74] On the other hand, he wrote that hams were treated with saltpetre, as was meat destined to be packed in barrels ('to give it a reddish colour').[75]

The *Dictionnaire général de la cuisine française* (1853) explains that saltpetre helps to redden salted meats ('that is why it is useful in curing rounds of beef, smoked tongues and hams'). In addition to 'scarlet' cuts, it also recommends saltpetre for *chorizos*, 'Lyon saveloys', 'so-called Bologna salami' and for '*Bayonne-style* hams'[76] – in other words, imitations of genuine Bayonne hams. On the other hand, the 'general curing recipe' given by *Le Cuisinier européen* (1863) doesn't mention saltpetre at all,[77] and other works recommend continuing to colour *chorizos* using paprika. The French vet Théodore Bourrier (a meat specialist and the chief inspector of the butchers and charcuteries of Paris) described numerous curing techniques, some involving saltpetre, some not. He wrote: 'Many people add 60 grams of saltpetre per kilogram of salt. This addition firms up the surface of the meat and makes it less susceptible to atmospheric influences and gives it a nice pink colour.'[78] But in 1897, Bourrier was still recommending that tongues should be coloured using carmine.[79]

Nitrate treatment was applied to meat products more often on the other side of the Rhine, for the German population traditionally liked their sausages to have a pink colour.[80] The French treatise *Charcuterie pratique* (1884) includes a chapter entitled 'German Charcuterie', which lists saltpetre in almost all the recipes.[81] The *Nouveau manuel complet* (1869) says that saltpetre is responsible for the 'fine bright-red hams of Westphalia',[82] and the *Charcuterie ancienne et moderne* (1869) mentions nitrate only in the context of certain types of ham and German salt beef: 'The reputation of Hamburg beef, which comes from its bright red colour, is only superficial; for this colour is due not to the quality of the meat but merely to the presence of a mix of saltpetre and salt used in its manufacture.'[83]

In the *Treatise on the Breeding of Swine and Curing of Bacon* published in Leith in 1811, saltpetre features as a common ingredient.[84] It was used in abundance by the small bacon curers that sprang up throughout the nineteenth century,[85] not only in Ireland and Britain[86] but also in the Danish workshops feeding the British market. The *Précis pratique de l'élevage du porc* [Practical Guide to the Raising of Pork] by Professor Alphonse Gobin (1882) is illustrative here: to make bacon, he proposes a French recipe that uses only salt, then points out that the British add sugar and saltpetre, so much so that it was use of nitrate that characterized what he called 'the English process'.[87] In 1898 a producer in Llandaff (Wales) was questioned by the Pontypridd police. To the question 'With regard to curing, do you use saltpetre?' he replied in the affirmative, but specified: 'It is possible to cure hams with saltpeter or with salt alone.'[88] A few years later, the American specialist Albert Fulton indicated that 'neither the sugar nor the saltpeter is absolutely necessary for the preservation of the meat, and they are often omitted'[89] – but he noted that US factories were systematically curing meat using saltpetre. In 1881, a newspaper in Missouri defended the salt-only option against the mixture 'salt + saltpetre': 'The *National Live Stock Journal* asserts that hams have a decidedly better flavor, and the meat retains a more natural color, when nothing but plain salt is used in the curing. *If the work of salting is carefully attended to*, the hams, when cured with salt alone, will be ready for smoking at from six weeks to two months, according to the size of the hams. Saltpeter has a tendency to harden and redden the meat. It undoubtedly hastens the curing process; but it does so at the expense of the flavor.'[90]

'ACCELERATED CURING' AND 'SALTPRUNELLE'

As well as the colouring effect, one essential advantage of saltpetre is that it shortens the production process, as we have seen. This was demonstrated in the book *Le Cuisinier moderne mis à la portée de tout le monde* [The Modern Cook Made Accessible to Everyone], published in 1836.[91] Several recipes (such as Bolognese or Milanese saveloys) do not involve saltpetre, but the author proposes using it to make 'Lyon sausages' with a drying period of only eight days and, even better, 'Bayonne hams' in three weeks (instead of the traditional nine months). The process produces an all-purpose, imitation Bayonne at a lower cost: a nitrate brine makes long maturing redundant. Moreover, a series of saltpetre-based treatments (the 'Sanson procedure') produces meats with a long shelf life in only seventeen days, with no need for cooking or even smoking.[92] Another author, Jules Gouffé, advises housewives that, thanks to nitrate, they can learn how to produce salt beef in fourteen days and 'Bayonne ham' in fifteen …[93]

These procedures anticipate certain aspects of modern food production: although curing was originally a technique of pres-ervation, 'saltpetre curing' seeks to create a product that, from a sensory point of view, *resembles* the product on which it is mod-elled. In other words, it is an imitation. A number of publications describe this 'accelerated curing' where the quantity of saltpetre is so large (a quarter of the volume of water) that salt can be dispensed with altogether.[94] In this salt-free 'curing', two days of preparation take the place of four weeks of work and waiting: '*Accelerated curing*. The meat is cooked on a low heat in a quan-tity of water and saltpetre […] By this method, sometimes used in France, one can start using the meat after forty-eight hours; it is as firm, red and tasty as Hamburg salted meat, which takes four weeks to prepare.'[95] These procedures are described in the historical literature as artificial methods, *short-cuts* to help the lady of the house prepare food that reproduces the character-istics (appearance mainly, but to a certain extent also the taste) of traditional products.

Another method has caught the attention of historians of food chemistry. As a pair of American producers put it: 'Many old recipes for curing stressed the use of a small amount of Sal Prunella to acquire a good colour in the meat.'[96] Manuals describe this 'sal prunella' or 'saltprunelle' as a 'concentrated form of saltpetre', and those who have looked into its history see it as an early version of nitrite.[97] Originally, saltprunelle was developed for medical purposes. A description of how it was produced can be found in pharmaceutical texts: sulphur was mixed with saltpetre until they fused together; then it was moulded into pills.[98] The etymology of the phrase is uncertain, but a number of apothecaries suggest that the name 'prunelle' derives from the fact it comes in the shape of small balls (like plums). Other chemists and pharmacists think it rather refers to the illnesses for which these pills offered treatment: 'burning fevers' (from the Latin *pruna*, red-hot coals).[99] It first appeared in a meat context in the eighteenth century in British recipes. *The Cook's and Confectioner's Dictionary* by John Nott (1723) and later *The Art of Cookery Made Plain and Easy* by Hannah Glasse (1747) contain a recipe showing how 'to pot beef like venison', which combines salt, saltpetre and saltprunelle.[100] In the quantities recommended, 'there are salts sufficient to colour a whole beef red'.[101] Saltprunelle appears throughout the nineteenth century in recipes for reddening and preserving meats: the *New System of Domestic Cookery* (1807) recommends saltprunelle to treat ox tongues and make hams; likewise *The Lady's Own Cookery Book* (1844).[102] In an 1847 manual, James Robinson offers a formula for 'spiced bacon', in which he completely forgoes saltpetre in favour of saltprunelle. He also suggests a saltprunelle/saltpetre mix to produce (specifically red) 'Hungarian beef' and 'British-American'-style ham.[103] One last book, from 1864, might mark the apogee of saltprunelle treatment: published in London by an author who identifies himself as a 'wholesale curer of comestibles', this book describes the then-fashionable techniques for the production of processed meats. Fifteen or so recipes describe treatment using saltprunelle; almost all the rest use saltpetre.[104]

In the 1890s, saltprunelle continued to be mentioned in publications written for Canadian meat-processors[105] and in some

British recipes, in particular for making bacon in warm weather and smoked ham using an accelerated curing.[106] In the twentieth century, 'sal prunella' still crops up occasionally in technical publications on the production of industrial curing agents.[107] But these mentions seem anachronistic, because by then a whole host of new nitro products had appeared which had put paid to the hit-and-miss approach of former times. At the end of the nineteenth century, makers of processed meats who wished to colour their products no longer needed to source their salts in apothecaries' stocks: the modern chemical industry was coming into existence, and its laboratories were producing dedicated and infallible preservative/reddening agents. Saltpetre curing was entering the scientific age.

CHAPTER 3

THE TRIUMPH OF MEATPACKING

At the beginning of the nineteenth century there was only one sort of saltpetre in Europe: potassium nitrate. After 1820, a new type appeared: sodium nitrate or 'nitrate of soda'. It came from immense deposits discovered in the deserts at the foot of the Andes in South America. Within a few years, whole convoys of ships were plying back and forth, supplying European ports with Chilean nitrate. This nitrate was used primarily as a fertilizer, but it had one particular feature that made it useful for meat coloration: it acted more quickly than potassium nitrate. As one technical text explained: 'Nitrate of soda (Chile saltpeter) is a little stronger, and 1.7 ounces of nitrate of soda will replace 2 ounces of saltpeter.'[1] Its use became more and more widespread throughout the nineteenth century. In 1895, an American publication recommended that both products could be used interchangeably: 'Meats for export are handled somewhat different in curing from meats of home consumption, color being one of the most essential points in this trade; consequently a considerable quantity of saltpetre is used, or nitrate of soda, either of which answers the same purpose.'[2]

A RED TIDAL WAVE

The last few decades of the nineteenth century saw an explosion of colorants designed for meat-processing. Published in 1899, the American professional manual *The Manufacture of Sausage* stated: 'The trade now insists upon having sausages of a color not otherwise obtainable than by the use of artificial coloring matter.'[3] One formula 'retains color', another gives 'a handsome lean-meat tint', another confers 'the much desired rich, deep red color'.[4] At the same time, on the other side of the Atlantic, experts in the German Health Ministry were indicating that the race to provide the reddest forms of processed meat was becoming more intense: 'Artificial colouring in meat products, which first made an appearance fifty years ago in certain parts of Germany, has become progressively more widespread, as a consequence of what we might call the *competition of appearance*.'[5] These colorants came in both powdered and bottled concentrate form. In one year alone, in Hanover, the food inspectorate recorded four new liquid formulae: one was composed of 25% salt, 30% boric acid* and 39% saltpetre, with the rest made up of water and starch; two others were a mixture of aluminium acetate, saltpetre and sugar; the fourth was a solution of salt, sugar and saltpetre.[6] In the space of a few years, health authorities witnessed the emergence of a global market, in which European preservative and colouring agents competed against those invented in the USA. In official reports, page after page was taken up with lists of such products.[7]

Among others, there were: *Viandol* (aluminium acetate + salt-petre), *Securo* (the same with sugar added), *Carniform* (saltpetre + phosphate), *Nadal* (salt + saltpetre + benzoic acid), *Lipsia* (ben-zoate + salt), *Sel Montégut* (salt + saltpetre), *Conservaline* (salt

---

* Like saltpetre treatments of meat, borate-based treatments underwent a rapid development in the nineteenth century. Borax (or sodium borate) is a salt of boric acid. Deposits of the mineral were often found next to deposits of sodium nitrate, to which it provided the perfect complement, as, although its colouring effect is negligible, its preservative power is very strong. Borax was banned because, like nitrate, it produces harmful compounds. Nowadays its use is confined to cleaning products and certain pharmacological preparations.

+ saltpetre + perchlorate). A large number of these preparations included sodium borate or boric acid (brands such as *Antiferment*, *Boroglycin*, *Antisepticum*); others were based on sodium sulphite (*Meat-Preserve Krystall*, *Excelsior*) or sodium bisulphite (*Double Cone*, *Phlordaritt*), or else sodium fluoride or aluminium sulphate. Others combined these ingredients: for example, *Carnit* (a solution of aluminium acetate and saltpetre) or *Eminent* (salt + saltpetre), *Cologne Salt* (salt + sodium benzoate + saltpetre) or *Enfin Trouvé* (salt + saltpetre + sodium phosphate). Nitrate/borax mixtures were very common, as were boric acid/sulphite. There were even meat preservatives whose principal active ingredient was chloroform:[8] no sooner was a chemical product shown to be capable of arresting the growth of cells than some enterprising chemist put it in a bottle in diluted form and sold it as a 'preservative' on the pretext that it was able to destroy bacteria and was harmless, given that it was highly diluted. Even formaldehyde and bleach (sodium hypochlorite) were sold for their 'preservative' and 'anti-bacterial' qualities.[9]

In 1907, a Canadian report[10] accused certain additive manufacturers of combining ingredients in complicated formulae in order to prevent them being detected (such as *Lakolin*, which combined boric acid, sodium sulphite and glycerine); others gave their products reassuring labels, along the lines of 'proven harmless' or 'undetectable in the final product'; others passed themselves off as 'condiments', 'spices' or 'spiced salt' (one *Spice Salt*, for example, was in fact a mixture of salt/sulphite/saltpetre with a vague sprinkling of spices, while one 'salt for spicing meat' was a liquid that contained as much nitrate of soda as pepper).

In Germany, right at the start of the twentieth century, certain products were very upfront about their function as reddening agents – thus the powders *Blutroth* ('blood red') and *Darmroth* ('sausage skin red') or the liquids *Blutrother Fleischsaft* ('blood red meat juice') or *Wurstroth* ('sausage red'), etc. – but colorants were starting to get a bad press: they were increasingly seen as disguise and artifice.[11] That is why they were often sold as 'preservatives', even though all they actually 'preserved' was the colour. In contemporary toxicology documents there are dozens of damning reports of these 'so-called preservatives'. Thus *Borolin Sausage Preservative Salt* or *Sausage Preservative Spice*, which were nothing more than mixtures of sugar, salt, sodium borate, saltpetre and sulphate of soda, designed to colour sausages, saveloys and salami.[12] Likewise, in 1907, a toxicologist of the German Health Ministry described in detail different 'pseudo-preservatives' (*sogenannter Konservierungsmittel*) that had appeared on the market and pointed out that the instructions for use spoke only of the colouring action and that the dealers themselves admitted that they had no other purpose than to give the meat a red colour.[13] The doctors noted that *uncoloured* meats were disappearing, because the consumers thought they were of poorer quality: driven by market forces, all meat products were turning pink or red, and the public didn't even notice any more. Unable to curb this red tide, one toxicologist sounded the alarm: 'By conducting these completely unjustified experiments we are playing with the health of the consumers.'[14]

From the end of the nineteenth century there was a consensus among observers that coloured meats were becoming the norm – even in towns with long-standing meat-curing traditions. In 1897, in Wrocław (Breslau), the Polish town where they make the venerable *kiełbasa* sausage, the director of the health inspectorate laboratories took samples from the markets and wrote: 'The preservative substances are not used only on fresh meat; latterly, they have also been applied to sausages.'[15] He talked about the artificial colouring of sausages and salamis as a real epidemic. Meats falsified and treated with sulphuric acid were so widespread that 'sausages made in the proper way are increasingly rare these days', to the extent that 'there are only a few regions in Germany where they still know how to make them'. And in conclusion:

'The only way is to inform the public. At the end of the day, it would help our producers to gradually wean ourselves off the considerable imports of foreign salamis and to devote as much attention as possible to production.'[16] Three years later, the same author stated that chemical colorants had made further advances: all the sausages and salamis that reached Wrocław in 1900 had been coloured using additives that were detectable only under a microscope.[17] In the scientific journals, there was a litany of articles on the same theme: researchers conducted experiments either to identify the additives and work out what they were made up of,[18] or to develop testing techniques,[19] or to denounce the ever-increasing list of reddening agents of all types.[20]

In 1908, the British medical journal *The Lancet* criticized the public authorities for not taking the necessary measures to protect consumers against producers who added preservatives even in sterilized products (which rather indicates that their true use was as colorants). *The Lancet* wondered whether the guilty parties were primarily food additives merchants, who never missed a chance to trumpet their products: 'these antiseptic materials are often sold under fancy names, with nothing to guide the purchaser as to their composition, and are often accompanied by a sort of guarantee to the effect that their use in accordance with the instructions given will not entail liability to prosecution. There can be little doubt that the great saving in trouble and material which is held out as an inducement to users of these substances and the harmless, not to say beneficial, effects which they are stated to have on those who consume them in food have had weight with a certain class of manufacturers of preserved meat.'[21]

THE RISE OF THE MEATPACKERS

Of all the historical causes behind the normalization of preservative and colouring agents, one phenomenon in particular played a major role: the nineteenth century saw the development of meat-processing on a large, industrial scale. In France the initiatives were at first quite modest (in 1829, one observer expressed regret that production was still rather under-developed),[22] but industrialization was beginning to happen elsewhere in Europe: in northern Germany (the great smokehouses of Hamburg) and in Ireland. In the second half of the seventeenth century, English farmers had been successful in getting a ban imposed on animal imports from Ireland.[23] Forced to seek new outlets, the Irish began to slaughter their own animals and export their beef and pork in processed form, especially in Munster – so much so, in fact, that Ireland very soon acquired a quasi-monopoly: these preserved proteins provided an essential food resource for privateers and mariners, and especially slaves in the European colonies.[24] The Irish producers conducted all sorts of experiments to speed up the production process and extend preservation: the victualling commissioners were 'desirous of leaving no experiment or method untryed'.[25]

Building on the success of Irish meat curing, a number of ham and bacon manufacturers sprang up in Cork, Waterford, Limerick and in several other towns.[26] But the real turbo-charge of development came from the other side of the Atlantic. In America, the meatpackers – essentially the first meat-processors of the industrial age – were on the rise. Strictly speaking, however, the meatpackers were not meat curers: they were 'conditioners', 'packers' and 'transporters'.[27] In the eighteenth century, a few small abattoirs in Massachusetts had started to pack salted meats in barrels, for use on long sea voyages, in isolated trading posts, and even for sending to a few local traders.[28] Progressively, this activity moved further west. In Cincinnati they were soon slaughtering so many pigs that the town was informally known as 'Porkopolis'. Hundreds of traders were packing pork in barrels, chests or enormous casks.[29]

Meat is a delicate, spoilable substance and so could only be processed in winter, when it was so cold that it turned the whole outdoors into a gigantic fridge: according to *Harper's Weekly*, the slaughtering season lasted from November until March.[30] Once the temperature touched zero, all roads into Cincinnati were filled with herds of beasts being brought in by farmers. The animals were slaughtered in dozens of sheds built along the Ohio River. There, the carcasses were piled onto carts to be sent to the curers. It was a race against time: as much meat as possible needed to be salted before the warm weather returned. If temperatures rose in the middle of the curing season, the workshops ground to a halt. In any case, the season ended in March, when the ice began to break up on the Ohio River. In summer, the abattoirs were deserted. The meatpackers then concentrated on getting the best price for the previous winter's stock.

In Europe too, production of cured meat was strictly limited to the winter months. In the Cantal region of France in 1826, for example, a report noted that the pork and sausage trade took place 'from the end of October to the end of February'.[31] Ham was produced using only salt.[32] In the Lyonnais mountains, where a few dry sausage (*saucisson*) factories appeared after 1860, it was also a strictly seasonal activity: it provided a supplementary income to farmers, who could not in any case be working in the fields in winter.[33] At the same time, in Ireland, the production of salt pork and salt beef also followed a strict seasonal pattern: pigs could only be slaughtered in months containing the letter R,[34] because, as one British expert put it, 'It is never wise nor safe to cure bacon from the end of March to the beginning of October. It is more often attempted than accomplished, as the weather conditions in summer are wholly against successful curing.'[35]

But in the far north of the USA, near the Great Lakes, the rules of the game were about to change. Shortly before 1850, in Milwaukee, a bold entrepreneur, John Plankinton, constructed the first plant – a wood and brick building – where it was possible to both slaughter pigs and pack them in barrels to feed the navy.[36]

Soon, others followed in his wake, such as the Englishman John Layton and the Cudahy brothers, born in Ireland, descendants of butchers who had fled a war-torn and famine-afflicted Europe. Within a few decades, they would completely revolutionize the ways in which pork was preserved and thus lay the foundations of a whole new methodology: they mechanized and reinvented 'the system which packs 15 bushels of corn into a pig, packs the pig into a barrel, and sends him over the mountains and over the ocean to feed mankind'.[37] During the American Civil War (1861–65), the demands of the military fuelled a phenomenal expansion of meatpacking. During 1860, Milwaukee processed 51,000 pigs. Barely two years later, that figure had risen to 182,000.[38]

Eighty miles south of Milwaukee, another city of abattoirs was starting to make its presence felt. Chicago stood at the cross-roads of the Great Plains, the salt mines, huge reserves of ice (the Great Lakes) and railways connecting it to large centres of population. In a few decades, its own population exploded: from 4,500 inhabitants in 1840, the city reached 112,000 in 1860; half a million by 1880. This figure had tripled by the end of the century. Chicago pushed its way to the forefront of the meat-processing towns, producing twice the output of Milwaukee and leaving Cincinnati trailing in its wake. The Armour company of Chicago alone produced more ham and bacon than the dozens of factories in Cincinnati.[39]

NO SEASONS IN CHICAGO

In 1870, more than 90% of Chicago's processed meats were still produced in winter and the very end of autumn.[40] It was in 1866 that an abattoir first found a way to control the cold.[41] The technology was still in its infancy: in winter, blocks of ice were cut on the frozen lakes and kept until the summer in warehouses insulated with straw or sawdust. The rooms in the abattoir were fitted with compartments (between the walls or the floors) in which blocks of ice were placed to lose those precious few degrees. A few decades later, one of the head technicians recalled: 'Some plants were built entirely of wood, while others had brick walls with interior structures of wood. Ice houses for the storage of natural ice in many cases occupied more than one-half the area of the plant. The insulation of the walls of refrigerated buildings was accomplished either by building parallel brick walls, each from one to two feet thick, with an eight- or ten-inch air space between them, or by furring the inside of the wall, sheathing with wood, and filling the space so formed with dry wood shavings or sawdust.'[42]

But ice melted and sawdust absorbed the water, so the meatpackers tried using cork instead – in the form of sheets, bricks and granules. These techniques turned the floors of the packing houses into unhygienic sponges, soaking up damp and oozing filthy water.[43] Nevertheless, such were the commercial pressures that all meatpackers with an eye to the future adopted this rudimentary form of refrigeration. Within a few years, production was all year round: instead of a window of around 100 days, cured meat could be produced for 365 days a year. This was the start of 'summer packing',[44] also known as 'ice-packing'[45] or 'all-the-year packing'.[46] Output increased at an almost unimaginable rate. In the Midwest alone, production quintupled in five years: the number of pigs slaughtered annually rose from 495,000 in 1872 to 2.5 million in 1877.[47] And this was only the beginning: as icing techniques improved, packing outside the cold season soon outstripped that done in winter. When the first refrigerators proper appeared, the producers acquired ammonia refrigeration units. The process was expensive but quickly became profitable – eventually massively so. Thus, from 1892, 70% of the meatpacking in Milwaukee took place from April to November, during the hot months.[48] Giant factories sprang up in other cities: Kansas City, Wichita, St Louis, Omaha, Sioux City ...

Curing techniques were transformed too in order to improve efficiency. Previously, hams and slabs of bacon were rubbed with salt and then heaped one on top of the other in alternating layers of salt and meat, either in barrels or in piles. Saltpetre enabled a much shorter production process: this was either rubbed directly on the flesh or inserted inside the ham by slicing it open and pushing it in using a rounded stick.[49] All-the-year packing required greater speed: the saltpetre mixture was no longer applied by rubbing, it was injected. From dawn to dusk, workers would cut the meat and 'treat' it as quickly as possible by inserting giant syringes attached to brine tanks.

MIRACLE SYRINGES

Injection (the technical term is 'pumping') came to be used all over – in the USA and in Europe. A French manual from the time suggests that this process reduced the length of production by two-thirds.[50] Injection is not strictly speaking an American invention. As early as 1805, the Dane Erik Viborg, a vet in Copenhagen, described a procedure whereby a saltpetre brine was injected with the aid of a funnel.[51] Then a Scottish surgeon introduced a new system that consisted of connecting a pump to the whole animal immediately after it was slaughtered.[52] This method underwent a number of subsequent variations,[53] but the injection process only really took off when the French mortician* Jean-Nicolas Gannal got involved.[54] The *Annales d'hygiène publique et de médecine légale* (Annals of Public Hygiene and Legal Medicine) recounted: 'In May 1841, the late Gannal proposed a method of

---

* He ran a business in Paris that embalmed bodies after death. In a famous book, *Histoire des embaumements* (History of Embalming), he described the experiments in which he tested a whole range of chemical products until he came up with the ideal formula for preserving corpses: a mixture of two parts alum, two parts cooking salt and one part saltpetre.

injecting meats to preserve them, the same injection method he applied to preserve corpses.'[55] In *The Preservation of Food*, the American Goodrich Smith confirmed that Gannal 'had already acquired a name by a method of embalming corpses. Afterwards, he employed himself in preserving meat.'[56] A shrewd businessman, Gannal used a separate company to promote the procedure he had invented for curing hams, so that the public didn't associate his meat-injection methods with his former activities.

Then, in 1877, the New Yorker John Alberger came up with a system where a saltpetre brine was prepared in a raised reservoir and injected into the whole pigs just moments after slaughter via a tube inserted into the heart.[57] According to its maker, this system reduced the production time by half. Another advantage: the product gained weight, whereas in traditional curing it loses weight. As it was designed to treat whole animals, this procedure was flawed because of its lack of flexibility, and it was superseded by equipment that allowed each piece of meat to be treated individually. All sorts of different models appeared, from the large industrial pump capable of injecting hams continuously, connected to a steam engine, to the small portable apparatus fitted with a pump that was worked by hand or foot.[58] At first, veterinarian hypodermic syringes were used, but later the manufacturers designed a range of specific syringes. Some still preferred injecting saltpetre in solid form using a spring mechanism, which they considered more efficient than the other 'instruments of vaccination'.[59]

THE 'DISASSEMBLY LINE' REVOLUTION

In 1868, abattoirs were already equipped with an overhead rail which was used to transport pig carcasses from one end of the plant to the other.[60] But once meatpacking ceased to be a seasonal activity, investment in equipment intensified.[61] Certain jobs in the abattoir were almost totally mechanized, while others were divided into separate, more specialized roles, which allowed increased production rates using less and less qualified staff. Meatpacking was inherently about speed and so it reinvented every stage of the process, adapting the older Irish procedures and rationalizing each separate operation. Anything that could be motorized was speeded up: the meatpackers invented the moving carcass line. At the start of the chain was the terrifying revolving pig wheel (or hog wheel), to which live pigs were attached when they entered the abattoir, and what followed was a veritable 'disassembly line', described thus by one French observer: the pig was 'fastened by the rear foot, hauled into the air, killed, bled dry, scalded, scraped, disembowelled, decapitated, split in two and deposited in a refrigerated chamber all in the space of between ten and fifteen minutes'.[62]

The trimming of the hams, the injection of the salt/nitrate/borate solution, the flattening of the bacon, the mincing of the flesh, the cleaning and stuffing of the sausage skins, the smoking, the slicing, the packing: in the course of two decades, all these manual tasks were mechanized. Within a short time of first entering the meatpacking plant, the pigs were, effectively, pulverized into hundreds of different products. One contemporary observer was highly enthused: 'From the slaughtering room to the refrigerator chamber, all the time, there is a continual stream of pork passing, no hog pausing more than a few seconds in any one place [...] So perfect are all the labour-saving appliances throughout this portion of the building that the carcass of the hog never once requires lifting or moving by manual labour.'[63] Though originally invented for pigs, this 'disassembly line' would later be adapted for cows (it required 157 men and 78 separate operations to 'disassemble' a cow).[64] Legend has it that it was from observing this process that Henry Ford had the idea of applying the principle in reverse to his automobile production line.[65]

The American example was followed in other countries. When in 1895 an official fact-finding mission was sent from Melbourne to examine options for increasing Australian agricultural production, the reporter was amazed by what he saw. 'The enormous number of hogs arriving daily at the larger stock-yards is a sight to witness. At Chicago, on a single day, 66,000 hogs have been delivered.'[66] With a mixture or fascination and terror, he described the Armour factory, the largest meat-processing plant in the world, from the constant flow of pigs coming in at one end to the barrels of ham and cartloads of bacon and sausages leaving at the other.

SALTPETRE, THE DNA OF MEATPACKING

Salted cabbage needs six weeks of fermentation to become sauerkraut, the time it takes for the flora of lactic bacteria, which transform its texture and flavour and ensure its preservation, to develop. Similarly, it needs a certain amount of time for the must of the grape to turn into wine, for milk to become Cheddar cheese, Gruyère or Gouda. Ditto for Parma ham and salami made without nitrate or nitrite: it takes months for the enzymes to complete their work. Months of close attention, with no guarantee of success, where bad weather could ruin everything if the artisan doesn't take the correct measures, or an error can lead to a lack of colour, a poor taste or unwelcome mould. The artisan has to keep an eye on the produce, examine it regularly, manipulate it, adjust the temperature, control the humidity ... and all the time that the products are maturing, there is no money coming in.

The Chicago system, on the other hand, was premised on a rapid turnaround and a permanent acceleration of the production process. A long maturation is well suited to the seasonal production cycle, not to continuous output. Short of filling the city with drying houses, how could the producers of Chicago possibly find the time for a proper maturation and a natural transformation of myoglobin? Low ambient temperature is imperative for traditional curing, but who would think of maintaining huge drying kilns at a temperature of 5°C or 6°C all year round on the plains of the Midwest, when from April onwards average temperatures rise above 15°C and then climb to 28°C, with peaks as high as 40°C? Who would contemplate keeping hams and sausages for nine months to a year in air-conditioned, ventilated rooms when you can get the same results using nitrate, in a fraction of the time, without worrying about keeping the place cold?

American publications suggest that nitro-additives were also used to combat certain bacteria which thrived in insanitary factories, identified as *Bacillus putrefaciens* and *Bacillus foedans*.[67] Virtually non-existent in traditional production, these bacteria were very common in industrially produced hams, where they were transmitted via the handling equipment, the thermometers that were inserted into the meat, and the syringes used to inject the brine. Badly maintained tubes, pumps and syringes were perfect breeding grounds for polluting micro-organisms.[68] In the jargon, these hams were known as 'sour hams' or, more often, 'stinkers': the bacteria produced a smell so repulsive that the meats were no longer fit for sale. These stinkers posed no threat to health, but they were uneatable and had to be rejected.

Until the closing decades of the nineteenth century, curing salts were rubbed into the meat by hand, without use of syringes. That is why the problem of stinkers was virtually unknown in the curing workshops that followed traditional methods: in 1911, a bacteriologist from the US Department of Agriculture showed that stinkers were caused by accelerated methods of industrial production, lack of due care, and non-compliance with asepsis regulations on the production line.[69] He asked the factories to improve their hygiene and recommended regular disinfecting of the syringe needles. A military report of 1926 made the same point: 'It is interesting to note that the percentage of sour meat found in the various establishments is almost in direct proportions to the care and sanitary methods used in all the processes of handling meat from the killing floor to the smoke-house. These establishments that observe the greatest sanitation in all of these processes suffer much less from sour meats than those establishments where sanitation is lax.'[70] Rather than improve hygiene, the meatpackers preferred to control the spread of microbes through recourse to antiseptics: 'it is an essential curing agent and [...] a large percentage of sour meats (in the trade sense) would result if its use were discontinued, it cannot be successfully controverted', wrote one meatpacker in favour of saltpetre curing in 1907.[71]

In short, nitro-additives were fundamental to the Chicago meatpacking model, they were part of the logistics; they enabled the meat to be 'treated as conventional inorganic raw materials', as *The Lancet* put it in 1905, in an article criticizing the insalubrity of meatpacking[72] – in other words, like an inert material such as wood, sand or metal. What would have become of these giant factories of 1880 or 1910 if they had been deprived of their bactericidal reddening agents? Without nitrate, we would have had nothing on the scale of the Armour factory in Chicago, where, 365 days a year, between 5,000 and 6,000 men, women and children from the poorest parts of town killed and emptied tens of thousands of pigs to transform them into hams and sausages and send them out as quickly as possible to all corners of the world.

CHAPTER 4

RAPID CURING TAKES OVER THE WORLD

To fully understand the history of modern processed meat, we need to remind ourselves of an often forgotten fact: at the start of the twentieth century, when automobile production was in its infancy, meatpacking was one of the most important industries in America. And it was an export industry. A second fact that is not so obvious to us these days: just like cotton or wheat, salted meats (hams, bacon, salami, etc.) were the stuff of intense international commercial competition.[1] That is why the invention of all-the-year packing revolutionized meat-processing – not only in the USA but in all the producing countries: everyone had to keep up with the pace of this revolution. How could you continue to take nine to twelve months to make a ham when your competitors were able to make them in 90 days? Thanks to its technological innovations, Chicago proclaimed itself the 'hog butcher for the world'.[2] In 1900, the number of pigs processed through American slaughterhouses each year passed 50 million, and 40% of the world's pigs were in the USA.[3] As well as being 'Europe's granary', the USA became the leader in processed meats.

MEATPACKING AS AN EXPORT INDUSTRY

At first, the American meatpackers concentrated on the British market, where they were in competition with the Irish curers.[4] But from 1865 onwards, American products were finding their way into markets across the world: they supplied the Antilles, Cuba, Haiti, South America, then the whole of Europe.[5] As the main producer of pork meat in Europe, France seemed at first sight to be more a competitor than a market. But very quickly France too was importing US bacon, ham and canned meat in particular,[6] and later salamis and sausages. This expansion of business was helped by the fact that, as time went on, American products improved in quality: in the first half of the nineteenth century it was mainly bottom-of-the-range salt pork aimed at the poorest in society, labourers and slaves in the southern plantations.[7] But the American meatpackers moved into higher-value products,[8] abandoning mess pork in favour of more refined fare, with the less favoured cuts disposed of in the domestic market.[9]

For the working classes in Europe, American bacon, hams and salamis were a godsend in a period when the price of meat was constantly increasing.[10] The wave of American processed meats was competing directly with local production, and between 1850 and 1880 American exports of bacon increased twentyfold. In a report presented a few years later, the head of markets at the US Department of Agriculture was stunned by the incredible volume of bacon coming out of the factories: 'By 1854 the exports amounted to 46,000,000 pounds; the quantity was over 200,000,000 pounds in 1863, nearly 400,000,000 pounds in 1873, nearly 600,000,000 pounds in 1878, and 760,000,000 pounds in 1880, the largest amount ever exported in one year.'[11] And the quantities continued to increase, as American bacon was about half the price of bacon produced locally in Europe.[12] In 1899, one Chicago meatpacker declared to a Senate hearing: 'That business has grown tremendously. The city of Liverpool alone will take from 18,000 to 20,000 boxes of our bacon weekly. Twenty-five years ago they wouldn't take that much in a year.'[13]

'American speculators can furnish the markets at much cheaper rate', the *Irish Farmer's Gazette* fretted in 1863.[14] Faced with this competition, British and Irish producers tried to shorten their production schedules and brought their procedures into line with those followed by the Americans. The Irish Bacon Curers' Association introduced use of the injector syringe,[15] and a company in Waterford obtained a patent for a device for cooling brine.[16] Two years later, in 1864, they developed a complete refrigeration system, which managed to 'make the curing house as much as possible unsusceptible to the action of atmospheric influences from without'.[17] A few decades later, the Scottish specialist Loudon Douglas took stock of the transformation: 'In times not long past, for one to talk of the summer being equally as favourable a time as winter for bacon curing, a lunatic asylum would have been suggested for a change of air and environments, and yet bacon curing is now carried on in many parts of the country all the year round.'[18] The use of chemical preservatives became widespread. In a piece entitled 'How to cure bacon in summer', Douglas in his *Receipt Book* (1896) listed three indispensable ingredients: 'dry antiseptic' (a mixture based on borax), potassium nitrate (saltpetre) and saltprunelle.[19] He went on to say: 'This brine will effectively prevent the bacon from decomposing, and will at the same time preserve the bright redness of lean desired when the sides are cut and exposed to the air.'[20] Artificial smoke was also a feature (in the form of 'smoke powder'),[21] as were brown colorants which mimicked the outer appearance of smoked meats; the professional manuals showed in great detail how these artifices should be applied.[22]

Bacon and ham produced by traditional methods started to become scarce in the London market; their distinctive taste began to acquire negative connotations.[23] In 1910, one journalist lamented the disappearance of traditional British curing, which 'consisted of packing cuts of meat into salt or immersing it in the brine produced by previous curings'. He went on to explain: 'This was how it was done in the farms of Cumberland, Westmorland and Yorkshire, and even now there is a preference for the products of these mountainous regions, and it is not unusual to find hams in Yorkshire with a special aroma that cannot be achieved in the modern industrial establishments. It is a matter of regret that curing on the farm is disappearing and that the old formulas are being replaced by the more scientific methods widely practised in the new factories, but which have one flaw, which is that they impart the same taste and the same aroma to all the products indiscriminately.'[24]

FRENCH ALARM

Statistically insignificant until the 1870s, American processed meats made deep inroads into the French market within a few years. In 1874, France tried to raise protectionist barriers by increasing tariffs. The question of American cured meats became a matter of lively argument in parliament: should they be banned in order to protect French producers? Was it wise to restrict imports and risk customs reprisals? Should free trade be curtailed at a time when France was short of meat and the workers were the main beneficiaries of meat products at knock-down prices? The historian Alessandro Stanziani describes how in 1876 a certain Guyot, a member of parliament from the Rhône region, tabled an amendment against protectionism, because cheap American cured products put 'meat' within the reach of all budgets.[25] In the end, cured meats from Chicago continued to arrive in increasing quantities, forcing French producers to respond to their cheaper prices. In 1882, a professor at the French College of Agriculture,

Alphonse Gobin, sounded the alarm: 'From 1856 to 1879, imports more or less quintupled: at the beginning of the period, they were the equivalent of around 70,000 heads of pig; in the end, the figure was nearer 360,000. In the same period, exports fell from approximately 35,000 to 17,000 heads of pig.'[26]

At the end of the 1870s, there were several cases of trichinosis in Europe. This is an illness caused by parasitic worms that cross from pigs to humans. Processed meat from Chicago was accused of being 'literally stuffed' with them.[27] Was this true, or was it rather a panicked reaction to the increasing influx of American products?[28] This alert was exploited by the supporters of protectionism,[29] to such an extent that from the beginning of the 1880s the import of American meat was banned in several European countries. American exports went into freefall. But the establishment of minimal norms and the creation of an official US inspectorate allowed the trichinosis problem to be brought under control.[30] After ten years of embargo, most barriers were lifted. France still maintained some protections through import tariffs,[31] but American exports continued their march towards world domination.[32] When France planned to ban certain American processed meats, America threatened to retaliate by blocking imports of French wine.[33]

Meat producers everywhere were forced to compete by making their products less and less expensive and more and more attractive, by reducing use of salt and always applying potassium nitrate, sodium nitrate or sodium borate. Even when its use was banned in the American domestic market, borate was allowed in products for the export market.[34] Caught up in a tangle of trade negotiations and reciprocal bans, the French authorities tolerated the import of hams treated with sodium nitrate and sodium borate (even though these were banned in France), which led French charcutiers to demand a level playing field.[35] One professional review complained in 1895: 'There is nothing but butcher's meat and

71

cured meat coming to us from New York and Chicago; even the army supplies itself with US tinned meat. Why? Why from America rather than France? The reason for this choice, the only excuse, is to do with budget: American tins cost 145 francs per quintal, while those produced in French factories cost 210 francs.'[36] A few years later, the official journal of the US meatpackers could confidently assert: 'The American packinghouse trade has been the pioneer in this wonderful movement, and American packinghouses always have been and are today the models for all the world. If foreign countries desire to compete with the United States in the meat trade they have to adopt American methods, they have to learn from American teachings. That is why America stands foremost in the meat trade.'[37]

EUROPE UNDER SIEGE

In 1888, the Frenchman Théodore Bourrier, chief veterinary inspector for meat, devoted a whole chapter of his book on pork to meat products from Chicago.[38] For Bourrier, their unique selling point was their cheap prices, and he laconically described US hams as 'monsters'. In 1892, another observer, Alfred Picard, described the ups and downs of tariff-based protectionism against transatlantic meat imports and stated that 'The United States sends to Europe enormous quantities of bacon, ham and lard'.[39] In the eight years between 1890 and 1898, the quantity of American exports tripled again.[40] They even managed to break into countries which had customs protections in place. The French consul in Gdansk, a major commercial port on the Baltic, described the ruses by which Americans infiltrated their products into the German market: for example, 'a meat curer in Amsterdam or Rotterdam presents to the relevant authorities 10,000 pounds of Dutch bacon and gets in return a formal document confirming that the merchandise is genuinely of domestic origin'. But instead of exporting it directly, he hands over the consignment to an importer of American cured meats, who 'substitutes his own merchandise in the cases originally containing the Dutch produce',[41] then sends it to Germany. Then the whole thing would start over: the Dutch producer would reclaim his products, obtain an export certificate, have them replaced by American merchandise, and so on. Even factoring in transportation costs, the margins were enormous.[42]

Similarly, the French consul in Belfast alerted his government to 'the considerable quantities of American pork and ham which are exported to France as Irish pork or ham'.[43] He quotes from the Belfast commercial press: 'Ninety-nine times out of a hundred, the hams prepared and sold here, and even more so exported as *genuine Irish* & *real Wiltshire*, are falsely labelled. The vast majority of these pork products and hams are sent to us under this labelling by the Americans.'[44] The consul concluded: 'It should be added that we were told by an Irish businessman of the utmost respectability that it was almost impossible for the people of the country themselves to procure real Irish ham.'[45] An American commercial agent said: 'Merchandise from all lands is stored on the English docks; from this store the continent is supplied. Many goods manufactured in America have been sold as English products and marked as such.'[46] He explained that in France, in a wholesale market, he was offered 'Hams, shoulders and bacon, called genuine "Yorkshire"', all of which, 'upon close examination, disclosed the word "Chicago" stamped on them. The mark was faint, having been almost effaced from the meat.'[47] In 1904, the *Manuel de l'épicier* ('Grocer's Manual', written by Léon Arnou, the former president of the Grocers' Association) stated that American hams also passed through Hamburg, from where they were re-exported to France under a German label: 'All the hams that come from America are prepared and sold as hams from Mainz, Westphalia or Hamburg.'[48] It was the same for hams sold as 'York hams'. According to the *Manuel de l'épicier*, these were often American

73

hams which were 'reworked' on their arrival at Liverpool, where they were 'prepared in England in the manner of York hams and sold to us as such'.[49] In British professional journals, there are even precise instructions on how to 'rework' American products. For example, certain manuals advise not subjecting American ham and bacon to normal smoking, because they would lose weight; the *Manual of Pork Trade* recommended artificial smoking.[50]

In the European market, only one country succeeded in competing effectively with the United States: Denmark. Throughout the nineteenth century, the Danes had been exporting pigs fed on the whey produced by the country's dairies. But in 1887, their main customer, Germany, closed its borders to live animals. Denmark instituted an emergency rescue plan for the pork industry. Instead of exporting pigs, the country equipped itself to export bacon. They had one single market in their sights: Britain. In the space of two years (1887–89), the Danes managed to build twenty industrial abattoirs, organized as cooperatives; the number would grow to around 30 by 1905, supplemented by 24 private abattoirs.[51] Like the American factories, these Danish plants were geared towards foreign markets: in 1900, half of their total production was exported, and more than three-quarters of this went to the UK.[52] The Danish industry used the most modern, efficient and economical methods – the ones developed in the USA. Something similar happened in the Baltic states: in the 1920s, a French journalist was surprised to discover a large abattoir in Lithuania that exactly reproduced the technology developed in Chicago.[53] Likewise in Holland: when the public abattoirs were modernized, it was the methods of the American meatpackers that were adopted.[54]

Isn't that how liberal capitalism works? The most economical techniques deliver a competitive advantage and the most attractive products sell the best. One example illustrates how the extraordinary competitiveness of the US meatpackers led all the other

countries to adopt American techniques. At the end of the nineteenth century, Germany was worried that it had fallen behind
in the manufacture of tinned meat. Easy to transport, store and
prepare, this was the ideal source of food for soldiers in the field.
Military experts were emphatic that, if war were to break out,
foreign enemies would have a strategic card up their sleeve since
they could call on food supplies that Germany lacked.[55] In 1907,
a long study published in Jena analysed the prospects for the meat
industry. The author, Carl Wagner, remarked: 'In Europe we so
often admire the extremely low price of American corned beef.
This price can be explained. The main factor is the cheapest meat
in the world, and production takes place in units which are integrated into massive factories, where the means of production are
tuned to optimal efficiency. There they use the most well-adapted
machines and the most highly tested equipment. In these conditions a satisfying and regular production can be obtained using a
relatively small, unqualified workforce on low wages.'[56]

Wagner observed that the use of chemical additives by the
American meatpackers extended the life of the goods but above
all simplified production to the maximum. He noted that food
additives gave corned beef a correct appearance even if the meat
had begun to degrade. Saltpetre, while not exactly essential,
served, according to him, to give the meat 'a fine colour all the
way through'.[57] He explained that in Germany, producers were
able to do without chemical preservatives, which were 'not at all
necessary'. But he continued: 'However, if you don't want preserved meat to go off rapidly, it requires a lot of effort, since
you have to carefully follow the rules. On the other hand, the
use of preservative salts minimizes the risk of deterioration.'[58]
Consequently, in view of the competitiveness of American products, Wagner reckoned in 1907 that it was economically unviable
to produce preserved food using German meat. And indeed, in the
years that followed, German imports of American processed meats

increased to record levels, and these imports were directed primarily to German regions with large working-class populations.[59]

THE AGE OF THE IMITATION

Originally, pork products were particular to their geographical regions. Whether it was 'Parma ham' or 'Bayonne ham', 'York ham', 'Suffolk ham' or 'Bradenham ham',[60] bacon from Westmorland or Cumberland, Nüremberger or Frankfurter sausages, each meat product was influenced by the peculiarities of its location, its culinary tradition, its unique agricultural heritage. But in 1908, the US meatpackers' press rapturously announced: 'Sausage to-day in many houses is made by the carload, day in and day out, not by the few hundred pounds, as in the early days. Popular brands of foreign sausage are being copied and surprisingly successful results obtained. There are factories in the United States to-day which turn out every brand of sausage and kindred products known to the trade in every part of the world, and which cannot be distinguished from the original. In fact, certain kinds of sausage are made here and sent abroad to be sold in competition with the native article and connoisseurs have been puzzled to distinguish any difference in appearance, taste or quality.'[61] One example out of dozens of manufacturers: in Chicago in 1895, Luetgert & Co. occupied a large, five-storey building. In their advertising they described themselves as: 'Manufacturer *of all kinds* – German, Italian and French Sausage'.[62] It was as if one single town-centre factory made Stilton, Brie, Gruyère, Swiss cheese, Roquefort, feta, etc. all under one roof. Additives were part of the box of tricks that made this magic possible: whether it was hams, sausages or salamis, the Chicago manuals described the processes for making products sufficiently close to the original that you couldn't tell the difference, unless you were a chemical expert. The *National Provisioner*, for example, described how to

make a 'Bayonne ham',[63] and as early as 1904 the French authorities were finding American 'Bayonne hams' even in the port of
Bayonne itself.

In Europe, a few larger companies were starting to adopt the
same techniques. The case of the French firm Olida is particularly illuminating. When it was founded in 1855 in Paris, it was a
supplier of gastronomic products for a well-off clientele that was
partial to fine dining. Every year, Ernest Olida, the founder, went
in person to York and Limerick to order the best hams of England
and Ireland.[64] Trading under the name 'The House of York Ham',
Olida quickly built up an excellent reputation. But from 1870, the
firm increasingly sold American hams, passing them off as British
products. In 1895, they changed strategy: the Olida family had a
factory built near Paris to produce meats in situ. Olida supplied
France with 'York ham', 'Westphalian ham' and 'Prague ham', all
produced at Levallois-Perret.[65]

Apart from fake York ham, the French adored American
'saucissons' (dry-cured sausages), which were systematically
treated with nitrate, borate and sulphite: from 1901, when the
category 'sausages' first appeared in records of the American
administration, France was by a long way the principal importer.
Whereas France bought six times less American ham than the
UK, it imported three times as many sausages and salamis.[66] In
the traditional meat-curing regions, workshops closed or went
into decline. Those that survived had little choice but to adopt the
quickest production techniques.

With the coming of the First World War, the conquest was
complete. Exports of American processed meats rose in leaps
and bounds year on year. The annual profits of the five largest
meatpackers rose from a little over $20 million in 1914 to nearly
$100 million in 1917.[67] After the war, national producers picked
up the baton. In France, this period saw the founding of a number
of industrial meat-processing companies. Taking its inspiration

from America, this 'charcuterie industry' set out to harmonize both the taste of the products and the techniques used to produce them: specialists in the 'American method' set up in Paris and popularized modes of production that had been honed on the other side of the Atlantic. French professional publications such as *L'Alimentation moderne et les industries annexes* [Modern Food Production and Allied Trades] reprinted recipes from the *National Provisioner* and adapted them to suit French tastes, sometimes indicating that a certain recipe was 'à l'américaine',[68] but often not mentioning it at all. For the first time, use of nitrate of soda was described in French publications, alongside potassium nitrate.

As for the US meatpackers, they were celebrating the victory: in 1924, the head chemist of one of the largest American factories claimed that 'The modern ham turned out under the guidance of chemical laboratory control is a standardized product of great uniformity. The myth of the old country-cured ham is rapidly passing away. That product, which was generally over-salted, not uniformly salted, over-dried, and over-smoked, could not compare in delicacy of flavor or succulence with the modern packing-house product.'[69]

A FORGOTTEN METAMORPHOSIS

The Chicago system was a forerunner of Fordism. It produced foodstuffs so homogeneous that it was possible to achieve standardization. The meatpacker Oscar Mayer said that his factories manufactured products 'of such uniformity and permanence' that nitro-salted meat could be traded daily on the Chicago commodity exchange.[70] A handful of intrepid entrepreneurs made vast fortunes out of this new industry. At the beginning of the 1920s, the Federal Trade Commission conducted an enquiry that showed that a small group of producers had become incredibly rich; they were known as the 'Big Five'. Their margins were tight, but volumes alone ensured gigantic profits.[71]

It was quite a transformation! In the middle of the nineteenth century, the Frenchman Denis-Placide Bouriat, a professor of pharmacy and an expert in public health, was lamenting that food preservation had not yet caught the attention of the new entrepreneurs of the industrial revolution. In 1854, he remarked: 'The strange thing is that speculation on the part of the great capitalists, which has invested in such a diverse range of objects, should remain a stranger to one that offers a way to double their capital, extend their markets and serve mankind.'[72]

In less than 70 years, Bouriat's call had been heard: the preservation of pork had effectively become a major 'modern' industry, and a highly profitable one to boot. This new and ultra-competitive branch of manufacturing industry fixed the rules of the game for the global processed meat industry. This 'pork remade' (to employ the expression of Roger Horowitz)[73] was so economical that it wiped out European production and consolidated the use of reddening agents. The meatpacker Oscar Mayer wasn't exaggerating when he said that 'Pork packing is uniquely an American industry. It is the product of American brains and energy, built without much precedent from Europe. The American pork packer, and especially the larger packer, has rendered a service which the public is just beginning to recognize. He has built the vast organization, and supplied the marvelous and original technique by which these highly perishable products are distributed in fabulous volume over land and sea through all climates, into every corner of the globe.'[74]

In Europe we pride ourselves on our gastronomy and cured meats, and so tend to forget about these decades of change, the nitro-sausages flooding into the ports of Le Havre, Hamburg and Liverpool, the flood of 'new hams' that swept through markets and kitchens. We like to think that the only impact of American meatpacking has been the novelty of corned beef on Italian or French tables.[75] In fact, European meat curing is now thoroughly retuned to the transatlantic model. Ultimately, the rules, tools, ingredients and machinery of our new industry were established in the USA between 1870 and 1900. It was in America that European products were reinterpreted, 'modernized' and accelerated before being returned to their countries of origin, where, by the contagion of competition, they forced the recipes to be rewritten. These nitro-meats were cheap and they were *beautiful*. In their deep-red homogeneity, the new raw hams were often thought to be more attractive than the old ones – as for cooked hams, you only need to browse the shelves of glistening pink meats in your local supermarket to appreciate what aesthetic masterpieces they are.

One can find oneself musing: what would have happened if Chicago had wanted to make wine? Would the 'winepackers' have come up with a system for making wine all year round? Would they have discovered a miracle powder that, once poured into a barrel of grape juice, would after a few weeks, a few days, produce what those backward Europeans still made following the rhythm of the seasons? Disguised as wine, would these 'processed grapes' have conquered the world in the way that 'processed meats' managed to take over the cured meat market?

CHAPTER 5

A CHEMICAL SUCCESS STORY

The development of nitro-additives was the work of a handful of specialist firms. Starting from scratch, they grew to become multinational companies producing chemicals for the processed food market. One of the most famous of these was the Heller company, which was founded in Chicago.

AN AMERICAN SAGA

In 1848, the year Europe was shaken by revolutions, Bernard Heller left his home near Prague, abandoning his thriving business as a livestock trader. At the age of 42, with his wife and family, he embarked for the United States and settled in Milwaukee, near the Great Lakes. The town was new; there were still Indians living in tents on the hill just behind the house. Bernard Heller opened a little butcher's stall on Third Street, then another outside the centre. There were many mouths to feed: from barely 1,700 inhabitants in 1840, Milwaukee's population had topped 20,000 by 1848. Heller thought big: aside from his shops, he became a sausage wholesaler. A few years later, his son Adolph married the daughter of another wholesaler, and the young couple concentrated on the production side of the trade. Success was not slow in coming: the Heller workshop grew into a proper factory, the Milwaukee Sausage Works.[1] The factory employed 40 staff and got its meat from nearby abattoirs.

Unfortunately, this was around the same time that the big meatpackers of Chicago were also starting to produce sausages – a market they hadn't paid much attention to hitherto.[2] A handful of them became very powerful; they used dumping tactics and secret agreements to corner the distribution networks. Most of the independent abattoirs were absorbed; the small producers were swallowed up or disappeared altogether. Even a major anti-trust law passed in 1890 didn't succeed in preventing the domination of the five firms (Armour, Swift, Morris, Cudahy and Schwarzschild & Sulzberger) who came to be known as the 'Big Five'. They knew how to get their way: when traders refused to follow the new rules of the game, the Big Five had the means to throttle them by cutting off their supplies and selling directly to the consumers at knockdown prices. In Milwaukee, Adolph Heller was struggling to fight back. The final blow came on a December night in 1883, when an arson attack reduced his factory to ashes. Heller tried to rebuild, but the Milwaukee Sausage Works soon went bankrupt: in July 1886, Heller chartered some carts and dismantled his plant. He loaded up his meat mincers and sausage stuffers and set off to find a more amenable place to do business. He took a punt on Nebraska, in the heart of cattle ranch country. The press announced: 'Adolph Heller, the great Milwaukee sausage manufacturer, has concluded a contract whereby he moves his works to Nebraska City at once. He will build in connection with the Nebraska and Iowa packing house and employ sixty men the year round.'[3] Even here, alas, the competition was fierce. The Big Five were everywhere, laying down the law, selling to whomever they pleased, using every advantage they had to crush the smaller traders. Heller packed his bags and moved on again, this time to Sioux City, Iowa. But as soon as he arrived, he had to lock horns with Cudahy Brothers Co., the behemoth meatpacker that controlled several factories in the region.[4] Adolph threw in the towel for good. The Hellers' great adventure in America seemed to have ended in ruin.

A VISIONARY

From the moment the sausage factory started to fail, Adolph's eldest son had realized that they needed to go in a new direction. His name was Benjamin. One day, when he was visiting the Chicago International Exposition of 1893, he was struck by the crowds of gawping people gathered around some of the stands. Benjamin Heller knew his stuff when it came to sausages, but he had never seen anything like this: it seemed that people were willingly handing over their ready cash for small sausages served warm inside bread buns. This would go down in family legend as the moment Benjamin had his intuition, his vision: these little sausages would become hugely popular because they could be cooked quickly and looked appetizing – plus, they cost next to nothing to produce.[5] Whether or not Benjamin Heller really had a premonition of the success of what would soon be called the hot dog, he worked out how to quickly produce these devilishly delicious little 'wieners', and in 1893, at the age of 24, he started selling chemical additives. He started off with just two products:[6] firstly, a brown colorant which would give the sausage skins that characteristic 'smoked' look; secondly, a pinking powder that just needed to be mixed in with the meat 'to color pork sausage so as to produce a nice appearance'.[7]

Heller had no training as a chemist but he had a remarkable talent for marketing. He christened his first invention 'Zanzibar Carbon'. The packaging suggested that this gummy substance was produced in the African bush by tribes who boiled up vegetable sap in large braziers in the open air. In fact, Benjamin obtained it from the chemical factories of Chicago: it was made from coal tar, a worthless by-product of the process of deriving coal gas from coal. The pinking agent itself was a mixture of salt, saltpetre, borax and aniline colouring.[8] Benjamin Heller was very unspecific about the ingredients on his packaging; instead he claimed that his miracle blend was made in Germany by a Berlin pork butcher with whom he had an exclusive import deal. In fact, Benjamin Heller made the mixture himself. As one of his descendants put it: 'One of the more colorful stories that has been passed down through the family tells of Benjamin Heller's early days in Chicago. In the early 1890s he didn't have very much and lived in a boarding house. Rumor goes that he would get up in the middle of the night [and] clean the bath tub where he mixed his product to sell to local merchants!'[9]

In a private letter of 1897, Benjamin Heller described how he would go about selling his products on one of his never-ending sales trips. 'When I walk into a shop and notice that a man is making his own bologna and uses no color, which you can always tell by the appearance of his goods, I walk up to him.'[10] Introducing himself as 'a friendly and very knowledgeable stranger in possession of a mystery', he would then present 'a new German chemical which is made in Germany'[11] and proceed to explain that the powder, called *Rosaline*, could bring about 'a fresh meat color, which does not fade'.[12] The label stated: 'Rosaline dissolves when it comes in contact with water or sausage meat, therefore it makes no spots in sausage. It does not fade when the sausage is cut or exposed to the light or air. Two to four ounces are sufficient to color 100 pounds of meat.'[13] 'Rich! Pure! Neat!'[14], it was 'the only thoroughly reliable coloring manufactured for coloring, curing and preserving Bologna, Frankforts, Summer Sausage, etc. It colors, cures and preserves all at the same time, and makes a natural, bright, fresh meat color.'[15]

Benjamin Heller published a book, *Secrets of Meat Curing*, which he sent as a freebie to thousands of potential customers. Lavishly illustrated, the book's scientific tone was intended to impress his unqualified readership. On every page, Heller subtly name-dropped his chemical additives as if he were talking about indispensable ingredients of all meat curing. Taking advantage of the lack of regulations on use of food adulterants,

Heller & Co. launched dozens of synthetic colouring agents (mostly created by the manipulation of petroleum and coal derivatives) and 'economical spices' (that is, artificial 'spices' obtained by chemical means). Alongside all this, Heller developed insecticides: a whole range of anti-cockroach powders ('Roach Killer', 'Roach Destroyer') as well as 'Moth Killer', 'Bed Bug Killer', 'Ant-Bane' and 'Fly Killer', the latter a liquid compound of phosphorous acid, fluoride and cyanide. These were made in the same workshops as the food additives, but to avoid any confusion, Heller marketed them under a different brand: the Chicago Insecticide Laboratory Inc. Within a few years, the Heller company had achieved stunning commercial success, and Benjamin and his brothers found themselves propelled into the upper crust of Chicago society.[16]

'NO ICE NEEDED'

After launching Rosaline, Heller developed a colouring powder based on sodium nitrate and sodium sulphite, which the firm marketed as 'Red Konservirungs Salz'.[17] One early piece of advertising claimed that this 'preserving salt' preserved meat 'the same as if it were frozen'.[18] Subsequently, the firm brought out a whole range of products with names ('Iceine', 'Cold-Storine') suggestive of freezing. This was particularly the case with another mixture based on sodium sulphite: 'Freeze-Em'. The label stated: 'Made expressly for keeping all kinds of fresh meat, chopped beef, and pork sausage in a fresh condition without ice from one to three weeks in just as fresh condition as if the meat were frozen.'[19] According to the sales brochure sent to retailers, 'the day is past when it is necessary to keep meats hidden in the ice-box. They can be kept fresh and sweet, bright in color, and as tempting as a ripe peach by the use of Freeze-Em.'[20]

Heller & Co. subsequently came up with another product: 'Freezine', a liquid that was 'more reliable than ice'.[21] It was a solution of formaldehyde – in other words, formalin. Heller launched a PR campaign to convince small dairies to adopt the 'Freezine process', rubbishing counter-claims by microbiologists and doctors that all that is required to keep milk fresh is good hygiene and control of temperature. In the free brochures Heller & Co. handed out, they listed the advantages of this liquid with its quasi-magical powers: 'Freezine produces the same effect on milk as freezing does, without the use of ice, or without lowering the temperature of the milk or cream',[22] 'it is much cheaper to use Freezine than ice', 'it is perfectly harmless', 'it cannot be detected when used'.[23] Printed in several languages, a booklet entitled *The Healthfulness of Freezine* insisted that if a dairy wished to avoid the costs associated with maintaining refrigeration equipment, all it needed to do was pour a few drops of Freezine into its milk churns. The milk would then be 'sterilized' for 36 hours, 'and by recharging each thirty-six hours this state may be extended indefinitely'.[24] The Massachusetts State Board denounced Heller publicly: 'The manufacturer of a largely used preservative, known as "Freezine" (which is a weak solution of formaldehyde) issues an attractive pamphlet in which he makes the following remarkable claims: "It is not an adulterant. It immediately evaporates, so that no trace of it can be found."'[25] The journal *Farmer's Voice* remarked: 'The name is a stroke of genius: it makes one think of ice, refrigeration and cold', and continued: 'The question is: how much of the above chemical compound does a man want to take into his stomach and how much of the claims for the innocence of the compound can be believed? Can a substance be a powerful poison outside of the body and an innocent compound in the stomach …?'[26]

In fact, it was already well known that formaldehyde is toxic even in low doses.[27] As sales of Freezine increased, so did the number of intoxications. But the Heller company feigned ignorance, not wanting to surrender any of its huge profits. Using images of suckling babies, the firm insisted: 'Freezine does not lessen the value of milk, but increases the value by adding to its keeping properties and by rendering it healthful to the human system.'[28] Heller & Co. talked about a 'milk poison' ('tyrotoxicon'), which, according to them, could occur if the milk wasn't treated with formaldehyde. In the booklet *The Healthfulness of Freezine*, Heller described the case of several hotels in New Jersey in which guests were afflicted with terrible indigestion caused by milk rendered unfit to drink by the August hot weather. Heller explained: 'During the same summer a similar epidemic occurred in Iowa. Four cases also appeared in Michigan, three of which proved fatal. Dozens of similar epidemics might be mentioned where the presence of tyrotoxicon has been proved by analysis of the milk used, and by post-mortem examination of the viscera of the unfortunate persons. Lack of space prevents mentioning one thousandth part of the fatalities occurring from causes directly due to tyrotoxicon poison.'[29] The conclusion, according to Heller, was that Freezine wasn't dangerous, quite the opposite: it helped to save lives. For: 'poisoned milk epidemics are frequent, but could be wholly prevented if Freezine were used in all milk and if the cans or vessels were washed in water containing a little of it.'[30]

PRESERVATIVES, THE GOOSE THAT LAYS THE GOLDEN EGG

In 1892 a US congressional report on food adulterations was alarmed at 'the grave and growing consequences of a greed for gain which is assailing the public health',[31] and noted that 'it has only been since the great opportunities for fraud provided by modern science, and especially modern chemistry, have presented themselves that the sophistication of articles of commerce has reached its present height'.[32] Another congressional report explained that the lack of food regulation 'amounts not only to a premium to dishonesty but is a threat to national health. Honest manufacturers and dealers are placed at a disadvantage or are forced into a reckless competition with fraud. Legitimate trade is handicapped and demoralized.'[33] One local board of health noted in 1901: 'It is not fiction that unhealthful adulterants are used in many food products. Under the labels of "Freezine," "Preservaline," "Liquid Sweet," "Liquid Smoke," "Rosaline," and other fanciful names, they are manufactured by hundreds of tons, placed in every market in the United States, shipped to foreign countries in immense quantities, and advertised with a skill and effectiveness that compels public attention.'[34]

In brochures aimed at housewives preparing jars of cherries and jam, a firm offered assurances that its formaldehyde-based antiseptic liquid was a better alternative than pasteurization using heat: 'It enables the busy housekeeper to put up fruits and vegetables without having to stand over a hot fire on a hot day, and without any loss, for the fruit thus prepared will not work [ferment] or mould, and the life, flavor, and vitality will not be cooked out of it. It saves broken jars; food cooked away; time, labor and money.'[35] One health bulletin tried to lecture farmers: 'Do not believe the statements which are made for any food preservative, and above all, do not use such substances in food products thinking they are harmless. Preservatives will not take the place of hygienic conditions in any manner or form and with proper care the use of such substances is not necessary in order that the product may reach the consumer in good condition.'[36] The journal *Wallaces' Farmer* alerted small producers that 'various companies are trying to persuade us to take out a licence to be dirty by using their compounds warranted to arrest decomposition and decay in almost everything we eat and drink [...] For example, one brand is warranted to keep sausage and all kinds of chopped meats perfectly sweet and moist, retaining its natural color and working like a charm.'[37] For this journal, it was 'simply an excuse to be dirty, or at least careless and slovenly'.[38]

'EMBALMED MEATS'

In 1898 a dispute broke out that crystallized the disquiet caused by the rapid increase in use of food additives. It all began when some US soldiers claimed that the General Staff had forced them to eat rotten meat crudely dosed with chemical preservatives to disguise the disgusting look, taste and smell. The Chicago meat-packers copped the blame: as the mother of one soldier wrote, 'thieving corporations will give the boys the worst'.[39] The suppliers were summoned to face questions in what soon came to be known as the 'embalmed meat' scandal.[40] An official inquiry was carried out, and dozens of soldiers and officers appeared before a Senate committee. Some of them thought that the meat had been treated with chloroform.[41] Many observed that it had an abnormal colour.[42] The soldiers testified: 'It was chemically treated';[43] 'the men would eat at it, and throw it on the ground';[44] 'it was thrown into the sink hole, it was buried. We didn't cook it, or make any attempt to cook it.'[45]

The inquiry concluded that only a few carcasses had been treated with unauthorized substances (by fumigation with sulphur). But the experts reckoned that overall the meat in question was not unfit for human consumption. The 'embalmed meat' affair nevertheless highlighted certain serious dysfunctions – in particular, the absence of any official controls over the production methods used by the meatpacking factories. True, an official of the veterinary services verified the state of health of the animals at the point of slaughter. But beyond this point, the processing was not subject to any oversight – except by the meatpackers themselves.[46] It was a black box in which the manufacturers were able to do whatever they wanted. Free from any scrutiny, they could use any food additive they wished.

According to the meatpackers, all these fears were unfounded: why be afraid of additives in such small doses? This was the line taken by National Provisioner (a publisher whose eponymous journal was the mouthpiece of the meatpackers' association). In 1899, National Provisioner brought out a thick volume entitled *The Manufacture of Sausage*. This was presented as a collection of 'the recipes and methods which are in daily use by the most successful sausagemakers and packers'.[47] The book stated: 'Saltpetre, or, chemically speaking, potassium nitrate, is of great value because it imparts the peculiar and desired natural ruddy color to meats, whether cured in dry salt or pickle. Without its assistance it would be almost impossible to prepare meats to suit many markets.'[48] A full-page advertisement announced: 'You *must* use Preservaline to produce good sausage. Progressive sausagemakers have discarded all the old antiquated methods and now depend entirely on Preservaline. Acts like a charm ... Never fails.'[49] The book suggested: 'Many objections have been urged against the use of these artificial coloring matters, on the improper supposition that they are a menace to health. As a rule, these materials are not as injurious as general opinion would have them.'[50] The same message is reiterated further on: 'The trouble lies in the fact that when there is a legitimate use in moderate quantities of such articles, abuses creep in which arouse antagonism and agitate the question of their entire prohibition; but those who use preservatives intelligently and in moderate quantities will always have beneficial results and need not fear any harmful effects there from.'[51]

PRESERVATIVES ON TRIAL

In 1899, the US Senate finally decided to tackle the issue of additives head-on and summoned the kingpins of the industry. When Albert Heller – brother of Benjamin Heller and co-director of Heller & Co. – was obliged to come and explain himself, a senator reminded him that he had been subpoenaed by a commission 'directed to inquire into what food products are deleterious to health and what are frauds upon the community'.[52] The journal *The Farmer's Voice* wrote: 'Of the meat preservatives, three of them, "Rosaline," "Freezine," and "Freeze-Em," are made in Chicago, and the expert chemist who had analyzed them declared every one of them to be deleterious to health. And yet there are those who object when it is intimated that the commercial spirit has seized upon the people and is leading them into all sorts of evil ways. It would be difficult to conceive of a more diabolical wretch than one who will deliberately feed his customers slow poison. We earnestly hope that the result of the senatorial investigation may be to break up and destroy this fraudulent business.'[53]

Albert Heller wasn't overawed. He had an answer for everything. Why did food processors not affix a label on their products that indicated whether a chemical additive had been used? Because that could cause cases of malnutrition among the poor: 'If they are marked, the poorer classes will be afraid to use the goods, or they might be backward in buying them, while they can't afford to buy the higher grade of the pure goods and pay the price.'[54] Why did Heller & Co. make out that Freezine was a 'natural product'? Because there was no reason not to: as formalin was obtained by a process involving the oxidation of wood alcohol, Albert Heller didn't consider that he was misleading anyone when he included the following inspirational description in his promotional literature: 'Freezine is a liquid gas made from vegetable materials, and is used as a substitute for ice.'[55]

When accused of selling an additive capable of poisoning foods, Albert Heller didn't turn a hair. He claimed that formaldehyde-treated milk was 'not only perfectly harmless, but it is positively healthful, and especially so with infants'.[56] He even maintained that it could protect children against the risk of cholera.[57] He reiterated: 'it improves the milk and makes it more healthful.' Likewise for Rosaline and all the 'preservative salts' that Heller & Co. manufactured: they were all harmless, they were all beneficial. 'Chemicals are used for curing the meat', Albert Heller asserted, before going on to say: 'Some use salt and saltpeter. Some use boric acid and salt.'[58] Adding provocation to arrogance, he exclaimed: 'I would like to make a statement that there has been much agitation in the papers in regard to the embalmed-meat question. I wish to say that every one of us eats embalmed meat – and we know it and like it and continue to like it – and that is, hams and bacon. That is actually embalmed meat.'[59] Nevertheless, he refused to reveal the exact make-up of Heller additives in public. The next day, a journalist reported: 'During the afternoon the committee went into secret session to hear the testimony of Albert Heller, manufacturer of Freezine and Rosaline, used for preserving milk and meat. Previous witnesses had testified that these were poisonous. Heller denied this emphatically, but asked to be heard in secret in order that his formulas might not become public property.'[60]

ONE FREEZE-EM AFTER ANOTHER

When the Senate concluded its hearings in December 1900, the press was up in arms against the 'falsifiers' and the 'poisoners'. For the *Los Angeles Times*, 'the adulteration of food is one of the crying evils of American commerce, and it is indeed about time that there was a stop put to the practice of feeding the populace something it hasn't ordered, and doesn't want'.[61] With no law forthcoming at a federal level, a number of local jurisdictions began bringing in their own restrictions. In Los Angeles, for example, seventeen butchers and pork butchers were prosecuted by the board of health for using a noxious additive. Heller sent his own lawyers to defend one of his customers. In 1903, the *Los Angeles Herald* warned against additives, which did 'for a slice of meat what rouge does to a woman's cheek',[62] and announced that 'experts of the "Freeze-Em" company will attempt to show that their preparation "improves meat in color and taste and aids in digestion"'.[63] From Heller's point of view, 'instead of prohibiting the uses of preservatives, their use should be encouraged and strictly enforced by every mandate of law'.[64] Further prosecutions took place two years later, first in Los Angeles (several users were arrested),[65] then in nearby Pasadena, where some traders were sentenced for using Freeze-Em. The newspaper commented: 'Pasadena is stirred by a health department discovery that meat is sold in that city which is doctored with a preservative usually known as "Freeze-Em"';[66] 'Pasadenans are satisfied with plain meats. When they want ice cream they order it.'[67]

In 1905, in one of its promotional publications, Heller & Co. reproduced a ditty on the mystery surrounding the exact composition of its additive:

'Freeze-Em, Freeze-Em, in a jar,/How I wonder what you are./Keeping all things sweet and fine,/How I wish your secret mine./Keeping meat so fresh and red,/Nothing does so well instead./Priceless crystals in a jar,/How I wonder what you are.'[68]

But Heller could see which way the wind was blowing. In the end, the firm took a radical decision: it withdrew Freeze-Em from its list of ingredients for use in food. They continued selling it, but as a disinfectant for equipment such as knives and pans: Heller explained that Freeze-Em 'prevents the corrupting influence of decomposing meat particles in meat block tops and other tools, utensils and machinery'.[69] Butchers could use it 'for sterilizing and deodorizing their utensils and other equipment',[70] 'added to the scrubbing water used for washing tools'[71] or 'for the purpose of disinfecting and deodorizing chopping blocks'.[72] In place of Freeze-Em, Heller & Co. launched a new product which had an almost identical name: Freeze-Em-Pickle. It was a mixture of salt and nitrate that turned meat 'as red as a cherry'.[73] Then Heller brought out a powder called 'Prepared Dry Pickle', then another called 'Improved Freeze-Em-Pickle', then 'Quick Action' (or 'QA Pickle') and a number of others with somewhat cryptic names ('I-X-L', 'Custom Cure', 'SchnellSalz', 'Microsized Cure'). And so it was that Heller became the world's biggest distributor of nitro-additives.

CHAPTER 6

PURE FOOD VERSUS NITRO-MEAT

In the closing decades of the nineteenth century, a popular coalition began to take shape in the United States. This movement went under the name of 'Pure Food'. It included doctors, women's groups, consumer organizations, and health board officials. Their common target was what were known at the time as the 'adulterators'. One activist painted a portrait of these modern-day sorcerers: 'The scientific and skillful food adulterator might well be called a modern alchymist with this difference, that while the alchymist of yore attempted to change baser metals into true gold, his modern successor is constantly trying to make from baser or cheaper materials mixtures that shall possess the appearance through not the quality of more expensive natural foods.'[1]

The Pure Food movement owed much to the work of one man, Harvey Wiley. A medical doctor and chemist, he ran the Bureau of Chemistry at the United States Department of Agriculture. He was the main inspiration behind the first significant American food laws that offered a way to limit the use of dangerous additives. Wiley's work has been the subject of several books. One question, however, has never been satisfactorily answered: what was Dr Wiley's attitude towards nitro-additives?

INCREASING CONCERNS

At the start of the Pure Food movement, Wiley and his colleagues weren't too bothered about saltpetre: they concentrated their attack on other additives they regarded as more noxious, such as formaldehyde in milk. In 1898, Wiley noted that in the case of corned beef, 'all recipes for its preparation published in encyclopedias and in the correspondence columns of our agricultural papers and similar publications, prescribe the use of both salt and saltpeter'.[2] He assumed that saltpetre treatment was so widespread that there was no particular reason to worry about its effect on the human organism, especially since the quantities of saltpetre used to treat the meat were fairly minimal. At a meeting of the American Public Health Association, Wiley said in 1899: 'The addition of nitrate of soda and nitrate of potash to sausages and meats tends to preserve and at the same time to intensify the red color of the meat. I believe that potassium nitrate is uniformly employed by all packers of corned beef so that it may be considered as a normal constituent thereof.'[3]

But around the same time, a number of biochemical investigations were under way in Europe, and they cast a more critical light on the issue.[4] Stopping short of explicitly condemning saltpetre treatment, they nevertheless started asking questions.[5] In 1890, when the German biochemist Karl Lehmann reported on new work on nitrate, he noted that: 'there has been observed a very variable degree of sensitiveness to poisoning by saltpetre'; he reckoned that 'the question as to the poisonous character of saltpetre seems to require a special experimental investigation'.[6] For his part, the British chemist Charles Ainsworth Mitchell noted: 'The effect on the human system of the continued use of meat containing nitre has not yet been determined.'[7] In a chapter titled 'Judgment of the saltpeter content of pickled meat', the *Manual of Meat Inspection* by the German scientist Robert von Ostertag made the same point: 'Nothing is known concerning saltpeter poisoning from the consumption of meat. The question of the hygienic judgment of saltpeter appears still to require a more thorough examination.'[8] In 1901, a report by a US commission of inquiry on food additives proposed in fact that 'saltpeter is regarded as more deleterious to health than either common salt, borax, or bicarbonate of soda, having a tendency to produce degeneration of muscle and an injurious effect upon the kidneys'.[9]

In the first years of the century, Wiley's position shifted perceptibly under the influence of the scientific publications he consulted. He pondered the impact that products treated with saltpetre could have on the health of consumers, especially the most vulnerable (those with medical conditions, young children, older people). At an official hearing at the House of Representatives, one congressman asked Wiley if he thought saltpetre was dangerous. He answered: 'I would not be afraid to eat a piece of corned beef, because the amount of injury would be immeasurably small.'[10] He said that he didn't consider saltpetre exactly harmless, but he wasn't sure enough of its deleterious effects to recommend banning it: 'Do not misunderstand me. I am not saying that it [saltpetre] should not be used in corned beef. I would be sorry to see it left out. But if you put it on the principle of harmlessness, it could not go in.'[11] Wiley wanted to wait and see more evidence: in order to make a definitive judgement he recommended to subject saltpetre to some real-life experiments.

THE 'POISON SQUAD'

In 1902, the American Congress asked the Department of Agriculture to 'investigate the character of food preservatives, coloring matters, and other substances added to foods, to determine their relation to digestion and to health, and to establish the principles which should guide their use'.[12] So the secretary of agriculture commissioned the Bureau of Chemistry to conduct a series of 'feeding experiments': Harvey Wiley's team would subject groups of young volunteers to a diet including regular doses of food additives.

Morning, noon and night these human guinea pigs recorded their weight, their temperature and their pulse. Once a week, they would undergo a battery of tests (physical examination, blood tests, analysis of stools and urine, etc.). Every participant signed two written pledges: 'one to take no food or drink outside that prescribed by the tests, and the other exempting the government officers from any liability for damages on account of sickness or injury to the subject on account of the tests'.[13] The investigation was set up in seven parts: the first study was dedicated to borax, the second to salicylic acid, the third to sulphurous acid, the fourth to benzoic acid, the fifth to formaldehyde, the sixth to sulphate of copper. The seventh and final series was exclusively devoted to the effects of saltpetre. For each of these additives the team from the Bureau of Chemistry had to produce a complete scientific report – 400 to 500 pages of statements, tables and analysis. After being submitted to the government, each report was then made available to manufacturers, political decision-makers and journalists.

These experiments, and the people involved, became known to posterity as the 'poison squad'.* From today's perspective, their methodology seems frankly unethical and archaic, as no one would now contemplate testing the toxicity of substances by feeding them to a group of people.[14] Nevertheless, the 'poison squad' marked an important moment in the history of food additives because it represented the first attempt to systematically evaluate their impact on the human organism. It allowed the

---

* Which lent its name to a recent book by Deborah Blum on the life of Harvey Wiley: *The Poison Squad: One Chemist's Single-Minded Crusade for Food Safety at the Turn of the Twentieth Century*, Penguin Press, New York, 2018.

Bureau of Chemistry to assemble experimental data in order to bring in laws to limit the use of dangerous substances. At a hearing in the House of Representatives, Dr Wiley explained that the observations made with the 'poison squad' had led him to revise his opinion on the majority of additives and to advocate stricter regulations: 'I formerly believed that certain preservatives could be used [...] simply by having the people notified on the label. I was strongly convinced of the truth of that proposition. I have, before committees in Congress and in public addresses, stated those sentiments. I was converted by my own investigations.'[15]

In parallel with the 'poison squad', Wiley led a team of experts who were seeking to establish 'standards of food purity'.[16] Their task was to define the ingredients that could legally be included in each individual foodstuff. In 1903 and 1904 the standards covering meat products were officially published. They stated that processed meat products could contain salt, spices, traces of smoke ... but they didn't mention saltpetre at all.[17] Wiley explained that the complete standards for additives would be published later, but this omission provoked fury among the Chicago meatpackers; in November 1905, one of their representatives interrupted a meeting of the standards committee to demand that saltpetre should immediately be included in the list of standard ingredients.[18] Meatpackers asserted that a restriction on saltpetre would constitute 'a complete revolution in the methods of curing meats'.[19] The *National Provisioner* argued that: 'In regard to colors, the trade declares that the public will not buy foods without the characteristic color to which consumers are accustomed. Corned beef without the red shade due to saltpeter is passed by in scorn by the average purchaser, and brown chopped meat or sausage which should have a natural red meat color is absolutely a drug on the market.'[20] For its part, the *Washington Times* outlined what would happen if the standards of food purity became law one day without permitting saltpetre: 'The beef may be corned in brine, that is all. No preservatives shall be used. The beef will be a dirty gray, instead of red.'[21]

THE SCANDAL OF *THE JUNGLE*

In 1905, the medical journal *The Lancet* published a long investigation into the meatpacking factories of Chicago. Supported by photographs, the articles, spread over four weeks, revealed that the meatpackers showed a complete disdain for cleanliness. The expert at *The Lancet* talked about 'the filthy ways prevalent in Chicago',[22] describing what he saw as an 'abomination': the walls of the workrooms were never washed down, grime accumulated in the corners and on the benches, the workers spat on the ground and dragged meat across the floor, the place was a breeding ground for germs. The chemical additives used in the processing of the meat compensated for the lack of hygiene: 'when meat that has been put in the brine for pickling is found to smell, hollow needles are driven into the flesh and brine is pumped into the body of the meat. This saves the time of penetration and men have assured me that when the meat still retained an unpleasant odour the operation was repeated.'[23] *The Lancet*'s verdict on the Chicago meat-processors: 'So long as a thing pays it does not matter what ignoble expedient may be employed to make it pay.'[24]

A few months later, the famous hyper-realist novel *The Jungle* corroborated *The Lancet*'s revelations, and its publication provoked an international scandal. *The Jungle* showed what was hidden behind the fences of Packingtown: a small group of industrialists ran everything and recruited the most vulnerable for their workforce – recent immigrants, wretched, expendable, often sick with tuberculosis – and worked them to death. As portrayed in *The Jungle*, an American meat-processing factory was nothing more than a monstrous machine for crushing the exploited – a machine for making money out of tears. It was also a gigantic poisoned cauldron: the killing floor swarmed with rats, the toilets overflowed in the work rooms, and, as *The Lancet* had said, rotten hams were sometimes injected with chemicals to 'restore' them – and if that didn't work, they were turned into sausages. The author of *The Jungle*, Upton Sinclair, declared in the press: 'I saw one of the [packer] trust's employees doctoring a spoiled ham in a cellar of one of the packing houses. The stench that arose from the ham was overpowering. The man was working a pump with one foot. Attached to the pump was a tube, on the end of which was a big hollow needle. He would jab the needle into the ham and then pump it full of chemical to take away the dreadful odor.'[25] A representative of the Armour company rejected all the accusations: 'There was no odor.'[26] He explained that the injection was part of the standard technique of nitrate curing: 'In preserving ham with the saltpeter mixture employed – curing it, some call it – it is impossible to distribute the preservative equally without using a needle.'[27]

The furore sparked by *The Jungle* gave fresh impetus to the reformist efforts of the Pure Food movement, which had been campaigning for years for tighter food laws. For the first time in history, the US Congress seemed determined to vote through measures to protect consumers against the abuses of industrial companies – in particular, against dangerous food additives. In an article headlined 'Proof of packer's use of chemicals', the *Evening World* newspaper described 'how chemicals are used on meat to cheat public'. The paper gave a detailed description of the substances used: the 'chemically-prepared salts', the magic powder that 'imparts a color to the sausage which makes it "sell faster"', the one that gave bacon 'that bright natural color with a golden mahogany finish, also gives them a fine flavor, prevents shrinkage and keeps off skippers [the larval stages of small flies]',[28] etc. The manufacturers' defence was that the market could regulate itself;

banning controversial additives would hamper free competition: according to them, 'the question of coloring matter can be adjusted by the consumer without the aid of the law'.[29] One senator refuted this line of argument: 'I cannot see that any person has a valid right to use any such packing on any meat as would be injurious to the public, and if depriving him of the right to so pack meats is injurious to him it is depriving him only of an opportunity to do a wrong to the public.'[30]

The meatpackers claimed that the proposed law was the fruit of a conspiracy orchestrated by Dr Wiley and his colleagues: according to the *National Provisioner*, 'this bill is advocated by the food crank, by the political food official who looks to its enactment to give him a job or increased power and prominence, and by the misguided individual who believes all he reads in the "horror sections" of the Sunday newspapers'.[31] If there was a rotten smell, lobbyists said, it wasn't coming from the meatpacking factories; rather, it emanated from the doctors, the food standards experts: 'it is time this matter of the slander of the food supply was ventilated and some of the high salaried officials who have done the slandering were cast out of office'.[32] On page after page the meatpackers' journal attacked Wiley as the 'food dictator',[33] the 'political chemist'[34] determined to 'throttle the legitimate food industry of the country':[35] 'When we read his speeches many of us believe that the Lord made a mistake in not taking him amongst the million infants he says have been murdered by impure food, and not let him grow up to mislead the people of this country.'[36] A report by the House of Representatives retorted that 'the penalties of the bill are aimed at cheats. That which is forbidden is the sale of goods under false pretenses, or the sale of poisonous articles as good food. No honest dealer need fear any provision in the bill. Legitimate trade should welcome its enactment into law. Only those wishing to deceive the public will object to its provisions.'[37]

SALTPETRE ON PROBATION

When in early June 1906 a leading figure in the meatpacking industry appeared before the House of Representatives, a congressman put it to him: 'I can not quite understand why you object to the prohibition of any chemical or dyes that might be deleterious to public health being put in your meat products.'[38] The meatpacker replied that the problem hinged essentially on the likely ban on saltpetre curing. He explained: 'We use it in our curing – that is the chief object in using it – to maintain the fresh color of the meat, and it helps in the curing, too. And it is used in such quantities that it is absolutely not harmful.'[39] Unfortunately, he continued, medical experts such as Harvey Wiley were likely to use the new law to ban saltpetre: 'Some chemists, and a chemist of high authority, might say that it is injurious';[40] 'Some inspectors might think that the very small quantity of saltpeter we use is deleterious, and yet in the small quantity in which it is used it is all right.'[41] An article in the *Washington Times* stressed that such a ban would be catastrophic for the meatpackers: 'People, they say, would much rather eat a piece of meat or sausage that has the deep red color preserved than a brown piece that had no saltpeter in it.'[42]

On 10 June 1906, the meatpackers held a council of war in a large Washington hotel. They wanted the putative ban on chemical preservatives to be struck out of the proposed legislation.[43] The arguments over additives led to some crazy speculation. On 11 June, for example, a daily newspaper wrote: 'It is denied tonight that secret service men have been gathering information to prove the extent to which the packers have used preservatives.'[44] In the weeks that followed, Washington became a battleground of lobbyists; Harvey Wiley described it as an enormous fistfight: 'Every man who used alum, coal tar dyes, salicylic acid, burning sulphur fumes, benzoic acid, copper sulphate, saltpeter, saccharin,

103

borax, or other nonfood ingredients in his products joined the solid phalanx that struggled to prevent the passage of a law which would interfere with these despicable means of making money.'[45]

At first, the industry believed it had achieved a result: in the final draft, the law that was adopted made no reference to the standards of purity.[46] The meatpackers were jubilant: 'The law passed was the law advocated by the meat industry', proclaimed the *National Provisioner* in block capitals.[47] The meatpackers' journal congratulated itself that the initial text had been 'emasculated',[48] that 'the features of the original bill [...] which were dangerous to the provision trade, viz., those giving Doctor Wiley exclusive jurisdiction over preservatives, were totally eliminated'.[49] Under the title 'No more food standards', the journal wrote: 'the work of the food standards commission appointed by the Secretary of Agriculture to report standards to him will be discontinued. No more meetings of that committee will be held for the purpose of establishing official government standards.'[50] The *Washington Times* described the meatpackers' satisfaction when they thought they had succeeded in abolishing the standards: 'When [...] this provision was stricken out there was great rejoicing in the food and drug lobby, because it was thought Dr Wiley had been eliminated.'[51]

But it was not long before disillusionment dawned: the meatpackers discovered that three days before the passing of the bill the government had just proclaimed the standards in a separate regulatory act.[52] As the *Washington Times* explained, 'the truth is that the opposition lobby has discovered that it won only a Pyrrhic victory when it succeeded in striking out of the law the provision for establishing food standards'.[53] Even worse: the government promulgated a regulation that authorized saltpetre only on a temporary basis, 'pending further enquiry'.[54] A weekly newspaper in Chicago explained that 'the law recently enacted by Congress prohibits the use of all preservatives with the exception of sugar, salt, spices, vinegar, wood smoke, and, *until further investigations* are made by the Bureau of Chemistry, saltpeter'.[55]

The *Washington Times* sounded a warning: the game was not yet over: 'The realization that the effort to eliminate standards was a failure has given an added bitterness to the food people's cup of gall, and the authorities are expecting them to make a long and hard fight.'[56] To illustrate the article, a caricature represented 'the bête noir of the food and drug makers': Wiley as Mephistopheles, complete with a devil's horns and tail. The caption read: 'Dr Harvey W. Wiley, leader in pure food fight, as he appears in the mind's eye to certain large interests in this country.'[57] The paper warned: 'The big interests that have been for years opposing the enactment of the law have not done with their work.'[58]

TENSIONS RUN HIGH: THE 'SALTPETER SQUAD'

In September 1906, the Wiley team set to work. 'Tests of effect of saltpeter to begin in October', announced the *Washington Times*.[59] The newspaper reported that 'Dr Harvey W. Wiley, chief chemist, is going to "try it on the dog" once more. This time it's saltpeter that will be fed to his class of self-sacrificing young men for the good of society and to the detriment of the makers of preservatives and foods cured with them.'[60] Another newspaper described what the groups of guinea pigs were undergoing: 'the eight members of the present squad were placed on a diet containing a certain proportion of saltpeter. The substance is so delicately disguised that it cannot be tasted in the food, and only such an amount is administered as would be eaten were the members to partake of corned beef or some such substance daily.'[61] In New York state, a journalist noted: 'The experiment this year is regarded as especially interesting because of the general use of saltpeter in preserving and coloring meats. The results will be surprising to a great many people if it is found that ordinary corned beef put up with saltpeter is not healthful. Nearly all the other preservatives of meats and canned foods have been ruled out by the secretary of Agriculture in the meat inspection regulations as a result of the tests made by Dr Wiley's poison squad.'[62] And the *Argus* newspaper offered this prognosis: 'It is regarded as probable that the work of the squad will clearly demonstrate that saltpeter is deleterious to the human system, in which case the regulation will be amended to prohibit its use.'[63]

The meatpackers themselves denounced this latest round of 'farcical poison-feeding experiments of Dr Wiley's bureau'.[64] The *National Provisioner* reproduced extracts from an interview in which Wiley declared: 'The law provides that nothing injurious shall be added to food products. Under that clause we have the power of deciding what is, and what is not, injurious, and by that is meant what might become injurious ultimately, in weeks, or months, or years.'[65] The following week, the paper tried to reassure the industry in a front-page article. Under the headline 'Saltpeter is not prohibited', it said: 'Some prominence has been given to a report that saltpeter is now prohibited under the federal meat and food laws in the preparation of meat and food products. Such is not the case. The regulations under both laws permit the use of saltpeter, pending investigation as to its effects. This investigation is now in progress, but until it is completed no change in the regulations will be made regarding saltpeter. The trade can use saltpeter as before, without fear of violating the regulations. It may not be declared a preservative at all, but a colorant.'[66]

The study by the Bureau of Chemistry came to an end in the spring of 1907. Each of the human guinea pigs had eaten 0.15 to 0.6 grams of saltpetre a day over a period of two months.[67] The saltpetre was not delivered in treated meat, it was ingested directly in the form of capsules. This was a significant methodological error, because administering it this way, while allowing an assessment of the toxicity of the nitrate itself, almost completely masks the effects of the nitroso compounds. No clear-cut toxic effects were recorded, but Wiley remained cautious: his team had detected abnormal levels of pain in the stomach region (in medical language he described this as 'pains in the epigastrum').[68] At the end of March, a Kentucky daily interviewed Wiley on the results of the 'saltpetre squad' experiments: 'He is unable to announce definitely what will be the conclusions as to its effect upon his squad, but thinks that he has already noticed material evidence of some effects. He believes these tests will prove it injurious, and, if so, he will endeavor to stop its use.'[69] A few weeks later, a newspaper in Illinois reported on the Bureau of Chemistry study under the heading 'To stop saltpetre sandwiches'.[70] In a conference the following year, Dr Wiley explained that the results hadn't allowed him to come to as cut-and-dried a conclusion as for the substances previously tested by the squad. Nevertheless, according to him, uncertainty concerning the effects of saltpetre was reason enough not to use it in food: 'While the data are in this case far less conclusive than those in any of the preceding cases, they are of a character to warrant the suggestion that so far as health and digestion are concerned it is safer to omit a body of this kind from the food.' He continued: 'While, therefore, the data which have been accumulated are not such as to warrant a sweeping condemnation of potassium nitrate in foods, they are sufficiently indicative to justify the conclusion that its presence in foods is undesirable and open to suspicion.'[71]

Today, we would say that Wiley was applying the precautionary principle: as he indicated a few years later, 'the Bureau of Chemistry was pledged to one very simple but most important principle, namely: "When in doubt protect the consumer."'[72] In a book that appeared shortly after the end of his experiments, Wiley placed saltpetre in the category of 'injurious preservatives and coloring matters'.[73] Given that he considered that saltpetre 'is not a proper substance to mix with foods',[74] he refused to approve its usage[75] and quickly recommended that suppliers to the armed forces desist from using saltpetre in products destined for the troops.[76]

MEATPACKER SCIENCE: THE 'MIRROR INVESTIGATION'

On 1 April 1907, there was a strange item in the American newspaper *The Sun*: an article announced that a new 'saltpeter squad' was about to be launched. This was news to Harvey Wiley, who had just completed his Department of Agriculture study and thought at first that this new squad was just a journalistic invention. An April Fool's trick perhaps? But the article gave actual details: this squad would be up and running 'in a week or two', its experiments would run for '6 to 12 months', and 'it is said that they will be the most thorough ever undertaken in the United States'.[77] According to the paper, 'This was settled yesterday, when what will be known as the National Commission for the Investigation of Nutrition Problems was formed at the Fifth Avenue Hotel'.[78] This so-called 'National Commission' announced that its guinea pigs would be students from the University of Illinois at Urbana (about 130 miles south of Chicago), and that the project would be led by one Harry Grindley, a professor of biochemistry. The journalists smelled a rat: barely six months earlier, this same Grindley had put his name to a report financed by the meatpackers.[79] This document maintained that there was virtually nothing to criticize in the Chicago factories, that levels of hygiene were fairly satisfactory and that 'the general procedures adopted give warrant of cleanliness'.[80] The industrial meat-processors had used the Grindley report to support their opposition to reform and to argue that official inquiries were 'fake and inconsistent'.[81] The journalists were quick to spot that the new saltpetre squad was a ruse by the meatpackers (then dubbed the 'Beef Trust'). The *New York Times* wrote: 'It came out incidentally that this is to be paid for – in part it was said – by the Beef Trust, with the idea of course of restoring confidence in meat as treated by the packers in Chicago.'[82] The paper described the press conference with Harry Grindley: 'When he was asked who was going to pay the cost of the investigation, which he had said would be very large, he replied that it was a University of Illinois project. "Isn't the Beef Trust going to contribute?" he was asked. "Well," said the Illinois professor, "the American Meat Packers' Association will give some money towards the work. How much I cannot say."'[83]

The background to this is that the day the law of 1906 had entered into effect, 70 meatpackers had got together and decided to create an organization to defend their interests. And so the American Meat Packers Association (AMPA) was born.* Its first mission was to prevent the ban on saltpetre.[84] To do this, the AMPA set up a 'mirror investigation' explicitly modelled on the official saltpetre squad investigation led by Wiley. In April 1907 a Washington newspaper described it in broad terms: 'The "saltpeter squad", as it might be called, will be boarded in a specially equipped house in such a way that the weight of all foods eaten by each man can be accurately determined and the food completely analyzed. A physician will keep a daily record of the physical condition and health of each member of the squad.'[85]

The meatpackers would admit later that the experiments were carried out 'at the request'[86] of the American Meat Packers Association, 'as a result of the threatened prohibition by the federal government of the use of saltpeter in curing meats'.[87] They had drawn up the programme and committed to covering all the expenses.[88] 'We obligated ourselves', one of them admitted a few years later. 'We made some sort of a promise that we would, when

---

* The American Meat Packers Association became the Institute of American Meat Packers in 1919, then the American Meat Institute (AMI) in 1940. In 2015, the AMI became the NAMI (North American Meat Institute).

the cost would be found – that we would all subscribe certain amounts to pay that because we thought it was a very necessary expense incurred.'[89]

ARBITRATION

Wiley had no doubt that this so-called 'investigation' was nothing but a plot to derail the work of the Bureau of Chemistry. He thought it was a clumsy tactic on the part of the packers: he felt that this 'alternative study' would have no scientific credibility, given that its sponsors were so easy to unmask. Knowing how important saltpetre was in the meatpackers' business model, who would believe that a study conducted under their control would reveal the truth about the possible deleterious effect of saltpetre on health? In a private letter, Wiley noted ironically: 'I should be interested to see the report of the commission which the devil would appoint to see whether or not the Bible is an inspired book.'[90] He concluded: 'My impression is that the results of this test under the patronage of the packers will not be received with very great consideration.'[91] He failed to understand that this manoeuvre of the packers could be successful even if it did not reach any scientific result.

During the summer of 1907, Wiley and his colleagues began to draw up their report presenting the results of the official saltpetre squad. The *Washington Times* stated that saltpetre would probably be banned, noting that 'saltpeter is very extensively used in connection with the packing and preserving of all sorts of meat products, and its prohibition would be especially a serious blow to that industry'.[92] The annual bulletin of the Washington health officer indicated: 'Saltpeter is not classed among the food preservatives, but its use in certain meat products is a common practice, and should be prohibited in view of its poisonous character.'[93]

On 30 October, the *Washington Herald* announced that the official report was finally 'ready for the printer'.[94] It had been proofread and typeset, and its title was announced: *Influence of Food Preservatives and Artificial Colors on Digestion and Health, Vol. VII, Potassium Nitrate*.[95] An article headlined 'Saltpeter sausage violation of the law?'[96] indicated that in Washington the local health authorities were waiting for the report in order, if necessary, to begin action against any offenders. But behind the scenes, the meatpackers asserted that saltpetre couldn't possibly be condemned on the basis of the Bureau of Chemistry study when their 'mirror study' had only just begun. The Department of Agriculture decided to block the publication of the report, under the pretext that the scientific findings proffered by Wiley were unconvincing.[97]

As a result, saltpetre remained authorized 'pending further enquiry'. This delay caused much confusion: in February 1908, the head of army supplies at the War Department complained that the quartermasters didn't know which regulation to apply – he stated that there was one rule that authorized saltpetre and another that forbade it: 'A decision with regard to this matter is necessary as this Department, because of the difference of views in these two publications and also because of an expressed opinion in writing received from Prof. Wiley stating that the use of saltpetre was considered deleterious, has prohibited its use in the curing of meat for the army, the result being that a number of the packers have declined to submit proposals unless the use of saltpetre were permitted, which results in the Department paying a higher price for the meat food products as competition is greatly reduced.'[98]

One month later, under pressure from the meatpackers, the Bureau of Animal Industry of the Department of Agriculture unilaterally decided to publish a new regulation.[99] It ignored the results of Wiley's Bureau of Chemistry and removed the obligation to run further tests. As the *National Provisioner* put it, 'Saltpeter is permitted. The words "pending investigation" have been dropped';[100] 'The revised regulations now announced supersede all old regulations and amendments.'[101]

Though there was nothing more at stake, the meatpackers' saltpetre squad continued, but its results only came out in dribs and drabs. The final volume of its report didn't ultimately appear until twenty years after the experiment and actually had nothing to say on the impact of saltpetre-treated meat on human health. In its introduction, a note indicated that Harry Grindley and his team had encountered a number of difficulties and hadn't managed to reach a consensus on basic facts: 'there is a distinct conflict between the two joint principal authors of this report. This conflict concerns personal viewpoint, methods of presentation and interpretation of scientific data, and apparently in some instances, even accuracy of record of observations.'[102] As for the report by the Bureau of Chemistry, that never saw the light of day. In a book published towards the end of his life, Dr Wiley insisted that the report was 'censored': he claimed that the US Department of Agriculture abandoned its publication because 'vigorous protests from those engaged in adulterating and misbranding foods were made to the Secretary of Agriculture against any further publicity in this direction'.[103]

CHAPTER 7

WONDER PRODUCT OF THE TWENTIETH CENTURY: SODIUM NITRITE

As we have seen, the main nitro-additive used before 1820 was potassium nitrate (saltpetre). During the nineteenth century, a fast-acting reddening agent made an appearance: sodium nitrate. At the beginning of the twentieth century, a new – and even more active – product started to come into use, illegally at first: sodium *nitrite*.

In 1925, the American government allowed its use in meat-packing factories.[1] Elsewhere, industrial meat-processors had an uphill task getting their health authorities to authorize it. In France, for example, permission wouldn't come until four decades later, in a decree of 8 December 1964.[2] The archives reveal the reason for this 40-year delay. The issue was quite simple: in France, sodium nitrite was included in the official 'list of poisonous substances' under the category of 'dangerous products'. In order for sodium nitrite to be used in meat-processing, the health authorities had to take an exceptional step: allow its use in low doses, exclusively in the form of a pre-mixed product called 'nitrited curing salt'. In this and the following chapter, we will see why this decision was taken: at the time, authorizing 'nitrited curing salt' in low doses seemed the lesser evil compared to the then anarchic and widespread use of pure sodium nitrite.

MEDICAL NITRITE: AN ABORTED EXPERIMENT

Even before it was understood that it made meat carcinogenic, sodium nitrite was identified as a dangerous substance because it causes an illness called methaemoglobinaemia. The nitrite releases nitric oxide (NO), which binds to iron in the body. The blood and the muscles lose their ability to transport and store oxygen. The flesh changes colour, the skin and mucous membranes take on a bluish tinge – the technical term is cyanosis (from 'cyan', blue) – and methaemoglobinaemia induces a lack of oxygen that is often fatal. This toxic effect of sodium nitrite has been known since the middle of the nineteenth century. 'Blood that has undergone action by nitrites loses virtually all power to fix oxygen', the Scottish biochemist Arthur Gamgee noted in 1863.[3] In a treatise on toxicology published in Paris ten years later, Dr Antoine Rabuteau described sodium nitrite and potassium nitrite as 'haematic poisons' (poisons of the blood), and explained how they 'powerfully modify the sanguinary fluid, altering not only the cells but the plasma too'.[4]

Because of its powerful effect on the blood, sodium nitrite was sometimes used in minute doses (0.065g) to treat cardiac disorders such as episodes of angina. This is when sharp chest pains occur as the arteries supplying the heart become restricted: the administration of nitrite can dilate the blood vessels.[5] Encouraged by the good results achieved in cardiology, a few British doctors had the idea of treating epileptic patients with tiny doses of nitrite. Early publications on this topic (1882) pointed out that in most cases seizures stopped immediately.[6] One enthusiastic experimenter wrote: 'The nitrites, on account of their remarkable physiological action, will, I venture to anticipate, in the form of their simple and more safely administrable compounds, soon have a wide and important application in the treatment of various forms of disease.'[7] The apparent success of this therapy created a sort of frenzy

114

in the medical community: even foreign journals encouraged practitioners to take a leaf out the British doctors' book and test the effect of sodium nitrite on epileptics.[8]

In 1883, however, the medical press was starting to ring alarm bells: at a conference, a doctor reported that, after giving sodium nitrite to an asthmatic patient and two epileptics, 'all three returned in a few hours with blue lips, in a state of semi-collapse, evidently poisoned by the drug'.[9] Another doctor said he had given sodium nitrite to an epileptic child – the seizures became worse. In three other cases, sodium nitrite treatment had 'given rise to alarming symptoms'.[10] *The Lancet* published an article in which two doctors described experiments they had conducted with sodium nitrite on mice, cats and human patients.[11] Nausea, vomiting, fainting, lips turned blue or violet; one patient 'said she thought she would have died after taking a dose; it threw her into a violent perspiration, and in less than five minutes her lips turned quite black and throbbed for hours; it upset her so much that she was afraid she would never get over it.'[12] Another woman 'said that ten minutes after taking the first dose – she did not try a second – she felt a trembling sensation all over her, and suddenly fell on the floor. Whilst lying there she perspired profusely, her face and head seemed swollen and throbbed violently, until she thought they would burst.'[13] One man 'felt as if [his head] would split in two'. A young woman vomited for two hours, 'she thought she was dying'. Another 'said the medicine upset her so much that she went off in hysterics, and could not hold a limb still'.[14] The *Lancet* article continued: 'They seemed to be pretty unanimous on one point – that it was about the worst medicine they had ever taken. They said if they ever took another dose they would expect to drop down dead, and it would serve them right. One man, a burly, strong fellow, suffering from a little rheumatism only, said that after taking the first dose he "felt giddy", as if he would "go off insensible". His lips, face, and hands turned blue, and he had to lie down for an hour and a half before he dared move. His heart fluttered, and he suffered from throbbing pains in the head. He was urged to try another dose, but declined on the ground that he had a wife and family.'[15] The article concluded with a typically British understatement: 'It must be admitted that our experiences have not been altogether satisfactory.'[16] One of the doctors explained: 'This was the first intimation I had had from any source that nitrite of sodium was a toxic agent.'[17]

One after another, the proponents of sodium nitrite therapy started backtracking. They published articles in which they explained why – contrary to what many had believed – this chemical should be handled with extreme caution.[18] They accused each other of having underestimated the risks, and even of having recommended incorrect, possibly fatal dosages.[19] By the end of 1883, most of the therapeutic applications of sodium nitrite had stopped. In cardiology, they were replaced by the use of less dangerous substances – such as 'glonoin' or trinitrin. A few years later, a German doctor wrote that nitro products 'have sometimes been recommended and used by American and English doctors for angina, asthma [...] and even epilepsy. But these therapies now seem to have been completely abandoned, and in view of the toxicity of these products, that is no bad thing.'[20]

SODIUM NITRITE, A POWERFUL – BUT REGULATED – TOXICANT

At the end of the nineteenth century, experimenters established the injected doses above which sodium nitrite presents a serious risk: above 0.2g per kilo of body weight, a frog would experience difficulties breathing, then would die.[21] In dogs, the lethal dose was established at 0.15g of sodium nitrite per kilo of body weight. The cat has lower resistance: ingestion of 0.035g per kilo is enough to kill it.[22] Nowadays, in the United States, formulae based on sodium nitrite are used to eradicate invasive local mammals, in particular certain species of wild boar. In New Zealand, a pesticide company offers Bait-Rite, a 250g bait pellet dosed with 10% sodium nitrite for killing feral pigs and marsupials.[23] The Australian company PetSmart markets an almost identical product under the brand name Hog-Gone. The irony of the situation was not lost on one journalist from *Chemical and Engineering News*, who wrote an article in 2014 entitled 'Counterattacking the wild pig invasion with bacon preservative sodium nitrite'.[24]

In humans too sodium nitrite is a formidable poison. Loss of consciousness occurs rapidly: 'The effect of sodium nitrite occurs within minutes after ingestion and after 30–60 minutes maximum',[25] explained the Dutch intensive care doctors who treated several victims in 2018. They went on: 'The oral intake of 1–2g of sodium nitrite will give severe symptoms; it is estimated that 4g of sodium nitrite is lethal.'[26] Toxicology studies confirm that the fatal dose lies somewhere between 2g and 4g for an adult weighing 75 kilos, in other words a teaspoonful.[27] The medical literature has described several cases of suicide – including, occasionally, by workers in a meat factory.[28]

In the Netherlands there has been a recent upsurge of suicides by sodium nitrite following press coverage of the 2018 suicide of a nineteen-year-old woman who managed to get hold of the substance via the internet.[29] At the start of 2019, a Dutch medical journal devoted a study to a man of 27: 'After resuscitation for a total of 40 minutes, the patient died due to cardiac arrest caused by hypoxia due to methaemoglobinaemia. Afterwards, it became clear that a few weeks before his death, the patient had ordered sodium nitrite over the internet.'[30] Another patient (who had also procured sodium nitrite online) was found barely fifteen minutes after ingesting it. He was administered a life-saving antidote to re-establish oxygen in the blood. (By a strange quirk of nature, the antidote is the same colour as the symptom:

methaemoglobinaemia is treated with an intravenous injection of *methylene blue*.)

The history of nitrite is peppered with criminal episodes. In China in 2020, a schoolteacher in Henan province was sentenced to death for having attempted to poison 23 children by pouring sodium nitrite into their food.[31] A few years earlier, the national radio ran a report on a certain Mrs Fan, who had thwarted a murder attempt by her husband (he had poured sodium nitrite into her drinking water and herbal tea).[32] In a similar case, American law enforcers arrested one Tina Vazquez. She had procured some sodium nitrite in a meat-processing plant where she worked and used it to fill a small capsule which she then gave to her neighbour, whom she wanted to kill, passing it off as aspirin.[33] In reality, the dose used would probably not have been sufficient to kill, as it was less than the necessary 2 grams. The court nevertheless sentenced Mrs Vazquez to twenty years in prison for attempted murder.

THE DANGER OF ACCIDENTAL INTOXICATION

Beyond these newspaper headlines, sodium nitrite has been the cause of a whole host of accidents.[34] In Toulouse in 1943, five people died and 47 were intoxicated when a baker made some biscuits to which sodium nitrite had been added instead of baking powder. When the police arrived at his bakery, he tried to prove his innocence by swallowing some of his 'baking powder'; he was immediately stricken with cyanosis and only just managed to escape with his life.[35] In *The Lancet*, one doctor described a tragic case that occurred in Salisbury in 1954: a two-year-old boy died 'three hours after licking an almost empty bottle of sodium nitrite'.[36] It subsequently appeared that 'the sodium nitrite had been bought at a chemist's shop by the child's elder brother, aged 9, who wanted it for a chemical experiment about which he had read in a book'.[37] In South Africa sodium nitrite, 'mistaken for table salt, had been sprinkled over a meat and cabbage meal in a hostel. The majority of approximately 30 diners became ill 1–2 hours after the meal.'[38] In India, the Food and Drug Toxicology Research Centre in Hyderabad described an accident in which fourteen people from a single family were poisoned by a tamarind soup 'because sodium nitrite, kept in a polyethylene packet in the kitchen, was used by the relatives, mistaking it as table salt, which was also kept in the same rack in another container'.[39]

In 1945, a team of New York doctors recorded three series of fatal accidents: the poisoning of three men who salted their stew with nitrite (three deaths); a family who used nitrite instead of salt (three deaths); and a group of eleven men who ingested nitrite while eating breakfast (one death).[40] This last event was the subject of a book that became a bestseller at the time: entitled *Eleven Blue Men*, it recounted how an employee of a cafeteria in Manhattan made a mistake when refilling the salt cellars. Instead of salt, he filled them with sodium nitrite 'that he used in corning beef and in making pastrami'.[41] Less than half an hour after eating their morning oatmeal and leaving the restaurant, the eleven customers collapsed. The book described how 'All were rigid, cyanotic, and in a state of shock. The entire body of one, a bony, seventy-three-year-old consumptive named John Mitchell, was blue.'[42] Another 'blue man' was discovered by a police officer 'slumped in the doorway of a condemned building', and yet another discovered at the end of the day in a hotel where, for nearly nine hours, he 'had been lying, too sick to ask for help, on his cot in a cubicle'. The doctor who tended to the sick men noted: 'Sodium nitrite isn't the most powerful poison in the world, but a little of it will do a lot of harm.'[43]

In the industrial domain, sodium nitrite is used to prevent metal corrosion. In refrigeration systems operating on a closed circuit, the application of nitrite prevents the gears rusting up (because when dissolved in a fluid, sodium nitrite acts as a strong oxidizing agent). Unfortunately, it sometimes happens that nitrited refrigerating fluid is absorbed accidentally. In the Netherlands, a fault in the refrigeration unit of a lorry transporting food led to two deaths.[44] In the USA, more than 40 children from one elementary school 'developed acute onset of blue lips and hands, vomiting, and headache after the school lunch periods':[45] some nitrited solution had accidentally been used in the preparation of the food.[46] In 1994, a 34-year-old secretary was admitted to Birmingham hospital after drinking water mixed with a sodium nitrite liquid 'intended for the central heating system'.[47] She survived after emergency care. In Hertfordshire, a young woman wasn't so lucky; the medical registrar at Watford General Hospital reported: 'Miss H., a 17 year old dental nurse, was admitted to the casualty department at 18.30 hours after being found at home in an hysterical condition and bright blue.'[48] Two hours later, she was dead. It transpired later that she had accidentally swallowed a small pellet of sodium nitrite used as anti-corrosion protection for surgical equipment. Describing this tragedy, the *British Journal of General Practice* stressed that in hospitals 'the label on a sodium nitrite container indicates that the material is toxic by ingestion both in text and using the standard symbol of a skull and crossbones in black on an orange background'.[49]

JOHN HALDANE DISCOVERS HOW NITRITE REDDENS

The idea of using sodium nitrite in processed meat came about when chemists discovered that at very low doses, this substance could achieve the same reddening effect as saltpetre, only much more quickly. A prominent figure in this story is the British biochemist John Haldane, a specialist in the respiratory functions of blood and haemoglobin. In 1897, at the Oxford Laboratory of Physiology, Haldane noted that the flesh of mice that had been injected with nitrite became more and more red after the animal died.[50] Then Haldane conducted a thorough investigation into the question of the reddening of the flesh: in a ground-breaking article ('The red colour of salted meat'),[51] he revealed the mechanism by which, when meat was cured with nitrate or nitrite, it became pink rather than grey. He proved that nitrate only acts on the meat when it is transformed into nitrite. Drawing inspiration from the work of a German specialist in the detection of adulterated foodstuffs[52] and a young chemist who had studied the action of nitrite and sulphuric acid on meat,[53] Haldane described in detail the transformations that gave rise to the appearance of the pink pigment in meat treated with nitrite. He noted: 'Nitrites are somewhat actively poisonous; and as has been recently shown they cause death by asphyxia through their action on the haemoglobin and the consequent interference with the oxygen supply to the tissues. Their presence in very appreciable quantity in raw salted meat might thus seem to be possibly harmful.'[54] Haldane noted that in the scraps of nitrited meat he analysed, he detected hardly any nitrite: the nitrite had changed into nitric oxide, which, when it interacted with the haemoglobin present in the meat, gave the nitrited flesh its distinctive colour. John Haldane was thus the first to identify the pigment in nitro-meat, which he named nitro-oxy-haemochromogen (NO-haemochromogen,[55] sometimes also called nitroso-haemochromogen, nitro-haemochromogen or nitroxy-haemochromogen).[56]

Almost simultaneously, the Russian chemist S. Orlow had embarked on a virtually identical project: seeking to explain the coloration of sausages and salamis that had been treated with saltpetre, Orlow conducted a series of experiments using nitrite. In 1903, he wrote: 'after fifteen to twenty minutes of boiling, the pieces of beef and pork had acquired a very fine pink coloration, exactly analogous to the shade of sausages. One can conclude from this that these commercially available sausages are boiled in the presence of nitrates, or even that the flesh itself is impregnated with these salts.'[57] By establishing that sausages treated with nitrite were identical to sausages treated with nitrate, Orlow could confirm Haldane's conclusions: meats cured with saltpetre are pink because, in fact, saltpetre is transformed into nitrite. Like Haldane, Orlow suggested this offered good reason to be wary because 'these compounds are dangerous for the organism'.[58] He asserted that 'nitrites act on the organism in a very harmful way, these are toxicants which induce the formation of methaemoglobin, dilation of the blood vessels, a weakening of the function of the brain and spinal cord and irritation of the digestive tract'.[59] However, he did note that the quantities of nitrite detected in the final products were quite low: in his view, 'the maximum content of nitrite observed in sausages is equal to the minimum dose prescribed by doctors for certain illnesses'.[60] After analysing strips of commercially produced processed meat, Orlow concluded: 'At most, we found 0.012% nitrite of soda, which is less than the medical dose. However, in certain cases, larger quantities can be found, which might be considered harmful.'[61] This was the point picked up by the technical press: the French review *Le Mois Scientifique et Industriel* [The Scientific and Industrial Month] stressed that the maximum content detected in sausages 'is at the most equivalent to the lowest therapeutic dose'.[62]

FIRST ATTEMPTS AT SODIUM NITRITE CURING

In the USA, the *National Provisioner* reported on John Haldane's results in its April 1903 edition under the title 'The red color in salted meats chemically explained'.[63] The head biochemist at the Department of Agriculture's Chicago office reproduced and verified Haldane's experiments. He confirmed that the coloration was derived from the nitrite that was produced in the chemical decomposition of the nitrate.[64] This led to the idea of using pure nitrite to skip a few steps. The very first experiments were secretly carried out in a few American meatpacking factories.[65]

In 1909, a German vet, Friedrich Glage, brought out a booklet for pork butchers which offered to reveal 'an easy method for giving meat and sausages an intense red colour'.[66] He showed how nitrite could be produced by heating nitrate to a temperature of 350°/400°C. He explained that anyone could do it in the back of their shop and procure a fast-acting powder. These amateur alchemists were warned, however, of the risk of explosion: saltpetre is a very unstable substance, and 'many have already lost their lives'.[67] Glage explained that it was a case of allowing producers to 'satisfy the wishes of the public, who are only interested in meats that taste good and are very red'.[68] He noted: 'The effect is accelerated because it is no longer necessary to wait until the meat has broken down the saltpetre. Because of this, the sausage is ready more promptly. This is especially useful for sausages meant to be eaten very quickly, because they often have to be made in very large quantities in very little time. What's more, the finished effect is always consistent, always satisfactory, and the colours obtained don't fade. The red colour is strong but nonetheless soft. This shade corresponds exactly with the colours we see on this type of merchandise today. The sausages appear more appetizing, the public is more willing to buy them, all to the benefit of the butcher.'[69] His techniques were widely adopted in professional publications.[70]

During the First World War, the German chemists Fritz Haber and Carl Bosch developed revolutionary procedures which allowed the artificial synthesis of alkaline nitrites for use in industry and armaments, particularly metallurgy factories.[71] Sodium nitrite became widely available and inexpensive, and German butchers began to use it more and more.[72] Various entrepreneurs submitted patent applications, hoping to cash in on royalties for this technology: the US chemist George Doran obtained a series of patents relating to nitrite curing, which he described as 'a process whereby a complete cure of packing-house meats may be effected in a more convenient and rapid manner than heretofore'.[73] The manufacturers of food additives played catch-up in a race to acquire licences: the managers of Heller & Co. pursued George Doran for months to get him to grant them the right to exploit this miracle recipe.[74] In Europe, several products based on sodium nitrite began to appear on the market. In Prague, a certain Ladislav Nachmüllner sold 'Praganda',[75] while a German company offered a nitrite mixture under the name 'Nitrosin Salpetre'.[76] The Berliner Georg Lebbin obtained German, French, Finnish and Swiss patents for a 'procedure for curing meats' with sodium nitrite,[77] and his 'Lebbin salt' was added to the list of fashionable additives.[78] Accidents soon followed: in 1916 in Leipzig, a communal meal with nitrite-treated meats ended in dozens of hospitalizations and one death.[79] Without delay, the German authorities passed an ordinance on 14 December 1916[80] that stated that the use of sodium nitrite (and all other products containing salts of nitrous acid) was 'forbidden in the commercial preparation of meats and sausages'.[81] The 'import, stocking and selling of meats containing' the substance were also banned.[82]

At the beginning of the 1920s, chemists developed new techniques which consisted of mixing sodium nitrite with cooking salt by a process of dissolution/evaporation. This was the invention of 'nitrited curing salt', which itself became the subject of a number of patents.[83] Formerly used in a pure form, sodium nitrite was now being thoroughly mixed together with salt. Poisoning by nitrite overdose was now impossible: if the butcher added too much of this mixture, he would end up with a product so salty that it would be inedible. Buoyed by this safeguard, the producers of chemical additives and the patent owners sought to have their 'nitrited curing salt' legalized, but at first most countries refused to authorize this method. For example, in 1922, the medical experts of the German health authority decided to uphold the complete prohibition of nitrite curing. Their justification was as firm as it was laconic: 'suffice it to recall the original reason for the ban currently in force: sodium nitrite already has toxic effects at doses less than one gram. Such a substance has no place in the food industry, in the restaurant trade or in the kitchen.'[84]

COMPETITIVE ADVANTAGE

At the start of the 1920s, the American meat-processing industry was facing a significant economic crisis because food exports to Europe had gone into steep decline at the end of the First World War. The Chicago factories saw their share of the international market shrink to a worrying degree. The US Department of Agriculture representative in London made a tour of European capitals. His 1922 report entitled 'Danish Bacon Displaces American Product on British Markets'[85] concluded that American bacon was too salty and not appetizing. Factories had to find a way of producing mild cured bacon that could withstand the Atlantic crossing, even without refrigeration. 'American exporters of meat operate under big handicap',[86] the association of US meatpackers complained. As their lobbyist put it: 'Danish, Swedish, Dutch and Irish all enjoy advantage of proximity to European markets.'[87] One meatpacker railed against this injustice: 'Denmark is much nearer to England than is the United States. A shipment from Denmark may reach England within forty-eight to seventy-two hours from the time it left the Danish port.'[88]

So the US meatpackers set out to regain their competitive edge. They lobbied the government for the authorization to run full-scale tests 'to determine the practicability of using nitrite directly'[89] in order to accelerate production and lower costs.[90] The experiments concentrated on the technical aspects: they wanted to check whether production using nitrite could be cost-effective; whether nitrite posed too great a danger to those handling it in the factories; whether the procedure gave satisfying results from a colour and taste point of view; and finally, whether the resulting processed meats contained more nitrite than when they were produced using nitrate. The tests were described in detail by officials from the US Department of Agriculture. The initial experiment consisted of producing 80 large hams, made from the rear thighs of 40 pigs. 'To eliminate differences in the meat due to age, feeding, conditions of the animals, etc., they were carried through the cure in pairs – a left ham, for example, being given the nitrite cure, the corresponding right the nitrate, and the two again brought together for final judging.'[91] Three criteria were used to assess the hams: colour, taste and texture. The results were declared excellent: 'All judges agreed that the nitrite-cured meat was fully equal in quality and flavor to that cured by the customary process. No deficiency in quality or flavor was noted by any of them, and none of them were able to distinguish the nitrite-cured meat from that cured by the customary process.'[92]

The report noted that 'one-fourth of an ounce, or less, of sodium nitrite appears to be sufficient to fix the color of 100 pounds of meat.'[93] From a technical point of view, there were no specific problems: 'No curing difficulties caused by the proper use of nitrites were observed. At no establishment was any increase in the proportion of spoilage observed.'[94] As a consequence, the tests were run again on a larger scale: seventeen meatpacking factories began using sodium nitrite on an experimental basis. They all found that the use of nitrite hugely reduced the time of production: 'In one establishment the shortening amounted to 60 per cent of the former curing period; in the other establishments the shortening amounted to from 10 to 40 per cent of the former curing period.'[95] As well as the advantage of speed, the process produced a more uniform pink colour. According to the scientific director of the meatpackers' association, 'The hams, bacon, beef tongues, and beef hams cured by nitrite were in every way as high-grade and desirable products as those cured by nitrate. The color produced by the nitrite was produced more rapidly and in general was a better and faster color.'[96] He stated that for certain products the nitrite procedure resulted in meats containing less nitrite than in the same products using nitrate.[97]

AMERICAN AUTHORIZATION

In October 1925, at the request of the meatpackers and with no further trials, the US Bureau of Animal Industry authorized the use of sodium nitrite in meat factories.[98] The meatpackers' representatives pointed out that this authorization would lead to considerable savings.[99] To justify its decision, the Bureau of Animal Industry explained that under this method the production time could be shortened and the dosage of additives controlled more strictly.[100] The official report was insistent that: 'Since sodium nitrite consists of the color fixative in active form, the quantity required is much less than is the case with the nitrates.'[101] The criterion used to determine the authorized dose was simple: achieving the right pink colour, that is, a homogeneous coloration that didn't fade.[102] Moreover, the Bureau of Animal Industry specified that the additive was used exclusively for its colouring effect: 'The function of sodium nitrite in the curing of meats, like that of sodium nitrate and potassium nitrate, is the fixation of the red color.'[103] The bureau stipulated: 'Neither the nitrates nor nitrites are of any particular value as preservatives in the quantities used.'[104]

When they heard that sodium nitrite had been authorized, many medical professionals were up in arms. The American Public Health Association pointed out that there had been no proper evaluation of the health impact of the additive. An editorial in the *American Journal of Public Health* warned that the decision gave the meat industry 'the privilege of using nitrite solely as a dye in imparting a natural red color to the preserved meats. On the basis of this claim it would seem that nitrite should be specifically forbidden.'[105] Doctors wrote: 'It is the opinion of many eminent experts that sodium nitrite, even in very small quantities, is a highly dangerous ingredient in our foods. We should look with suspicion upon any further inroads on the purity of our food supply in the way of permission to use substances therein which are recognized as inimical to health.'[106] With one voice, the meatpackers and the chemists at the Bureau of Animal Industry responded that the nitrite levels in nitrite-treated products were acceptable because they were no higher than in products treated with nitrate.[107] Consequently, they couldn't be *more* harmful ... So why ban them?

CONTAGION

Once the USA had authorized nitrite curing, curers in other countries pointed out that they were at a competitive disadvantage. One chemist noted for example that 'the New Zealand meat packers are being undersold by American packers who use this process',[108] and attributed this disadvantage 'to the economies resulting from the use of nitrite'.[109] To protect their own industries, some countries adopted exactly the same standards as the USA. This was the case in South Africa, for example: Cape Town authorized nitrite curing in 1929.

In Europe, nitrite use was authorized in only one country: Austria. Elsewhere, sodium nitrite circulated as contraband, in liquid, powder or tablet form.[110] The prohibition was most weakly enforced in Germany, because traffickers could easily lay their hands on nitrite in Austria. The German veterinary expert Raphael Koller wrote about how, in the 1920s, 'the ban on sodium nitrite was not respected in Germany. The pork butchers would rather risk punishment than surrender the advantages that this new product gave them.'[111] Koller explained that dishonest druggists peddled pure sodium nitrite to the butchers: 'The reputation of cured meats is compromised by this trafficking. In Germany (as no doubt abroad), sodium nitrite and potassium nitrite are sold under the counter, under various fictitious names such as "Saltpetre Bar" (in fact potassium nitrite) or "Extract of Saltpetre", "Saltpetre Concentrate", "Salpetrine", "Scouring Salt", "Purifying Salt", "Reddening Bar", etc. And also, sometimes, under their real name.'[112] In 1935, a medical journal thus presented the case of a German meat curer who for several years had been able to buy a ready-made nitrited mixture. From 1928 he acquired barrels of sodium nitrite directly from a chemical factory and had one of his workers mix it with salt.[113]

Because of these widespread violations of the ban, there was an upsurge in cases of nitrite-related cyanosis in the German Reich. Cases of poisoning were recorded almost every year.[114] In 1929, for example, 25 people suffered serious intoxication from nitrite-treated mince. In 1931, twelve people were hospitalized in Fürth after consuming a sausage soup (*Wurstsuppe*) treated with pure nitrite.[115] After another case in 1934 linked to nitrited 'boiled sausages' (*Brühwürstchen*), the German authorities decided to put an end to this illicit activity by adopting the 'law on nitrited curing salt' (*Nitrit-Pokelsälz-Gesetz*, also known as the *Nitritgesetz*).[116] The law forbade meat-processors from possessing pure nitrite but permitted use of a salt + nitrite mixture manufactured by authorized establishments who could guarantee that the proportion of sodium nitrite was never in excess of 0.6%. The reference dose was set at 0.5%. If a producer mixed 20g of this compound with a kilogram of meat they would be adding just 100mg of sodium nitrite.[117] With this limit, German authorities were trying to protect the population against the risks of immediate intoxication. They didn't know that the chemistry of nitrite had other cruel tricks up its sleeve.

CHAPTER 8

NITRITE IN THE UNITED KINGDOM AND IN FRANCE

A new law on food additives was passed in the UK in 1925. For meat products, it authorized the use of nitrate, and even, in the case of sausages, of sulphur dioxide.[1] The total ban on *nitrite*, on the other hand, remained. But the importers, distributors and producers lobbied incessantly for a relaxation of the rules. Their French counterparts were no less determined: they too argued for the legalization of nitrite curing on the basis that it would be a boost for business. But for decades the health authorities resisted. In the end, the founding of the European Common Market in 1964 made most industrial meat curers happy: in the name of commercial harmonization, the processing of meat using sodium nitrite was legalized everywhere. It was a triumph for what was still known at the time as 'the chemical method' or 'the American cure'.[2]

BRITISH BACON AND SODIUM NITRITE

In early 1926, a distributor of chemical products in London wrote to the British Ministry of Health to ask whether, as he hoped, they would follow the American example: 'We are advised that the use of nitrite of soda in the curing of meat has received the final sanction of the U.S. Bureau of Animal Industry [...] We understand that the function of nitrite is the fixation of the red colour and shall be glad to know whether the use of the chemical for this purpose has received or would receive the sanction of your department.'[3] The chief food officer at the Ministry of Health replied in the negative. He noted that 'sodium nitrite is a dangerous drug, with a powerful action on the heart' and considered that 'its use for the purpose suggested is therefore open to great objection'.[4] The Ministry of Health then set out its official position, which would remain virtually unchanged until the Second World War: as the toxicologist of the ministry wrote, 'nitrite has always been regarded as one of the more poisonous preservatives'.[5] As a consequence, 'It does not seem to be a proper course to use the Preservatives Regulations as a means of facilitating the extended use of a toxic substance, especially when it is admitted that the substance is not used for purposes of preservation but for fixing the red colour in cured meat.'[6] At the same time, HM Revenue and Customs reported that they were picking up an increasing number of unusual cargoes. The central laboratory in Liverpool (the major port of entry for American processed meats) alerted Whitehall: analyses were often showing up excessive levels of nitrite in the processed meats they examined.[7]

For their part, British manufacturers were quick to complain to the authorities: 'If you won't allow us to use nitrites are you proposing any action to prevent the Americans doing so?'[8] After a meeting with the representative of E.M. Denny and Co. Ltd, the Ministry of Health noted: 'He stated that his firm considered that they were at a great disadvantage compared with America. Nitrite acts much more quickly in curing and it would be a great service to them if they could use a limited amount of nitrite.'[9] And after a further meeting: 'if American products to which nitrites have been added are admitted into this country, it would only be equitable that they should be permitted to adopt the same practice as the Americans.'[10]

The British Food Manufacturers' Research Association pressed the industry's demands. Its director wrote letters and had several long meetings at the ministry. He never failed to mention that the nitro-additive 'is used only to give a colour to the meat'[11] and that it is thanks to nitrite that 'the necessary colour compound is immediately produced on penetration of the pickle'.[12] He emphasized that saltpetre worked by being transformed into nitrite; as a consequence, saltpetre-treated hams were no less problematic: 'if there is any question of the injurious nature of nitrites in these products, the permission to use sodium nitrite would enable most curers to produce goods containing the minimum amount of this substance, as the quantity present can readily be regulated, which is not the case when nitrates are used.'[13] Finally, he explained: 'it should reduce costs. The smaller quantity of nitrite required, together with the shorter time of curing and the additional number of times the brine can be used, should result in a considerable saving in the course of a year.'[14] From the point of view of the manufacturers, the economic argument was perhaps the only one that really counted. This was evident, for example, in the minutes of a meeting held at the Ministry of Health with a delegation of industry leaders: 'We asked whether there was any advantage in using nitrite other than the saving of time. Did it produce a better or sounder bacon? The answer was no; saving of time was the only consideration.'[15]

A TRAGEDY IN MIDDLESBROUGH

At the end of the First World War, the British government had initiated the construction of a new chemical factory at Billingham in the borough of Stockton-on-Tees in County Durham. The Billingham plant used the 'Haber-Bosch' technology perfected by the Germans during the war: it captured the nitrogen content in air and used it to produce nitrate, nitrite and other derivatives of synthetic ammonia. The nitrate made in Billingham (sodium nitrate, ammonium nitrate, barium nitrate, potassium nitrate, etc.)

was used in heavy industry and for the manufacture of explosives and fertilizer. As for the nitrite, the Billingham factory predominantly manufactured the cheapest version: sodium nitrite. It was used for the treatment of metal and production of pigment for dyes (known as 'azo-dyes'). Before the Billingham factory was built, it was very expensive to produce sodium nitrite in Britain: it was obtained by transforming nitrate imported by sea from South America, where it could be found in mineral form. The Haber-Bosch technology, on the other hand, allowed the fabrication of synthetic nitrite at very little cost.[16]

In 1926, Synthetic Ammonia and Nitrates, Ltd of Billingham was acquired by a new group, Imperial Chemical Industries (ICI), which grew rapidly, quickly becoming one of the main employers of the nearby town of Middlesbrough. One Sunday in August 1936, Raymond Cooper, a 44-year-old employee of ICI, was at home having lunch with his family when he started to display the characteristic symptoms of cyanosis: first, his skin took on a slate-grey tinge, then turned a distinct blue colour. At the table with him were his wife Madeline, 42, and their daughter Dulcie, five. The doctor arrived too late: 'When I reached the house both the man and his wife were dead, and the child had been removed to hospital. I was told by neighbours who occupied adjacent rooms that the family had complained of abdominal pains immediately after the meal, that they had vomited, and their faces "went blue".'[17] A few days later, the newspapers explained that 'A fourth member of the family, Maureen, 10, had left for Sunday school before the meal began, and is the only survivor.'[18] In the throes of her agony, her mother had had enough time to alert a neighbour: 'Mrs Cooper was terribly ill and told her to prevent Maureen from getting her dinner out of the oven when she returned from Sunday school.'[19] According to the chemist who conducted the forensic examination, 'so far as could be ascertained, the man and woman died approximately within an hour after partaking of the meal. The child was removed to the local infirmary and its stomach washed out, but it died after an illness of about 3 hours. The symptoms observed previous to death were characteristic of this type of poisoning, namely, difficulty in breathing, marked cyanosis, vomiting, and finally stupor and collapse. The dinner consisted of meat, potatoes, cabbage, Yorkshire pudding, rhubarb tart and custard.'[20]

Sodium nitrite was found in their kitchen. The evidence suggested that Raymond Cooper had brought it home from his work at Billingham. His wife had confused it for salt and used it in the cooking. Analysis revealed that the most deadly parts of the meal were the cabbage and the pudding: they contained 6.5% and 4.5% sodium nitrite respectively. In an article headlined 'Three deaths from sodium nitrite', *The Times* reported the coroner's remarks: 'when the deaths were discovered, people suspected the possibility of foul play or that their own food might be contaminated […] It is satisfactory to the public to know that the poison was accidentally introduced into the food.'[21] He concluded by urging for more caution with sodium nitrite: 'it may be a warning to manufacturers that it is dangerous and that they might take steps to educate their workmen that it should not be taken away from their works.'[22]

'COLOURED PRESERVED PORK'

A few months after the tragic events in Billingham, a man in Hull committed suicide using sodium nitrite.[23] There had already been other poisonings: 'Please attach Manchester food poisoning case in which a family was poisoned 2 or 3 years ago owing to the use of sodium nitrite in mistake for common salt,' said a note in a 1926 file entitled 'Use of nitrite in curing of bacon etc.' from the Ministry of Health.[24] The British authorities were concerned that the risk of accidents would greatly increase if they authorized nitrite curing, as this would have meant that nitrite could be freely traded among butchers. When representatives of ICI insisted that sodium nitrite should be widely authorized in the manufacture of bacon and ham, the ministry replied: 'As you are, however, no doubt aware, very few bacon curers in this country have had experience in the use of nitrites',[25] and so this substance 'in the hands of inexperienced persons may lead to disastrous results'.[26]

It was one thing banning sodium nitrite, quite another actually preventing it from being used. In 1935, Dr Monier-Williams, head of the Food Research Laboratory at the Ministry of Health, reported on an exchange he had had with an industry insider: 'He said the conditions here are chaotic. All the canners are using nitrite, most of them without scientific control, and the whole process is haphazard in the extreme. German travellers are pushing it everywhere and also *R-Salt*, which is a β-naphthol sulphonate, and which is apparently an admirable agent for getting a good colour in ham and bacon.'[27] According to this insider, 'Of course the meat can be cured much more quickly if nitrite is used, but the flavour suffers and much of the stuff now being produced in this country is no better than coloured preserved pork.'[28] Dr Monier-Williams concluded his memo by noting: 'I told him that the use of nitrite and of *R-Salt* was illegal and that local authorities were empowered to enter factories and enforce the Regulations by inspection.'[29] But one of his colleagues reckoned that 'the prohibition is unenforceable by the normal machinery of sampling and analysis, since, although analysis will discover if nitrite is present in meat, it cannot discover exactly how it was put there.'[30]

In 1937, the UK tightened up its regulations on food additives and imports of processed meats.[31] In order to be allowed to export processed meats to the UK, producing countries had to obtain an official certificate proving that the goods conformed to UK regulations. An official circular was distributed to the American meatpackers which stated: 'The addition of nitrite of soda as such to food, including bacon, imported into England and Wales is prohibited under the Public Health Regulations of England and Wales.'[32] But the Americans went on the offensive.[33] The head of the US Bureau of Animal Industry, a strong ally of the Chicago meatpackers, travelled to London. He accused British decision-makers of 'unfair discrimination' against the USA.[34] The dispute over nitrite reached as high as the office of the foreign secretary, Anthony Eden.[35] After months of consultation, the Foreign Office, however, had to inform the American authorities that the Ministry of Health was not willing to allow the import of nitrited ham and bacon: 'the Minister is not prepared, on the information before him, to allow the use of nitrites in the curing of meat for reasons of practical expediency'.[36] The Board of Trade sounded a warning: 'the USA are very disturbed about this matter and will not let it rest';[37] 'if they were dissatisfied they might be expected to take reprisals and from the Board of Trade's point of view the matter would be serious'.[38]

THE PRETEXT OF WAR

In June 1939, as war with Germany grew ever closer, it became clear that the UK was in danger of running short of bacon, because supplies from Denmark would be interrupted: according to the calculations of the importers, 'in about the third week after the out-break of war there might be a break in supplies'.[39] They went on to say that the Americans would be able to supply the required quan-tities at short notice – but that this bacon could not be imported unless Britain lifted its ban on nitrite-treated bacon.[40] They then asked the minister of health if, as an emergency measure, he would consider temporarily suspending the law. The minister answered that if the circumstances demanded it, he would effectively give a green light to the import of bacon treated with nitrite and with borax,[41] 'as a temporary and emergency measure'[42] and only in order that the importers could 'fill the gap'[43] left by the lack of non-nitrited bacon. On 3 September, war was declared; the ban on nitrited bacon was lifted in October.[44] In the space of a few weeks its scope was widened to apply not only to bacon but to ham and all types of 'cured and pickled meats', and then even to pâtés.[45] As the historian Derek Oddy put it: 'prewar regulations banning food preservatives were relaxed. Borax and sulphur dioxide were once more permitted in margarine, bacon, jam and dehydrated vegetables; sodium nitrite was permitted in bacon and pickled meat.'[46]

Initially intended to cover only bacon imported from the USA, the authorization of sodium nitrite was quickly extended to Canadian products too. Naturally, local British meat-processors wanted to enjoy the same advantages,[47] and ICI lobbied intensively for nitrite to be available for use by British curers.[48] But the Ministry of Health remained fearful of the prospect of seeing sodium nitrite in general use. In the end, only seventeen British manufacturers were authorized to use it.[49] The American exporters were delighted, British producers were livid.[50] The regulation was confirmed a few years later: on 1 March 1944 the Bacon (Control and Price) Order[51] came into force; it specified that 'the use of sodium nitrite is prohibited except under licence'.[52] To obtain a permit, British curers had to submit to a whole range of very strict safety obligations. The regulation specified that every meat-processing plant should appoint a designated qualified chemist who was personally responsible for the stock of sodium nitrite. As an essential precaution, the magic powder had to be kept in 'a suitable locked receptacle in containers clearly marked "POISON"'.[53]

THE FRENCH CASE

In France, nitrate had been officially authorized since 1912. As in the UK, nitrite remained banned – even after being authorized in the USA. As soon as the *Nitritgesetz* (Nitrite Law) was passed in

Germany, the Federation of Charcutiers of Alsace-Lorraine asked the French government to allow the use of sodium nitrite. This request was submitted to the main French authority on food additives, the Conseil supérieur d'hygiène publique de France (CSHPF – French High Council of Public Hygiene), which commissioned the toxicologist Frédéric Bordas to examine the issue.[54] He was much feared by those producers who employed underhand methods: according to Bordas, the mission of the expert toxicologist was to 'make sure that the health of the consumer was not sacrificed to the rapaciousness of unscrupulous manufacturers'.[55] He had fought against the use of sulphurous acid in sausages, which certain meat curers maintained was essential for preservation.[56] Even in the early years of the century, he had been talking about 'the many diseases attributed to other causes, but certainly due to the daily consumption of foodstuffs preserved with an antiseptic'.[57]

Bordas' 1934 report on the use of sodium nitrite was published the next year in the *Annales d'hygiène publique* under the title 'Les nitrites dans les saumures' (Nitrites in pickling brine). He wrote: 'The question in front of us is not a new one. We have known for a long time that beef and pork treated with nitrate eventually takes on a pink colour, an appearance much sought after by customers, as they see this pink hue as a sign of well-preserved meat.'[58] Dr Bordas rejected the arguments presented by the producers: they claimed that the nitrite-based procedure would be advantageous for public health, 'but what the meat-processors omit to tell us is that if you add nitrites to the brine, the desired effect (the pink coloration of the meat) is obtained in less than twenty-four hours, instead of thirty days'. Bordas concluded: 'Should we facilitate operations which consist in reality in fooling customers, with the aggravating circumstance that nitrite itself is toxic?'[59] So the French High Council of Public Hygiene refused to change the law: as in the UK, the use of nitrite remained banned in France. This merely led to more illegal use: reports noted that in France, 'nitrite remains banned in theory but is in fact used in secret, quite widely even.'[60]

FRENCH MANUFACTURERS FACE COMPETITION

After the Second World War, a few French companies received a special dispensation for products destined for export. But the curers and food additive merchants agitated for a general authorization. At the start of the 1950s, twenty years after its first decision to ban nitrite, the High Council of Public Hygiene was once again called in. Dr Bordas and his associates were long dead by now; two toxicologists, Henri Cheftel and Louis Truffert, were commissioned to write a report. Both men were well attuned to the industry's priorities (Cheftel worked for a tin can manufacturer). Being fervent advocates of the Chicago model and its amazing productivity,[61] they had previously campaigned for the adoption of nitrite curing. They explained that nitrite was better than nitrate when it came to dosage control. When doctors questioned the very principle of nitro-additives, Truffert replied that 'that isn't the issue':[62] since saltpetre curing was authorized, there was no reason to ban nitrite. Cheftel and Truffert developed an argument that would be trotted out in all subsequent discussions: because it was banned, 'nitrite is used covertly, with all the potential dangers to the consumer that that entails'.[63] They explained that manufacturers tended to cheat by enriching their brine with nitrite 'bought from druggists',[64] and that some were already 'directly adding pure nitrite to "regenerate" meats',[65] which proved that 'nitrite is being used illegally'.[66] Some experts attempted to stem this tendency. The *Manuel Pratique du Charcutier Moderne* (Practical Manual for the Modern Pork Butcher) reminded readers that nitrite 'is a violent poison':[67] the Manual acknowledged that nitrite does indeed produce a nice pink colour, but stressed that it was far too dangerous to handle.[68] Rather than maintain the ban and continue with inspections, Cheftel and Truffert recommended legalizing nitrite in order to exercise some control over the practice: they claimed that, once it was legalized, the authorities would be able to define maximum levels, and any producers exceeding those levels would then be punishable.

The clinching argument in favour of nitrite was, however, the economic one. Some experts stressed that the attitude of Cheftel and Truffert 'was mostly due to the preoccupation with bringing French regulations into alignment with foreign legislation – principally American and German'.[69] Cheftel and Truffert pointed out that France was one of the few Western countries that still continued to ban nitrite use.[70] Others made the point that any curer who used nitrite 'derives a substantial advantage from the process: rapid fixing of the desired colour, hence a time saving, which increases productivity, lowers the cost price and facilitates competition in foreign markets.'[71] Consequently: 'Foreigners bring their products, which they produce much more rapidly than we do, on to the international market. They can sell them more cheaply, which limits our exports.'[72] Another industry chemist declared: 'It is clear that foreign countries can compete with us in the international market thanks to their use of nitrite. For example, the federal government of West Germany recently gave instructions to curers on the application of rapid curing (six days) to hams destined for England.'[73]

Despite all these arguments from the industry, the doctors on the High Council of Public Hygiene stuck to their guns: in their view, by authorizing nitrite, 'we would risk setting a very serious precedent, since a substance that is included in the list of poisonous substances and whose toxicity is well known would then legally be added to foods'.[74] In 1953, the Council reiterated its position of 1934 and refused to authorize the use of sodium nitrite in processed meats.[75] The French Ministry of Agriculture concluded: 'This issue has from the start encountered a marked reticence on the part of the High Council of Hygiene, because nitrite is a Class C poisonous substance.'[76] The ministry explained that nitrite curing could not be authorized as it would require 'admitting the addition to food of a substance classified as toxic, whereas it is normally a hard and fast rule that we keep hazardous matter well apart from edible goods'.[77]

CAPITULATION

Sodium nitrite was finally authorized in France in 1964. There were a number of reasons behind this decision. After the signing of the Treaty of Rome (1957), trade in pork was one of the very first areas in which the European Common Market came into play.[78] Nitrite-treated meats were able to circulate more freely, which exposed French products to sterner competition: in 1962, the Ministry of Agriculture deplored 'the import of hams of foreign origin prepared in a more economical manner',[79] and a few months later the newspaper *Le Figaro* expressed the difficulties faced by French producers: 'Meats of foreign origin were recently sold in the French market, principally hams offered at prices that defy all competition.'[80] French producers started again to intensively lobby the administration. In a letter of June 1963 to the minister of agriculture, the head of the association of industrial meat-processors emphasized the huge profits that the use of sodium nitrite would unlock: 'It would allow us with the present production capacity alone to *quadruple tonnages*. It would reduce the volume of stock. It would favour the rapid turnover of capital – all elements that would have an impact on price and would help us resist foreign pressures.'[81]

There was another argument in favour of legalization: at the end of the 1950s, there were many cases of illegal nitrite use in the news, and several meat producers were prosecuted.[82] Toxicology reviews listed whole strings of accidents, in both Europe and the USA: overdosed salami and knockwursts which caused a series of poisonings in Louisiana in 1955 (ten people);[83] overdosed sausages which caused deaths in Florida in 1956;[84] a couple of pork butchers poisoned in 1957 by nitrite obtained on the black market,[85] etc. In South Africa, a family of seven poisoned themselves by eating sausages prepared with too much sodium nitrite.[86] Even in Germany – where nitrited curing salt had been authorized for twenty years – some producers were employing underhand methods to lower their cost price by using pure nitrite. In 1946, for example, two accidents occurred a few weeks apart in one town. The first intoxication involved around 100 people: without fully understanding the instructions, a butcher had used a new product (a 'nitrite concentrate' – *Nitritkonzentrat* – of which he had thrown 'two large fistfuls' into the preparation). Three children died. Barely had they been buried than a fresh tragedy threw the community once more into mourning: another pork butcher had used an overdose of nitrite, intoxicating 33 people and killing a woman.[87] A few months later, another poisoning affected 146 people in another German town.[88] These were the latest in a series of accidents involving 'sausage soups' (*Wurstsuppe*) and all sorts of German pâtés.[89]

In January 1958, the newspaper *Die Zeit* published a report entitled 'Poison in sausages':[90] a huge contraband trade in sodium nitrite was dismantled, more than 500 manufacturers were prosecuted, a wholesaler was indicted for having dispensed tonnes of pure sodium nitrite to butchers. Jail sentences were handed down. The pharmacists' journal condemned the butchers but defended those who had sold the chemicals: 'The accused druggists were unaware of the law on nitrite.'[91] The popular press was less subtle, *Neues Deutschland*, for example, announcing: 'Nitrite scandal grows – a poison used by the barrel load – countless arrests'.[92] The affair caused such a fright that honest tradesmen put signs up in their shop windows. The magazine *Der Spiegel* described an artisan in Stuttgart who had put up a sign saying: 'We have not abused the trust of our customers. Our products contain no harmful substances.' The magazine also showed a pork butcher who had placed a large notice in the centre of his counter saying: 'In our establishment we do not use sodium nitrite!'[93] The French press picked up on these incidents,[94] and the Fraud Prevention Office suggested that identical scandals could erupt in France too. Describing what he called 'the sensational trials currently being staged in Germany', the director of the municipal laboratory in Metz alerted the French Ministry of Agriculture: 'I am convinced that the same reactions will soon manifest themselves here, it's inevitable.'[95]

Anxious to defend the interests of French curers, prevent illegal activity and avoid intoxications by overdose, the Ministry of Agriculture decided at the beginning of the 1960s to reopen the case to legalize the use of sodium nitrite. Up until the last moment, several departments of the Ministry of Health were reluctant to agree – to the ire of French cured meat industry.[96] The administration decided to settle the issue and applied the option first proposed in 1953 which aimed to find a solution that was 'satisfying from a health point of view and favourable towards industry':[97] the ministry set an upper limit of 200mg of detectable nitrite per kilogram of finished product. According to the experts, 'you would need to eat 1.5 kilograms of this meat to absorb 300 milligrams of nitrite'.[98] On the basis of a report drawn up by Louis Truffert and Henri Cheftel, the High Council of Hygiene finally gave its consent.[99]

'NOT UNFAVOURABLE', NOT FAVOURABLE: RELUCTANT AUTHORIZATION

On 9 June 1964, Edgard Pisani, the French minister of agriculture, sent his draft orders to the Ministry of Health. He indicated that he had just referred the matter to the Académie de médecine and that a favourable recommendation was a mere formality; he therefore requested the minister of health sign the orders without even waiting for the academy's verdict.[100] A few months later, the orders were effectively published in the *Journal officiel*.[101] It stated that the table of 'poisonous substances' had been modified: sodium nitrite was still officially a poison, but it was authorized *exceptionally* in cured meats and solely in the form of 'nitrited curing salt', that is, as salt with 0.6% nitrite additive.

Eventually, the Académie de médecine gave its agreement, but with considerable reluctance. The academy explained: 'The method of curing using nitrate is relatively slow: it takes about three weeks; that is why many foreign countries have allowed a more rapid method – one week – which involves use of nitrited curing salt.'[102] The academy gave in because it felt it was important, 'as soon as possible, to harmonize our regulations with those of the other countries in the EEC so as not to put French meat curers and butchers in a disadvantageous position'.[103] Sodium nitrite was seen as a replacement for 'the more or less blind nitrate method which requires, it seems, more products than with nitrite'.[104] Because nitrite curing appeared *less dangerous* than saltpetre curing, the academy thought it represented 'a certain progress', while at the same time making it clear that they endorsed neither method. The Food Commission of the academy explained: 'It is of course a matter of regret that use of nitrate should have become commonplace in the practice of meat curing. It is a method that the commission accepts only with reticence.'[105] The academy declared: 'The absence of refusal does not imply that the academy is in favour of such process.'[106]

CHAPTER 9

THE NITROSO SURPRISE

In the very first patent for sodium nitrite curing in 1917, the inventor, the American George Doran, explains that this substance shortens each stage of the production process.[1] That is why methods employing nitrite were designated as 'quick cure' or 'accelerated cure', and why many additives based on nitrite had names that evoked speed. Heller & Co. offered a mixture of salt and nitrite under the label 'SchnellSalz' (Fast Salt) and another called 'Quick-Action Pickle' ('QA') with the claim: 'Quick action is assured when using this product'.[2] Griffith Laboratories used to sell their nitrited curing salt 'Prague Salt' by presenting it as 'the safe, fast cure' with the accompanying slogan: 'there is a shorter road to perfect curing'.[3] Even today, many manufacturers of nitrited curing salt still emphasize speed. One company sells a product known as 'Tender-Quick'. Other brands include 'Insta-Cure', 'Speed Cure', 'Kwikcurit', and 'Holly Quick'. The distributor butcher-packer.com suggests that, by using its nitrited curing salt, 'your sausage will be ready to cook or smoke as soon as you have it stuffed (there is no need to wait)'.[4] And when he was lobbying the British authorities to authorize the use of nitrite in the 1930s, the director of the British Food Manufacturers' Research Association explained: 'We have now carried out a lengthy series of experiments on the curing of bacon and hams using nitrite to speed up the cure. We have been able to reduce the time of curing considerably in each case, in fact we have cured hams sufficiently well for sale as cooked hams in from two to three days.'[5]

THE SAUSAGE RACE

After the invention of sodium nitrite curing, the manufacturers developed techniques to speed up the production process even more. First, they came up with new forms of nitrited brine injection ('arterial injection' and 'stitch pumping'). One patent explained how the combined use of nitrite and new pumping methods could shorten the production time because it did away with having to macerate in brine.[6] The 'pumps' then became 'multi-needle injectors', and the needles themselves became 'multi-hole' (instead of injecting just from the tip, they injected along their whole length). Eventually, automatic injection machines with an integrated conveyor belt were developed. The historian Roger Horowitz explained: 'Bacon finally began to move more quickly through packinghouses in the 1950s as firms developed equipment injecting curing solution through dozens of small needles [...] The "PerMEATor" built by the Cincinnati Boss Company allowed a continuous flow of graded bellies to automatically slide into the machine. In one rhythmic, elliptical motion the needle-laden header completed its task of penetrating, injecting the cure and moving the belly along.'[7] Today, the machines are called 'ValueJector' (GEA Group), 'Injector-Tenderizer' (Fomaco), 'Imax Injector' (Schröder), 'Multi Needle Injector' (Belam Wolfking), etc. They continuously inject salt + nitrate or salt + nitrite brine while moving the meat along the production line. Sometimes they even cook or steam the product to help the nitrite work more quickly.

By shortening the production time and reducing the time of maturation (or dispensing with it altogether), nitrite curing obviously brought about astronomical savings. Less manpower was needed, and also less space: shortening the production time meant that output could be increased without having to enlarge factories. The Heller company claimed that their nitrited curing salt 'is a time-saver and holds storage space at a minimum'.[8] In 1954, an economic study conducted at the University of Chicago pointed out: 'curing and smoking processes have changed rapidly in the 15 years since quick curing of hams by arterial pumping was developed. This process reduces the time required for curing from 6 to 8 weeks to a few days [...] It has been estimated that quick-curing techniques have reduced by a third the space required to cure a given quantity of meat.'[9] The figures given by Roger Horowitz show the amazing acceleration made possible by nitrite injection in the 1950s: for a standard ham, the production time was reduced from 90 days to five. In 1965, the French chemist René Pallu pointed out that manufacturers could choose between 'slow curing' (28 days) and 'ultra-fast curing' (twelve hours) made possible by nitrite.[10]

The same thing was observed when it came to bacon ('Twenty-four hours after passing through the machine, the curing solution had permeated the bellies and they were ready for the smoke house', according to Horowitz).[11] Likewise for smoked sausages: at the start of the century, the production process began with a maceration phase that extended over several weeks. Then came the mincing, the stuffing into skins, the smoking, the drying, the cooking. In all, the process could take as long as 30 days. The American historian Bruce Kraig wrote: 'The final "fast cure" method means chopping the meat while adding cure, spices and other ingredients at the same time, followed by smoking and cooking. These processes allow the sausage to be made in a matter of hours.'[12] This description chimes in with technical specifications found in professional publications: for example, *The Packers' Encyclopedia* (1938), the reference manual for entrepreneurs of 'modern meat-processing', explains that 'quick cure' methods allow shop-ready sausages to be obtained immediately. Thanks to nitrite curing the chemical transformation of the meat takes place during the time the ingredients are being minced, mixed together and stuffed in skins. This accelerated process plays a crucial role when it comes to 'frankfurter' sausages: thanks to nitrite, they can be produced in a continuous flow.[13] As Kraig spelled out, 'speed in processing means money for the processor',[14] especially when it comes to products for mass consumption: that frankfurters and wieners could be produced in the blink of an eye greatly aided their mass production.

The whole apparatus of industrial meat-processing is now predicated on the miraculous chemical reaction between nitric oxide and myoglobin. Traditional methods cannot compete, and unfortunately non-carcinogenic alternatives are not as cost-efficient.

FROM MIRACLE TO RED ALERT

The cancer problem began to become evident when it was discovered that meats treated with nitrate or nitrite gave rise to 'nitroso' compounds. The first to be identified – in the mid 1950s – were the *nitrosamines* (or *nitroso-amines*). In Britain, John Barnes and Peter Magee, scientists at the Unit for Research in Toxicology of the Medical Research Council (MRC), discovered that these hitherto unknown molecules were strongly carcinogenic.[15] Almost from the start, Barnes and Magee had realized that nitrosamines were likely to be detected in cigarette smoke, as tobacco leaves are rich in nitrate.[16] But they never imagined that nitrosamines would one day be identified in meats treated with nitrate or nitrite.

Concerns were first raised in 1960: in Norway, vets noticed that farm animals whose feed had been treated with sodium nitrite were dying.[17] At first, biochemists believed these cases were due to methaemoglobinaemia (intoxication by cyanosis). But tests soon revealed that the feed contained hardly any nitrite: not enough to cause an intoxication. On the other hand, it contained one of the most fierce forms of nitrosamine, dimethylnitrosamine (also known as N-nitrosodimethylamine). The scientists realized that it was the sodium nitrite that had brought it about, by reacting with the proteins contained in the feed.[18] After this initial discovery, the biochemists noticed that nitrite could give rise to another type of neo-formed compound, called *nitrosamide*. Like nitrosamines, nitrosamides emerge from a reaction between nitrous acid and compounds present in meat (amides). Later, the carcinogenic potential of a third nitroso compound, *nitrosyl-haem*, was discovered. It results from the reaction between nitrite derivatives and haem, that is, the iron found in the meat. The role of nitrosyl-haem in cancer began to be studied from 1975 onwards,[19] but it was only twenty years later that its role started to become clearer through the work of Sheila Bingham and her team in the 'Diet and Cancer' group at the Medical Research Council. Today, we know that nitrosyl-haem plays a crucial part in the mechanisms that cause tumour growth.[20]

By the end of the 1960s, the growing scientific evidence on the carcinogenic effects of nitroso compounds was sending industrial meat-processors into something of a panic. Their problem was that these compounds didn't seem to act in the same way as the toxic substances they were used to. First, a crucial nuance: it is not the nitrite itself that is carcinogenic, but the compounds that nitrite generates by reacting with constituents of meat. Second difference, the dose: whereas the fatal dose of nitrite is somewhere between 2 and 4 grams, nitrosamines can act at a dose 4,000 times smaller. At a meeting between the industry and the UK Ministry of Health, one of the pathologists on the Committee on Medical Aspects of Food Policy pointed out that 'a test dose so small that it has no effect on say 200 animals may nevertheless be the cause of many cases of cancer in a population of 50 million people.'[21] As another cancer specialist would later explain, nitrosamines 'induce tumors in all species of animals where they have been tested, and man is unlikely to be resistant to their action.'[22] Describing his experiments to the Medical Research Council, Peter Magee pointed out that 'the compounds are remarkably consistent in their action, and it is easy to cause tumours with them in most of a group of animals tested'.[23]

In 1968, a large meeting was organized involving John Barnes, Peter Magee and experts from the industry and from the UK Ministry of Health. Peter Magee explained that even 'an incidence of 0.1% or 0.01% cancer in 50 million people would give rise to a very large number of cases'.[24] He also pointed out that, contrary to substances that provoked acute poisoning, the impact of the carcinogenic compounds only became evident in the long term: for example, when experimenting with diethylnitrosamine on monkeys, scientists of the Medical Research Centre observed that tumours only developed *two years after the start of the test*. It was impossible to know exactly how *Homo sapiens* would react. John Barnes asked: 'Would cancer develop if one lived long enough? With diethylnitrosamines tumours developed in 10% of rats after 2 years, and in 50% of rats after 3 years.'[25]

'CHRONIC' ALSO MEANS 'THAT COMES WITH TIME'

At a 1969 symposium in Johannesburg which brought together all the big names in toxicology, the French expert René Truhaut insisted on the distinction between *acute* and *long-term* poisoning, which he called 'chronic poisoning'. Professor Truhaut knew his subject well: the original president of the Joint Expert Committee on Food Additives (JECFA) of the World Health Organization (WHO), he was the one who invented the notion of 'Acceptable Daily Intake' (ADI) used by almost every country to regulate food additives.[26] Truhaut explained: 'There is a regrettable tendency, still far too prevalent, to be only concerned with cases of acute or subacute poisoning as caused by the intake, in one dose or several doses in quick succession, of certain chemicals. The poisoning is revealed by spectacular symptoms, often immediate, and sometimes followed by death. Such is the case, for example, with the inhalation of toxic gases or vapours, such as chlorine, phosgene, carbon monoxide, or hydrogen cyanide, or with the ingestion of sufficiently large amounts of some chemicals to cause serious accidental, or deliberate, poisoning.'[27] Truhaut emphasized that certain substances had a second aspect hidden behind the first, as it were: 'the effects of long-term poisoning, often called chronic poisoning, which may be caused by repeated absorption of sometimes minute quantities of certain chemicals over long periods of time. Such cases may be no less serious than the former, however, for they are often irreversible. They are particularly insidious, as they very often give no noticeable signs of warning.'[28] He also noted that in certain cases 'it is not the additive itself which is toxic, but the product it generates by reacting with a normal and even essential constituent of food';[29] and he went on to explain that the identification of nitrosamines caused 'considerable concern about the use of nitrites as food additives'.[30] He declared that 'if thoroughgoing analytical studies, now being carried out in various laboratories, confirm that carcinogenic and mutagenic nitrosamines can form in certain foods to which nitrite has been added, then it is likely that a whole part of current food policy will need to be revised.'[31]

At the Johannesburg symposium, Professor Leon Golberg, former director of the British Industrial Biological Research Association (BIBRA), explained that, when it came to food-borne carcinogens, 'pride of place must be given to nitrosamines and related compounds. Many members of this group are not only highly potent but also multipotential in their ability to induce tumours in a variety of tissues.'[32] Golberg declared: 'in view of the carcinogenic potency of so many nitrosamines, particularly their capacity for transplacental tumour induction in the unborn child, some immediate action seems necessary.'[33] He thought it was necessary that the authorities take measures aimed 'at replacing nitrite and nitrite-nitrate by other preservatives, as far as possible. Foods known or suspected to contain nitrosamines should be listed and pregnant mothers advised to avoid them.'[34]

A few years before, the biochemist Richard Morton had been appointed head of the Food Additives and Contaminants Committee (FACC) of the Ministry of Agriculture.[35] Professor Morton warned the agriculture minister: 'In several respects nitrosamines are more important than anything the Committee has yet had to deal with.'[36] In a report that remained confidential, he wrote: 'Many persons will wish to argue that long experience in human diets of cured meats implies that any risks are small, but we must treat this argument with considerable caution.'[37] According to him, 'to support the use of nitrite simply because traditional procedures acquire a certain presumptive patina of safety is now seriously in question.'[38] His prognosis was clear: 'A full look at the worst outcome raises the possibility that the use of nitrite in food may have to be discontinued.'[39] In an internal memo, the head of the UK Food Standards Division wrote a little later: 'it could be argued that since bacon and cured meats have been eaten for many years the problem should be treated like smoking and virtually ignored; alternatively, it might be held that the area of risk should be reduced by doing all that is possible in the circumstances, e.g. reduce the possibility of hazard by prohibiting the addition of nitrates and nitrites to food.'[40]

The main object of concern was bacon. In 1968, British manufacturers put the size of their market at '622,000 tons of bacon (equivalent to 12 million pigs)'.[41] At a meeting with representatives of the Ministry of Agriculture, one company director warned: 'unless a common sense attitude could be taken towards the problem of nitrosamines, it would probably be the end of the bacon curing industry.'[42] The industry set up a 'high powered industry committee',[43] the Project Policy Steering Committee on Nitrosamines, headed by James Sainsbury (who was director of production of processed meats for the family supermarket firm) and Lord Trenchard, director of Unilever (the group then owned several brands of processed meats such as Mattessons, Wall's, etc.). After a visit by Alastair Frazer, president of the British Food Manufacturing Industries Research Association (BFMIRA, today Leatherhead RA), an official of the Food Standards Division noted in the Ministry of Agriculture minutes: 'The implications of nitrosamines being considered to be dangerous substances to the consumer are very important because of the widespread consumption of bacon and canned meat and the implications for the economies of various countries if anything was done to damage their production and consumption.'[44]

METHAEMOGLOBINAEMIA AND CANCER: NITRO-ADDITIVES CAUGHT IN THE CROSSFIRE

To fully understand how disruptive the discovery of the carcinogenic action of nitroso compounds actually was, it is worth stressing here one crucial point which has given rise to serious (often disingenuous) misunderstandings.

As we saw in the previous chapters, sodium nitrite was subjected to very strict regulations long before the 1960s, that is, long before the carcinogenic properties of nitroso compounds were discovered. Sales of sodium nitrite were rigorously controlled, and the authorities had permitted its use only in very diluted form – in 'nitrited curing salt', which contained less than 0.6% nitrite. This regulation was driven by a central preoccupation, virtually an obsession: to prevent consumers being exposed to doses of nitrite capable of provoking intoxication by cyanosis (methaemoglobinaemia).

The regulations had been conceived to minimize the risks of poisoning by nitrite overdose. But with the discovery of nitroso compounds, everything changed: it was suddenly understood that, alongside the risks of methaemoglobinaemia, nitrate and nitrite were capable of giving rise to agents that hitherto had not been taken into account. When they had defined the limits for the use of nitrited curing salt in meat, the health authorities had only been able to consider the risk that they knew about: the immediate toxic action of nitrite on blood. At the end of the 1960s, they were amazed to discover that nitro-additives could provoke the appearance of compounds that no one, it seems, had contemplated.

In May 1968, *The Lancet* discussed the problem in a milestone article headlined 'Nitrites, nitrosamines, and cancer'. The journal recounted how a hitherto unknown aspect of nitro-additives had been uncovered when 'the unsuspected formation of carcinogenic nitrosamine was dramatically demonstrated'.[45] The article pointed out that the discovery of the carcinogenic properties of nitroso compounds presented an enormous challenge to the authorities, because until then only the risks of methaemoglobinaemia had been taken into consideration: 'These aspects of nitrite toxicity have been taken into account in regulating the use of nitrites in food, but the situation has become more complicated since the discovery of the remarkable toxic, carcinogenic, and other biological actions of the nitrosamines, because, under certain conditions, these compounds may be formed by interaction between nitrites and secondary or tertiary amines.'[46] *The Lancet* stressed that 'the possible occurrence of these biologically highly active compounds in food or other environmental situations is a matter of the gravest concern'.[47]

A POISONED CHALICE

When nitrite was initially adopted as a curing ingredient, the chemists knew that it was a very potent substance, but they believed that nitro-additives would have no deleterious effects if the dosages were small enough to prevent poisoning by methaemoglobinaemia. The fear of acute poisoning was so much at the forefront that other potential risks were not seriously examined. Thus, in the 1950s, when a French expert was agitating for the legalization of sodium nitrite, he claimed that he fully understood all the risks the substance posed, since, he said, he had participated in the post-mortems of persons who had died of methaemoglobinaemia.[48] He was convinced that the health of consumers would not be put in danger if precautions were taken to avoid doses high enough to cause cyanosis.

Everywhere, fear of death by cyanosis meant that more complex, more insidious risks, risks that were slower to manifest themselves and had a less obvious chain of causality, were overlooked. Even allowing for the benefit of hindsight, some remarks made nearly 100 years ago have a disturbing ring to them. When, in 1925, the US Department of Agriculture decided to allow curing with nitrite, the *American Journal of Public Health* wrote the following forewarning: 'Little consideration has been given to gradual and cumulative injury requiring years to bring about the final disastrous result. Since there is no proof that in the past damage to health has not been caused by the use of nitrates, if the use of either nitrites or nitrates is to be permitted, it would seem that experiments should be immediately undertaken to thoroughly study the long time effect of the ingestion of these chemicals on health. Sole dependence should not be placed on experiments which show the value of nitrites to the meat industry.'[49]

In the 1950s, the great oncologist Hermann Druckrey had also warned: 'In dealing with toxic substances, we have been accustomed to think in terms of safe limits of exposure, safe "threshold" doses, amounts which will not cause injury or amounts from whose effects the body can recover. In dealing with substances that can produce cancer, however, we are concerned with substances that have a very unusual pharmacological action.'[50] As for the UK, Druckrey noted that British regulations permitted 'the addition to foods of an unlimited number of different colouring matters without control until there is evidence that harm may occur'.[51] Similarly, when sodium nitrite was examined by the French High Council of Public Hygiene in 1953, one of the members of the commission stressed that what should be considered was 'not merely the lethal dose, but the dose that is harmful through repeated introduction into the organism'.[52] A doctor upped the ante when he pointed out that sodium nitrite was an extremely potent chemical: 'We must be very cautious in matters of food health: sodium nitrite is already dangerous at a dose of ten centigrams.'[53] His advice was to be careful: 'We should only authorize a product if we are certain it is harmless.'[54] A few months later, at another meeting, scientists working for an industrial meatpacking factory noted: 'Experimental ingestion in laboratory animals for a period of months, even years, and over the course of several successive generations would be the only way to resolve this principal problem of physiological toxicity [...]'[55] And they added: 'In fact, certain toxicities have only manifested themselves in the long term (DDT introduced into grains after disinsectization of granaries and mills, food colorants, potassium ferrocyanide in wine).'[56]

By the time these fears were confirmed, it was too late: nitrited meats were deeply embedded in a complex and sophisticated system of production, and huge investments were at stake. In the face of growing indictments of nitro-additives and the accumulation of damning evidence, what should the industrial meat-processors do? Some curers (those in Parma, for example) had a long heritage, and they were able to return to traditional methods. But many other manufacturers had no such heritage: their success was founded on the development of a highly efficient system based on the use of chemicals. To give up nitro-additives, they would have had to readapt their equipment and declare a halt to the 'miracle' of instant processed meat – visually perfect, easy to make, easy to store, easy to sell. It is hardly surprising that they didn't admit this. This is the point where the real scandal begins: the moment when, to justify their use of nitrate and nitrite, the global processed meat industry had to begin telling lies …

PART 2

HOW TO MAKE CANCER ACCEPTABLE: BRING ON BOTULISM

CHAPTER 10

BOTULISM, THE 'BLOOD SAUSAGE POISON'

Nitro-additives are no longer used as colorants – according to the processed meat industry, at least. Nowadays, they are supposed to be there to protect the public: nitro-meat-processors claim that nitrate and nitrite are the only effective means to combat *Clostridium botulinum*, a bacterium that can develop in meat. This micro-organism produces a toxin that binds to the nerve terminals and causes muscular paralysis. It is one of the most potent poisons that exist, if not *the* most potent: swallowing just a few grams of food containing the botulism toxin would be enough to kill you. The meat-processors therefore insist that it is essential to include a powerful anti-bacterial agent in their meat. Industry advocates claim that there is no reliable alternative, that the risk of cancer is the price to pay for this peace of mind. An article in a journal published in 2006 by the Society of Chemical Industry, for example, says: 'Next time you visit the supermarket, take a look at the lists of contents on packets of bacon and preserved meats. I shall be surprised, alarmed in fact, if you do not find that they contain nitrite. The nitrite is there for an excellent reason. Our forebears discovered empirically that it was about the only thing that stood between them and a particularly unpleasant form of death, which we now know to be caused by the botulinum toxin.'[1]

According to nitro-additives advocates, a German doctor realized around 1820 that botulism (from the Latin *botulus*: sausage, stuffed skin) was linked to the lack of nitrate in processed meat.[2] The archives, however, reveal a very different story.

AN EPIDEMIC IN SWABIA

Set amidst a landscape of vines, Württemberg in southwest Germany is today a prosperous and peaceful region. But around 1815, it was the epicentre of a major health crisis. The population had already been greatly impoverished by the Napoleonic Wars, and several bad harvests in a row had brought about a harsh degradation in living conditions.[3] Then fear set in when several dozen peasants died after eating smoked blood sausages that had been undercooked, kept without due care, or made using rotten meat and improperly cleaned casings.

The deaths were clustered in Swabia. With its Protestant, almost mystical tradition, this part of Germany was markedly poorer than Bavaria, its neighbour to the east. Doctors had identified a few similar cases of poisoning in the immediately adjacent regions, but they were surprised that this 'blood sausage poison' was almost completely unknown abroad.[4] When the deaths started to accumulate, the government appealed to any scientist who could work out the cause of the intoxications and find a remedy. The progress of the investigation can be followed through the pages of the weekly medical bulletin in which doctors in the region communicated their observations and hypotheses on the ongoing poisonings, for which they had no treatment. The doctors noticed that the first symptoms often affected the eyes. In one family of wine-growers, a 28-year-old father began complaining at first of double vision, with 'mist in front of his eyes'.[5] Another man said he was seeing double, 'then triple', and then found he was unable to distinguish colours or read numbers when the doctor showed him his pocket watch to test his eyesight.[6]

Still today, diplopia is one of the first indicators of botulism, as the toxin causes paralysis of the optic muscles. This initial symptom is often followed by difficulties in speaking and swallowing (paralysis of muscles of the mouth); then the paralysis affects the arms, the legs, the muscles of the thorax. When the toxin blocks the respiratory muscles, the victims die of asphyxia. Nowadays, cases of botulism require weeks of intensive care and respiratory support; and in spite of modern technology, around 10% of cases still result in death.[7] Reading the reports in the local medical bulletin, it is easy to imagine how the illness ravaged the countryside of Württemberg at a time when doctors travelled by cart from one patient to the next, arriving in the middle of the night at tumbledown cottages lit by candlelight, finding whole families at death's door. Page after page, these dramas unfold; the doctors describe their desperate attempts to save lives, trying all sorts of treatments based on camphor, phosphorus, mercury, opium, the application of leeches, and even arsenic.[8]

DR KERNER INVESTIGATES

The poisonings seemed to occur randomly, striking one village or another for no obvious reason. A young doctor, Justinus Kerner, started to take an avid interest in these deaths. Though he never managed to identify the exact nature of the poison, Kerner managed at least to detail its symptoms very precisely. He wrote three books on the subject, in which he outlined a methodical approach to this strange epidemic. He described all the cases that had been spotted and was alarmed at their frequency: he said that in Württemberg, sausages appeared 'to kill as many people as snakes in the tropics'.[9]

Like his predecessors, Kerner was surprised at how geographically concentrated the illness was. He learned that similar poisonings had occurred in the same region 80 years earlier. Even more significantly, the epidemic seemed to be clustered in certain specific locations. For example, he found that 29 cases had been recorded around one single small village in the space of eight months.[10] As far as Kerner was concerned, there was no doubt: these poisonings did not occur by chance, there was a cause! He gathered samples of blood sausages and distilled extracts from them; he tested these on frogs, snails, birds and small mammals. He noticed that the majority of human deaths occurred in the spring – in April in particular, when the peasants ate up the very last of their stores of blood sausages made from pigs slaughtered before the previous winter. At this time of year, the days are mild, but the temperature can dip below freezing at night. The sausages were stored at the ambient temperature and so were frozen and defrosted many times over, which speeded up the alteration of the meat. Kerner wondered whether that was the cause of the poisonings.

Inquests into the deaths consistently accused a local speciality known as *Saumagen* or *Blunzen*, enormous 'blood sausages' which, rather than being made using intestinal skins, were encased in pigs' stomachs.[11] These larger-sized sausages still exist today in Württemberg. Nowadays they are produced under strict hygiene conditions, but 200 years ago they posed a serious health risk, because of both their size (impossible to cook them right through) and the primitive conditions in which they were made. What is more, several doctors noted that, to add flavour, these *Blunzen* often had decomposing beef blood added to them.[12] Sometimes, these bizarre concoctions caused extremely risky fermentations. In a paper entitled 'On humans and artificial food', Kerner criticized recipes that mixed all sorts of ingredients (tripe, milk, blood ...) to create ever more unusual sausages. Citing many historical examples, he pointed out that humans had learned from experience to steer clear of any substance that was likely to pose a danger by rapid decomposition – in particular, blood and viscera. Kerner quoted a Byzantine emperor who had banned 'blood sausages' and recalled that traditionally such products had been approached with much caution and suspicion: 'Let none sell blood sausage, for 'tis perilous meat,' read an injunction by the provost of Paris in 1258.[13]

Writers in the nitro-meat camp claim that Kerner was the first to establish a link between botulism and the absence of nitrate/nitrite.[14] As they don't give their source, this claim is hard to verify. In any case, these writers miss the essential point. In his writings, Kerner doesn't appear to mention saltpetre. Rather, he demonstrates that the cause of 'blood sausage poisoning' is the lack of care and the use of rotten meat.[15] Kerner was anything but an apologist for bactericidal chemicals; on the contrary, his writings are a firm denunciation of meat products which have been rendered unhealthy by use of inappropriate ingredients and a negligent production process.[16] For Kerner, the sausages became toxic because the people who made them did not respect the basic rules of hygiene required when handling meat. His conclusion was that strict controls should be imposed on sausage-makers and consumers should be educated. Since the victims were mainly the poor and illiterate, Kerner recommended running information campaigns relying on easy-to-understand engravings, printed almanacs illustrated with stories of disease, depicting the figure of death travelling from village to village.[17]

DEFECTIVE PRODUCTION ... AND MALFEASANCE

At the beginning of the 1850s, Württemberg suffered another economic depression. The number of cases of botulism spiked once again.[18] Julius Schlossberger, a professor at the University of Tübingen, was commissioned to coordinate the efforts of the public health services. The doctors ran up the red flag: the incriminated sausages had all been made during warm weather, the blood and fat used were not fresh, and the sausages had not been sufficiently protected by smoking.[19]

The Museum of the Meat Trade in Böblingen, in the very heart of the zone affected by the epidemics, with its collection of meat mincers, gives a good idea of the primitive nature of the meat-processing techniques used at the time. Documents from that era show the procedures followed in the making of 'all sorts of black puddings, chitterling sausages, blood sausages, liver and pork sausages, peppered blood sausages, saveloys, mortadellas':[20] the meat was chopped up with a knife on a wooden block, then the casings were stuffed using a funnel. In one illustration, a woman chops the meat which she drops into a basket, while two men do the stuffing. One of them holds the casing while the other uses the funnel and a large spoon to force the mixture through.[21]

A painting in a museum in Frankfurt shows an old couple in the back room of their butcher's shop, the woman preparing the casings from intestines in a wooden basin. Her husband has just slaughtered the animal.[22] In another painting, a young woman has set up her table in a barn under an enormous quarter of meat. Children are playing to one side, a cat is slinking under the table. Giblets hang from a hook, a snout pokes out of a basket, while other pieces of meat are lying on the bare ground. Another engraving represents a peasant woman cleaning intestines while a pig is being gutted. A dog is feasting on the morsels that fall on the floor.[23]

In such conditions, the making of sausages and black puddings involved a considerable risk of infection, and so butchers had to respect the rules of hygiene and traditional production methods that had been tried and tested. Processing of the meat should only take place immediately after slaughter, and exclusively in cold weather. The sausages should be cooked for a long time (cooking was a well-known way to preserve meat)[24] – the paintings and engravings of the time are full of scenes of cooking for the purpose of preservation.[25] The doctors who were investigating the Württemberg poisonings discovered that the epidemic always sprang up in places where such precautions were ignored: the fatal sausages were made with complete disregard for the traditional rules and then kept in conditions that were conducive to putrefaction.[26] For example, in 1854 the local newspaper *Schwäbischer Merkur* reported the case of a young woman who died after eating a piece of spoiled blood sausage; it was so thick it couldn't possibly have been properly cooked through when it was made. Then the smoking had been carried out incorrectly, probably interrupted each night and started again the next morning. And finally, the product had been kept without due care for several weeks, without refrigeration, exposed to the extreme fluctuations of the ambient temperature.[27]

A few weeks later, in mid-July, a penniless labourer received two pieces of liver sausage by way of wages. This *Leberwurst* seemed rotten, but the labourer was so poor that he ate it anyway with his wife and their fourteen-year-old son. 'They were so hungry that they wolfed down the sausage, ignoring its unpleasant taste, which they disguised with the acidity of a salad.'[28] All three of them died. A few dozen miles away, a shoemaker's whole household fell ill: parents, children, apprentice and servant – twelve people in total. The sausage had been stored at more than 25°C and tasted strongly of garlic. The local doctor conducted inquiries in the neighbourhood and reported: 'I was told that the meat wasn't fresh, that the butcher had mixed leftovers of an old sausage with new meat in order to sell it and pass it off as fresh.'[29] Had the butcher added garlic to mask the repugnant taste of this recycled sausage? The butcher denied it, but a subsequent inquest confirmed this suspicion.[30]

One of the local doctors sounded a fatalistic note: 'Given the enormous amounts of sausage of all types that the population devours; given the abominable lack of care that one sees in the making and smoking of meat products; given that the ingredients used are prone to rot very easily; given the negligence of the public, who seem quite prepared to eat these products even when they already have a rotten taste, it is a surprise in the end that such cases don't occur even more frequently.'[31]

PREVENTABLE POISONINGS

Eventually, Justinus Kerner's hypotheses were vindicated: out of a total of 55 cases of poisoning studied by Professor Schlossberger in 1852, 21 were found to have taken place in April. And still there were virtually no cases recorded outside of Swabia.[32] The statistics confirmed that a certain type of sausage was repeatedly involved: one to which milk, breadcrumbs and other perishable by-products had been added, in particular bits of brain. Preserved without due care, exposed to the variations of the climate, they had been frozen and defrosted dozens of times over.[33] Even more markedly, the sausages hadn't been made by qualified butchers, but always by small producers in the countryside who didn't know, or didn't bother, about the most basic rules.[34] The cause of botulism was neither bad luck nor chance: it was the absence of hygiene and a lack of professionalism – that is, when it wasn't simply about filthy lucre.

Official appeals made no difference: the poisonings continued, to the anger of the doctors. In 1860, Dr Schroter, an army physician, wrote an article in the Württemberg medical bulletin to denounce this litany of deaths: 'In the pages of this journal we find a long, uninterrupted sequence of poisonings. It is bewildering that such tragedies continue to occur despite all the warnings.'[35] He was dismayed that new cases were occurring in hamlets where there had been deaths only three years earlier. 'Intoxications are occurring even in villages which already appear in the annals of botulism, in capital letters!'[36] Those deaths hadn't been sufficient to daunt the local butchers, who quite knowingly continued to sell rotten meat products. For Schroter, the source of the problem was quite obvious: the poisonings were caused by thick, over-sized sausages, made with 'disgusting skins' filled with 'old blood', which hadn't been smoked long enough for the preservative power of the smoke to take effect.[37] In his official report, Professor Schlossberger reached the same conclusions: without exception, the incriminated sausages had been produced by under-qualified people who had made a succession of technical mistakes.[38]

The severest criticism focused on the heating. In the first official reports, Professor Autenrieth, dean of the University of Tübingen, had noticed that the incriminated sausages had not been cooked using customary methods.[39] Several doctors validated this theory and showed that the sausages hadn't been cooked correctly because they were too thick,[40] and because the population thought that cooking them too long would ruin their appearance: under the effect of heat, the sausages opened up and lost their shape, their skins split. In report after report the doctors tried to educate the poor peasants: 'Women, cook your sausages! Don't worry about them splitting open! Worry instead that, if you don't cook them, your sausages will poison your husband and children!'[41]

SOLVING THE BOTULISM PUZZLE

In 1855, a Belgian pharmacologist, Édouard Van den Corput, published an essay entitled *Du poison qui se développe dans les viandes et dans les boudins fumés* (On Poison That Develops in Meats and Smoked Blood Sausages). Like the doctors who went before him, the author noted that half the poisonings he had been able to identify had taken place in April.[42] He wondered why this mysterious organism proliferated in Germany, 'almost exclusively within the circumscription of a special part of the country or rather a bounded region'. He asked: 'what is the agent responsible for these "poisonous blood sausages" and "toxiferous" meats?'[43] Reiterating the warnings of his predecessors, Van den Corput reckoned that these intoxications should encourage consumers to be wary about processed meats which, 'infested with mould or half ruined, have undergone twenty metamorphoses before catching the eye of a consumer, and which, reworked to give them a more enticing appearance, escape the random inspections of the authorities with impunity.'[44] His study was the first publication to expound the theory of a toxic microscopic organism: the author imagined an 'elemental plant (fungus or algae)', invisible to the naked eye, which he called *Sarcina botulina*. A few years later, Louis Pasteur effectively described microscopic organisms capable of developing in the absence of air: this was the discovery of anaerobic bacteria. For his part, Robert Koch discovered the bacillus responsible for anthrax, subsequently identified the bacterium that caused tuberculosis, and finally confirmed that cholera was due to *Vibrio cholerae*. It was the big bang of microbiology: one by one, infectious diseases such as typhus, diphtheria, tetanus and pneumonia gave up their secrets.

The next stage occurred in Belgium, in a little village 40 miles from Brussels: in December 1895, in the church at Ellezelles, the funeral was taking place of Antoine Créteur, who had died aged 87. The municipal band played a few melodies to accompany the funeral procession. Then, following local tradition, the musicians met up for a bibulous meal. The raw ham that was served tasted so bad that many avoided it altogether. 'Some applied liberal amounts of mustard to help them swallow their portion but admitted that they still had great difficulty getting it down.'[45] A few hours later, the symptoms began to appear: the musicians had blurred vision, as if they were 'looking through a mist'.[46] A dozen men fell ill, and three of them died.[47]

A biologist, Émile Van Ermengem, was asked to do a medical assessment. He collected the results of the post-mortems and procured samples of the incriminated ham. By using new techniques in bacterial culture that he had learned from Koch, he managed to identify the agent of botulism: he described 'large bacilli that sporulate rapidly'.[48] Van Ermengem synthesized a century of research in a long report that he published in 1897, accompanied by the very first photographs of the elusive bacterium. Throughout the 246 pages of his final account, Van Ermengem emphasized one crucial point: botulism is avoidable through good standards of hygiene and sound production practices. He showed that the pork in Ellezelles had come from an animal slaughtered in August without any recourse to refrigeration. The animal hadn't been bled properly.[49] After slaughter, the meat had got dirty and then was treated with a defective brine.[50] It was kept in a barrel for months, immersed in a liquid that didn't have a sufficient concentration of salt to ensure efficient preservation. Van Ermengem indicated that it had often been the same deficiency in salting that had made Württemberg sausages so dangerous. By a series of experiments on the bacteria that he had identified, he proved that pure salt (without nitrate or nitrite) completely prevented the development of the bacillus. All that would have been needed to avert the accident in Ellezelles was a sufficiently concentrated brine, 'the sort that is normally used'.[51] Again, Van Ermengem underlined the role of heat. If the rules of hygiene can't be followed, if the curing can't be carried out with due care, it is enough that the producer properly cooks the hams or the meat of the sausages: 'A proper cooking would have made the Ellezelles ham perfectly harmless.'[52]

It was because butchers respected these rules that botulism was so rare outside of Württemberg, and was virtually unheard of in the UK and France.[53] Contrary to what many lobbyists for nitrite claim today, the cause of botulism was not the lack of nitro-additives: it was either ignorance or deceit.

CHAPTER 11

THE BOTULISM PRETEXT

What value should we place on claims so often put forward by the nitro-meat lobby that the use of additives is 'indispensable'? On its website, a Danish industrial meat-processing company which has not used nitro-additives since 1992 explains that modern production methods do not justify the use of nitrate or nitrite as 'anti-bacterial' agents: 'This is completely irrelevant in modern meat-processing. The main reason for the widespread use of nitrite is that it gives a red colour to deli meats and sausages.'[1] In its adverts for bacon and sausages, the company stresses: 'Appearance: natural colour of cooked meat, lightly marbled. The product is nitrite-free, and consequently its colour differs from that of other products on the market.'[2] In a similar way, a Canadian company that has given up using nitro-additives explains that 'many companies still use them as a way to extend the shelf life of their products, and to cut down on production costs'.[3]

BOTULISM IS WHEN YOU CUT CORNERS

In France, many nitro-meat producers claim that nitrite curing is necessary, 'because botulism hasn't gone away' or 'because cases of botulism continue to occur'.[4] Yet when we look closely at French botulism statistics, it is striking that most botulism cases are due to raw hams that have been home-produced in a cellar or garage, using whatever is to hand, by unqualified individuals.[5]

175

History is quite informative on this. France suffered a serious epidemic of botulism during the Second World War. Previously, the number of cases had been negligible.[6] 'The disease was rare until 1939', remarked a doctor who wrote his doctoral thesis on the subject in 1944.[7] In the four years running up to the war, the Pasteur Institute of Paris had identified only three outbreaks of botulism in the whole country.[8] Professor René Legroux, head of the microbiology service at the Institute, noted in 1945: 'During the four years of the Occupation, on the other hand, we know of 417 outbreaks, each generally affecting between 2 and 10 people', which puts the number of cases of poisoning at 'more than 1,000' in the space of four years.[9]

At the end of the war, the French Académie de médecine released a report entitled 'Botulism and salted hams',[10] which showed that this sudden epidemic was due to a disregard for the traditional rules of production. The bulletin of the academy offered a reminder that traditionally hams should only be pre-pared under proper conditions of refrigeration and be taken from animals slaughtered with their stomach empty – otherwise the meat goes off. But in 1944, because of food shortages, salted meats 'could be sold at a very high price, and as a result hams were prepared in the spring as well as the summer'.[11] To feed the black market, animals were slaughtered clandestinely with no regard to rules of hygiene and refrigeration. The animals arrived with full bellies, still digesting food, and the killing took place in deplorable conditions. The curing was poorly done because 'today salt is rare and farmers have to use it sparingly'.[12]

Thanks to an experimental study involving 300 pigs, Professor Legroux and his team confirmed that botulism was entirely avoid-able: they proved that the epidemic was due to the conditions of slaughter of the animals and to a contempt for basic sani-tary principles.[13] The faults were the same as those described by Van Ermengem at the end of the nineteenth century: botched slaughtering, loose hygiene, rough-and-ready refrigeration, incomplete and badly managed curing. Legroux's results were unanimously confirmed by several medical researchers who investigated the epidemic of 1940–44: they concluded that when meat is processed by qualified professionals taking due care, 'botulism can easily be avoided and cases of this food-borne disease should no longer occur'.[14]

In the 1980s, there was a resurgence of botulism in Poland, then under the Soviet yoke. The poisonings were linked to the preparation of preserves according to a primitive method known as 'meat in a jar', an expedient that was rendered necessary by the economic restraints of the time.[15] Today, in an age of ultra-modern production lines, where industrial meat-processors can work in conditions of perfect hygiene (including 'white rooms' and even, in the case of some firms, automatized production), it seems almost comical that some large manufacturers, these champions of innovation, should hide behind the tragic accidents caused by poor amateur butchers operating out of their cellars. That some neophyte working alone from a makeshift kitchen in his garage – or in a hypothetical farm in rural Scotland, or a shack in Sardinia – might have need to 'secure' a few rustic hams when he is unable to master basic hygiene is not to be denied: that is when a bactericide comes in useful – whether it is a nitro product or another, less dangerous antiseptic. However, can it be seriously argued that the state-of-the art processing plants of some of the world's largest food groups are unable to guarantee adequate refrigeration in their plants and follow correct slaughtering procedures?

Nowadays, several giants of industrial food processing are fighting to retain the right to nitrite-cure bacon, ham, corned beef and smoked sausages of all sorts, and even meats used in frozen foods such as pizzas. Who would believe that any of these flagship companies of the agri-food sector are unable to guarantee that the meat they use is clean, that their curing and cooking processes are up to scratch, that their workers wash their hands and use clean equipment?

FROM A (ASPARAGUS) TO Y (YOGHURT)

The bacteria responsible for botulism are abundant in nature: they are found in the soil (earth, humus), in fields and gardens, in lake and river sediment, on the banks of estuaries, etc. That is why they are often found on the scales of fish, in the digestive tracts of mammals (humans included) and on the skins of fruits and vegetables. The microbe, or its dormant spore, only needs a favourable environment in which to proliferate. At the beginning of the twentieth century, most microbiologists still believed that botulism was restricted to products made from animal flesh. That all changed in 1904, when there was an outbreak in Germany caused by a bean salad.[16] The same scenario occurred a few years later in California, where twelve people fell ill after eating a tin of beans. Since then, botulism caused by beans has resulted in hundreds of poisonings across all continents.[17] Similarly, cases of botulism have been caused by badly preserved mushrooms, by carrots, shallots, garlic, by jars of peppers, tins of spinach and peas, by 'canned roasted eggplants in oil' ... In the summer of 2019 it was a bottle of industrially produced vegetable soup that intoxicated a French woman when she consumed it after its best-before date.

In the UK, botulism cases have been linked to the consumption of a vegetarian nut brawn (carrots, peas and yeast gravy), home-prepared bottled mushrooms, hummus stored unchilled, and even a pre-packed rice and vegetable salad eaten as part of an airline kosher meal.[18] In Scotland, an industrially produced korma sauce was responsible for the poisoning of several children in 2011.[19] Previously, in June 1989, the UK experienced the worst outbreak of botulism in its history, caused by some hazelnut yoghurt: 27 people in northwest England and Wales were affected, and one woman died.[20] The accident was due to faulty preservation of the mix of pre-roasted hazelnuts used in making the yoghurts.

The risk has markedly increased with the development of vacuum packaging, as *Clostridium botulinum* is anaerobic. Unless they have a high degree of acidity, almost all prepared foods present a potential risk of botulism *if their production was defective*. A recent article gives a good idea of the diversity of foods that might be affected: *Clostridium botulinum* has been detected in chicken enchiladas, prepared prawns, peanuts, mortadella, mascarpone cheese, raw ham, sautéed onions, chili sauce, cheese sauce, pesto, tinned tuna and sardines, bamboo shoots, carrot juice, and so on.[21] Certain vegetables are particularly problematic: since *Clostridium botulinum* lives in the earth, root vegetables are often carriers, especially if the soil has been composted with manure. Tinned asparagus regularly causes poisonings in the USA,[22] as well as in Australia[23] and in Europe.[24] These intoxications have generally been linked to home-made preserves, but cases of botulism have also occurred after ingestion of industrially produced tinned asparagus.[25]

Microbiologists also point out that 'potatoes and potato products have been identified as presenting a significant risk':[26] cases of botulism have been caused by potato salads,[27] vacuum-packed potatoes, cooked potatoes, potato-based dip prepared from baked potato, cheese sauce with baked potato,[28] potato soups,[29] etc. In 1994, at a community meal in Texas, 30 people fell victim to a dish made with potatoes that had been kept in aluminium foil; unfortunately, this was not an isolated incident.[30]

In view of these poisonings, would it be reasonable to expect that the use of nitrite would be authorized in hazelnut yoghurt, korma sauces, tins of beans or asparagus, pre-cooked mushroom dishes, even in cooked potatoes?

NITRITE ON OLIVES?

Riskier even than potatoes, apparently, are olives. Three early outbreaks were recorded in the USA in 1919 (seven dead after a banquet in Ohio,[31] five dead in Detroit, seven in Tennessee).[32] The following year, in New York, a family of five was wiped out.[33] Accidents continued to happen, and the investigations showed that the cause was always defective processing: the brine used was insufficiently concentrated and minimum temperatures for processing were not observed. Olives that are less cooked have a better visual appearance, therefore producers baulked at heating them at a high temperature, claiming they were worried it would cause the skins to split. One year, when prospects for the olive harvest in California's San Joaquin valley were exceptionally good, a specialist from the local university conducted a campaign to remind amateur producers that if the olives were not heated long enough before being canned they risked becoming deadly killers. He explained that 'some people are concerned that so much processing will make olives mushy'.[34]

In Europe, numerous cases of botulism have occurred because of olives in the course of the last twenty years: seven people falling ill after a picnic in Italy in 1999;[35] a party of sixteen poisoned in a restaurant in Italy in 2004;[36] four Dutch tourists intoxicated in 2008; two people made ill in Finland after eating olives stuffed with almonds in 2011. That same year, a dish of puréed olives intoxicated five people in the south of France, while four others were struck down in the north of the country.[37] Following the example of the industrial meat-processors, olive producers could have cited these accidents to argue that they should have the right to treat their olives with nitrite. This solution would offer several advantages: the olives would be more attractive, they would keep for longer, and the levels of salt could be lowered. Manufacturers of pre-cooked dishes that include olives, such as pizzas, could mount the same argument and request authorisation to add nitrite to their products 'to protect the consumer'.

Fortunately for the public, olive producers do not benefit from the same privileges as industrial meat-processors. When it comes to olives, the view is that the accidents that regularly occur because of defective procedures in artisanal production are not enough to justify the use of dangerous bactericides in industrial production. Rather than authorize carcinogenic additives, the authorities choose to subject the industry to tighter controls, modify equipment where necessary, eventually revise hygiene norms – even if it means forcing technicians to retrain.[38] When faults in processing occur, the health authorities sound the alarm and withdraw the products from the market.[39]

The same goes for all other risky products: when it was discovered that several cases of botulism had been caused by commercial carrot juice, the US health authorities revised the rules that the producers had to follow.[40] When an epidemic of botulism was caused by poor maintenance of equipment used in the manufacture of sauces, one large American manufacturer was forced to comply with rules of hygiene and preparation.[41] In Europe, *Clostridia* in milk products has been tackled by improving hygiene[42] and developing detection techniques.[43] In Iran, when it became clear that certain traditional milk products were associated with outbreaks of botulism, the government demanded that they be pasteurized.[44] Everywhere, for all these products, methods were applied that meant it was not necessary to have recourse to nitrate or nitrite – or any other equally efficient antiseptic (bleach, ethanol, glutaraldehyde, mercury chloride or platinum chloride, etc.).[45] The processed meat industry is the only one where the false claim of 'there is no alternative' is taken at face value ... even though alternatives have been practised in the industry itself, in some cases for decades!

THE IRONY OF PARMA

The European meat-processing industry managed to get one key institution, the European Food Safety Authority (EFSA), on its side. The links between some influential members of EFSA's committees of experts and prominent players in the agri-food and agri-chemical industries have been criticized on more than one occasion.[46] The Brussels-based organization Corporate Europe Observatory has shown that in the period 2012–2018 close to half the members of the EFSA expert panel on food additives had a financial conflict of interest.[47] EFSA has proved to be a faithful ally of business, reluctant to compel the meat industry to reform itself, systematically defending the status quo, making only the minor concession of symbolically lowering the levels of 'residual nitrite', an adjustment which has no bearing on the carcinogenicity of processed meats.

In June 2017, EFSA published two thick reports on the use of nitrate and nitrite which followed word for word some of the arguments articulated by the nitro-meat industry.[48] Instead of giving a ruling on nitrate-treated meat and nitrite-treated meat, the agency has chosen to rule on the additives themselves (nitrate or nitrite), and not to take into full account how these react with the meat. This essential issue has been aired many times before, in particular in a long legal battle between Denmark and the European Commission that ran from 2003 to 2015: Danish authorities have stressed that 'regulating the nitrite amounts used on the basis of whether they were considered to involve an intake in excess of the established relevant acceptable daily intake (ADI) *does not provide* the necessary protection of human health',[49] because 'the ADI established for nitrites *does not take into account* the formation of nitrosamines'.[50] The 2017 EFSA reports simply omitted any reference to nitrosyl-haem, and in fixing its 'maximum nitrite levels' EFSA has acknowledged that it was not taking into account the effects of nitroso compounds. EFSA admitted that nitro-additives were a source of carcinogenic nitroso compounds, but merely recommended that 'further large-scale prospective observational studies' should be carried out.[51] In 2020, the European Commission requested that EFSA conduct a new evaluation. But in March 2021, the agency made it known that because of 'lack of time', the expert panel would still not take nitrosyl-haem into account. Once again, EFSA announced that it would merely 'provide recommendations regarding the need for further assessment in this area'.

Going back to 2017, the EFSA reports concluded that the use of nitro-additives represented no risk to health. As if hidden in the depths of the reports there is one postulate: the agency asserts that, if the quantities of nitrite present in processed meat are well below the doses that would cause cyanosis, there is no reason why nitro-meats would cause cancers.[52] The European agency deliberately conflates questions of acute toxicity with issues of carcinogenicity.* Hence press reports suggesting that EFSA has established that the doses of nitro-additives used in processed meats pose no risk, 'except for a slight exceedance in children whose diet is high in foods containing these additives'.[53] The public doesn't get the fact that this evaluation concerns the risk of methaemoglobinaemia, *not the risk of cancer*.

Conversely, and again in tune with the meat industry, the agency never stops going on about the dangers of botulism: 'In most cured meat products, the addition of nitrites (or nitrates) is *necessary* to prevent the growth and toxin production by *C. Botulinum*',[54] EFSA asserted; according to European regulation, nitro-additives are necessary 'as a preservative in meat

---

* As explained in Chapter 7, when you ingest pure sodium nitrite, your blood loses its capacity to transport oxygen. This results in cyanosis (the skin turns a blue colour), which is caused by methaemoglobinaemia. This intoxication has nothing whatsoever to do with colorectal cancer.

products to control the possible growth of harmful bacteria, in particular *Clostridium botulinum*'.[55] The dictionary defines 'necessary' as 'needed to achieve a particular result', 'that is absolutely required', 'that cannot be absent'. There can be no doubt that nitrite is *useful* to the industry. But *necessary*? *Indispensable*? By deliberately ignoring the real reasons for using nitrite, EFSA acts as if it were impossible to cure meats without recourse to carcinogenic additives.

The agency's head office is in Italy – in Parma, in fact, the Mecca of traditional meat curing. In the immediate vicinity of the EFSA headquarters, the outskirts of the town teem with curing plants. According to the agency's communiqués, the evidence is clear: nitrate and nitrite have to be maintained 'to ensure microbiological safety'[56] in order to protect the population from the perpetual 'threat of botulism'. The agency even insists that the most exposed products are raw hams.[57]

And therein lies the paradox: in the town of Parma itself, nitro-additives have been banned from raw hams since the 1990s! As we have seen, the Parma Ham Consortium decided in 1993 to revert to its traditional recipe and formally banned use of nitrate and nitrite, choosing instead to apply strict rules of production and hygiene, beginning at the slaughterhouse.[58] The 150 firms in the consortium produce between them 8 to 9 million raw hams every year; they are known around the world for their quality and taste. And in the 27 years since the change, not a single case of botulism has been detected![59]

CHAPTER 12

A CASE STUDY: SODIUM NITRITE AND SMOKED FISH

In the UK, the use of nitro-additives is strictly forbidden in all fish products.[1] Experience and tradition have taught those in the trade preservation techniques that are capable of dealing with risks from bacteria using simply salt, smoke and heat. The *Clostridium botulinum* bacteria are nevertheless present in aquatic sediments, and fish-based products pose a risk of botulism *when they are incorrectly manufactured*.[2] Producers of smoked salmon could argue – as industrial meat-processors do – that smoked salmon, like ham, would be 'safer' if they were allowed to inject the fish with a powerful anti-bacterial agent such as nitrite (tests on fish-based products have demonstrated that nitrite can effectively destroy bacteria[3]). Lobbyists of nitrite-cured fish could claim that such treatment would also help to minimize other bacterial risks (in particular *Listeria monocytogenes*). The health authorities could authorize the use of sodium nitrite in smoked fish as they have done for cured meats, because it simplifies production and offers, in theory, an additional 'security' in case deficient processing has led the fish to be contaminated with bacteria.

FISH AND BOTULISM

Citing statistics obtained by Scottish researchers who studied botulism during the 1960s, a document distributed by the Food and Agriculture Organization (FAO) indicated that 'tests on 646 samples of vacuum packed fish on sale in British shops as far apart as Aberdeen and Bristol, Liverpool and Hull, showed that toxin developed in 5 packages after improper storage'.[4] More recently there have been reports in the press of the UK Food Standards Agency ordering recalls of smoked fish,[5] and the scientific literature contains detailed descriptions of botulism intoxications that have happened in Europe, for example when some smoked salmon was eaten in Germany just three days after its use-by date.[6] In France, botulism bacteria and spores were found in scallops, defrosted king prawns and fish soup.[7] A young French girl and her parents died of botulism after eating tuna kept in a broth for a few days.[8] In 1978, two people died in the UK after eating tinned salmon.[9] A recent summary recorded botulism toxins in badly preserved smoked fish (Finland, Germany), salmon roe (Canada), sardines (South Africa), and clam chowder (USA).[10] In 1991, in Egypt, 91 people were hospitalized because of grey mullets that had been incorrectly salted – eighteen people died. Surimi can potentially lead to botulism poisoning if it is not produced carefully.[11] Tests have revealed traces of the botulism bacteria in haddock fillets, carp, rainbow trout, salmon, etc.[12] There are reports of cases of botulism caused by tinned tuna, crab meat, herrings, roach, anchovies, lumpfish roe, prawns and smoked trout.[13] A law firm specializing in food poisonings has a 'botulism blog' which refers to dozens of recent cases.[14]

That is why modern manufacturers have to respect certain rules of hygiene and apply tried and tested production techniques. This comes at some cost. But in the USA, producers of smoked salmon have found another solution: as with processed meats, most use sodium nitrite to maximize their profits and to achieve a more attractive appearance at little expense.

NITRITED SALMON?

Injecting sodium nitrite into salmon flesh considerably enhances the intensity of its colour; as one scientific article on the impact of nitrite curing on salmon put it, 'fillet colour is held to be an important quality parameter for salmonid fishes. Redness contributes significantly to overall enjoyment of cooked salmonid flesh, and has a signalling value as an indicator of product quality.'[15] In France, the National Archives contain a lot of correspondence from manufacturers of nitro-additives and fish producers lobbying for the use of nitrate and nitrite in order to improve the colour of the fish, to simplify the production process and to extend shelf life. The Fraud Prevention Service has consistently refused to allow this.[16] In the 1970s, some businesses that used nitro-additives to prolong the preservation of herrings and cod roe were investigated, but it was salmon production that was most prone to illegal activity. The perpetrators tried to justify themselves by claiming that the use of nitro-additives was the only way to make the products safe.

As it is often kept anaerobically (that is, in vacuum-sealed packs) and generally eaten without being cooked, smoked salmon provides ideal conditions for the development of bacteria.[17] A report on producers in Brittany showed that, in fact, improvements in microbiological safety could be obtained by 'modifications at the level of personal hygiene, disinfection of the equipment and temperature control',[18] but unscrupulous producers would always be tempted to talk up the risk of bacteria to make nitrite curing seem acceptable. Tests have shown that this treatment would extend the period of microbiological safety by three to four weeks. In smoked fish, the addition of nitrite would allow the shelf life to be extended from 35 to 56 days;[19] this means in effect that adding nitrite to a packet of smoked trout would extend its 'sell by' date by three weeks.

However, like nitrite-cured meats, salmon treated with nitrite becomes a source of carcinogenic nitroso compounds. So, in Europe, the health authorities demand that producers of smoked salmon apply rules of hygiene and respect standards of production that remove the need for chemical bactericides. Occasionally, producers or importers use nitrite despite the ban; when they do, they are heavily sanctioned.[20] Honest processors manage perfectly well without using additives: over the last few decades, the large increase in consumption of smoked salmon has not led to an increase in cases of botulism. Europeans eat smoked salmon in abundance, without botulism and without nitrite. Yet, as we said above, there is one country that authorizes use of nitrite in smoked salmon: in the USA, representatives of the industry have managed to convince the authorities that it is *indispensable* to use nitro-additives in order to ensure consumer safety. The tactics they used were those that some meat-processors employ even today. That is why the story of nitrite-treated salmon needs to be told.

THE AMERICAN APPROACH

Like meat, fish deteriorates very quickly. In the Middle Ages, the preservation of fish was developed to a high degree of sophistication: herring, abundant through three months of the year, could be protected against putrefaction by a combination of salting and smoking without having to be cooked.[21] It could thereafter be kept for months, without refrigeration, and transported long distances. This technology was refined over the centuries. As America was colonized, fish smoking plants sprang up along the east coast, and subsequently in the interior, especially around the Great Lakes.

From 1870, advances in European and American chemistry led to the development of new bactericides. These could take the place of traditional salting and smoking and so simplify the production process: the producer, by sprinkling preservative onto the fish or immersing it for a few moments in a disinfectant, could bypass the need for a complete smoking and end up with a less salty product, which would keep for a long time. As with meat, the market for fish preservatives exploded at the end of the nineteenth century: many producers adopted sodium borate, boric acid, potassium nitrate and sodium nitrate, sodium hypochlorite, aluminium acetate, aluminium sulphate, sodium bicarbonate, etc.[22] Because of the antiseptic effect, nitro-additives simplified handling and facilitated transport and storage, thus extending the period of commercial viability and reducing the cost price. For salmon, they offered a considerable advantage compared with other disinfectants in that they improved the colour and provided excellent resistance to oxidation.[23]

In 1925, the Bureau of Animal Industry of the US Department of Agriculture authorized the use of sodium nitrite in meat-processing. It was not allowed in fish products, but was used illegally. The smoked fish producers banded together in a pressure group called the National Fisheries Institute and campaigned to be allowed to use nitro-additives on the same footing as the meatpackers. To substantiate their request, they explained that 'the nitrates and nitrites reduce the "brashness" of the salt and improve color stability, stabilize the protein and have an antibacterial action. All of this leads to a longer shelf life.'[24] But the health authorities demurred: the Food and Drug Administration (FDA) considered that the use of sodium nitrite was not justified in the preparation of smoked fish, because it involved employing a chemical product in place of traditional methods which were entirely adequate.[25] Like other artificial preservatives (boric acid on fish and eggs, calcium fluoride in beer or wine, formaldehyde in milk), the FDA reckoned that the use of sodium nitrite 'would disguise faulty processing and handling techniques'.[26] In particular, the FDA indicated that nitrite should not be tolerated on fillets of fish because it would be employed 'in lieu of good manufacturing and handling practices'.[27] During a hearing before a congressional committee, the commissioner of the FDA pointed out: 'It seems evident to us that the public welfare demands some supervision of a manufacturer's unilateral determination that there is a reason for using a poisonous chemical additive. The play of the marketplace must not be the sole criterion in this field where the public health is at stake. We must have a law that will cope with the type of manufacturer who does not appear before your committee, the ruthless producer who will stop at nothing to make a fast dollar. Although they are distinctly in the minority, they do have to be controlled.'[28]

NITRITE, A SUBSTITUTE FOR 'GOOD MANUFACTURING PRACTICES'

Throughout the 1950s, the FDA remained implacably opposed to the use of nitro-additives for the preservation of fish and sea-food.* When one producer asked for authorization to treat whale meat, the agency refused, explaining: 'Since we regard nitrates and nitrites as poisonous or deleterious substances not required in the production of foods generally, we have never sanctioned their use in food.'[29] In another letter (this time to a manufacturer of nitrited

---

* The Bureau of Chemistry (headed from 1882 to 1912 by Dr Harvey Wiley) never approved the use of nitrate and nitrite in meat-processing. In 1927, the Bureau of Chemistry became the Food, Drug, and Insecticide Administration, whose name was shortened in 1930 to the Food and Drug Administration (FDA). At this time, the FDA had authority on additives used in food (including fish products) but, due to special dispositions, had no authority on additives used in meat products, which were regulated exclusively by the Bureau of Animal Industry.

curing salt) the FDA wrote that it considered that 'the harmlessness and suitability for food use of nitrates and nitrites has not been established to the satisfaction of our pharmacologists'.[30] This is even more explicit in a letter that the health agency sent to the US Department of Agriculture in 1957, indicating that 'our Division of Pharmacology has serious reservations about the safety of nitrates and nitrites as ingredients of food'.[31] Their opposition was unambiguous, and indeed in 1959 one producer who used sodium nitrate on fluke, a type of flat fish, was convicted and sent to prison.[32] Likewise, a written exchange between an officer and the chief counsel of the FDA describes the prosecutions undertaken against a New York firm, regarding a large shipment of salmon: 'Examination shows that the article contains sodium nitrite, a poisonous and deleterious substance. Investigation showed the sodium nitrite was added by the dealer in curing the article.'[33] The FDA seized the shipment and passed the matter on to the judicial authorities. The FDA stated that sodium nitrite 'is unsafe'[34] and 'is a substance not required in the production of this food and can be avoided by good manufacturing practice'.[35]

But approximately at the same moment, the American Congress decided to loosen the rules on additives. A few years later, at an official hearing, the commissioner of the FDA summed up this change of tack somewhat laconically: 'The congressional committees explained that the usefulness of an additive should be determined in the marketplace and not by FDA.'[36] The indefatigable fish curers' association then restarted campaigning for the legalization of sodium nitrite. Their first request concerned tuna, as cooking gives the flesh a chalky colour that the producers adjudged to be rather unappetizing. In 1961, fish curers succeeded in obtaining the authorization to add sodium nitrite to tuna for colouring purposes (the regulations specified that the additive was used as a colorant).[37] Then they turned their attention to smoked fish: first salmon (addition of nitrite authorized in

September 1961, 'as a preservative and color fixative'), then shad, a type of herring (nitrite authorized in July 1963), then sablefish (nitrite authorized in November 1964).[38]

In 1965, after the deaths of some farm animals in Norway that had been given feed treated with sodium nitrite,* the FDA sounded the alarm: the agency was worried because the feed was made out of nitrite-treated fish, which had been milled into flour.[39] The FDA then refused to grant further authorizations and began to re-evaluate those it had already passed. The agency reiterated that nitrite curing was not justified insofar as norms of hygiene and good manufacturing practices were sufficient to guarantee that the products were completely safe. But some manufacturers did not apply these standards.[40] There was an outbreak of botulism, caused by portions of smoked chub which had been prepared with no regard to norms of hygiene, refrigeration and cooking. Several people died. The industry managed to seize the moment and asked for sodium nitrite to be authorized to prevent botulism, in place of minimum standards of heating and smoking. To justify their request, they claimed that biochemists who had studied rotten fish had already proved that nitrite was able to destroy a whole range of bacteria, including *Clostridia*.[41] The FDA was opposed, pointing out that sodium nitrite was not justified because observing the rules of production would have been enough to prevent botulism.† But after months of lobbying, the agency finally gave way: in August 1969, the FDA allowed manufacturers to employ less heat in the preparation of chub and to replace 'good manufacturing practices' with nitrite treatment.[42]

---

* See Chapter 9: The Nitroso Surprise.
† FDA scientist Jacqueline Verrett analysed these events in a book published in the early years of the 'nitrite war'. See Jacqueline Verrett and Jean Carper, *Eating May be Hazardous to Your Health: How your government fails to protect you from the dangers in your food*. Simon & Schuster, New York, 1974, pp. 136–56.

SCANDAL

In 1970, an NGO called the Environmental Defense Fund (EDF) requested to be told the basis on which these authorizations had been granted, but the FDA refused to reveal the data that the manufacturers had supplied. The EDF took the administration to court.[43] The FDA was ordered to open up its files, and what they contained unleashed a scandal: it was discovered that the agency had authorized sodium nitrite even though its own toxicologists considered the additive was potentially carcinogenic and that it was never *necessary* to use it in smoked fish.[44]

The press seized on the affair. Journalists unpicked the arguments put forward by the industry. American manufacturers around the Great Lakes, for example, maintained that the bacterial risk was more pronounced there than in Europe or Asia, since the water in the region was significantly more infested with pathogens. This might well be the case, but the argument really didn't stand up: the proponents of nitrite asserted that the chemical treatment was indispensable ... whereas, on the other side of the border, Canadian producers – fishing in the same lakes or even the same streams – were able to deliver perfectly safe products by respecting the rules of hygiene during processing.[45]

Other arguments were surprisingly trivial: one representative of the fish curers justified nitrite treatment on the basis that they were reluctant to heat their products because they were worried their fish would end up twisted or lumpy; as their customers liked their fish smooth and straight, the producers preferred to smoke at a lower temperature and substitute nitrite treatment for heat.[46] The verdict of the EDF on this was: 'The industry and FDA portrayal of the choice as being between nitrite and botulism is a self-serving deception.'[47] They went on to say: 'Since alternate procedures exist for protecting people against botulism, nitrite cannot be regarded as "required".'[48]

The heads of the FDA were summoned to face a senatorial committee in 1971. Under questioning, the agency ultimately admitted that the use of sodium nitrite suited the manufacturers but was in no way 'necessary' or 'indispensable'. Investigations revealed that the producers used nitrite so as not to have to invest in more modern refrigeration and smoking equipment. A Senate scientific adviser summed up the result of the hearing as follows: 'The impression I am getting is that there are marginal producers in the industry whose capability and willingness to abide by good manufacturing practices is so relatively poor that, rather than enforce high standards, FDA is going in the opposite direction by suggesting what alternative methods can be made available in order to enable such producers to compete in the marketplace.'[49]

In response, the heads of the FDA explained that the real guilty parties were in another agency altogether: those advocates of laissez-faire and proponents of the light touch, the Bureau of Fisheries in the Department of Agriculture.[50] Thanks to the Senate hearings, the public became aware that nitrite treatment had been used on fish illegally long before it was authorized. Worse: it was shown that certain manufacturers obtained authorization to treat fish with nitrite in order to have a use-by date longer than two weeks. Internal documents from the FDA revealed that even when the agency became aware of growing alarm over nitroso compounds, it did not have sufficient political clout to stand up to commercial interests: instead of compelling the industry to apply rigorous production methods, the FDA first turned a blind eye to violations,[51] then had to commit itself to legalizing nitrite usage, on the basis that this sector was 'not highly sophisticated'[52] and that 'its controls are not the best'.[53]

Following the Senate hearings, the FDA renewed its efforts to restrain the use of nitro-additives on smoked fish. But it was a case of shutting the stable door. The industry had already mobilized to oppose a ban. The battle would last for five years. In 1976, the

Marine Fisheries Review summarized the corporate position. The industry refused to replace nitrite with extended heating, and their reason could be summed up in one sentence: 'The complaint of industry is, however, that they are not equipped to use that alternative and, moreover, use of this alternative results in a product showing excessive thermal damage.'[54] When the FDA carried on regardless and regulated the length of the heating process, the industry took the administration to court. And they won: industry lawyers managed to obtain the concession that fish prepared following the FDA's norms might end up less appealing due to the extended heat exposure, to the extent that the products could be 'commercially unsaleable'.[55]

And that was that. Forty-five years on, smoked salmon is still treated with sodium nitrite in the USA: today, only a minority of US manufacturers eschew the use of nitro-additives. The FDA never managed to revoke the authorizations: strangled by legal red tape, the agency finally threw in the towel at the end of the 1970s, to the massive benefit of the nitrite curers.

But this was nothing more than a dry run for a much bigger conflict to come. The decisive battle was not over smoked salmon, but meat ...

CHAPTER 13

WARNINGS
AND DENIALS

In October 1969, the US Department of Agriculture held an emergency meeting with the meat-processing companies to discuss nitrite and cancer. The problems that lay ahead were already flagged up in the first public pronouncements: in an article entitled 'Do we make carcinogens from nitrites we eat?', the journal *Medical World News* reported one of the scientists of the FDA exclaiming: 'The truth is we don't need the nitrites nearly as much as we once did. But color is now an important part of a processed meat product. If corned beef didn't look the way it does, it would start tasting different.'[1] The US Senate summoned oncologists and biochemists to appear before it between 16 and 30 March 1971. In an article entitled 'Meat color additive linked to cancer', the *Washington Post* reported: 'A medical researcher said on Capitol Hill yesterday that a needlessly large cancer toll is the possible price Americans pay for the chemical treatment of ham, other cured meat products, and smoked fish to keep them pink, red or otherwise appetizingly colored.'[2] The popular press picked up the story: 'Health warning on meat cosmetic' (*Daily News*);[3] 'Cured meats yield cancer causatives' (*Evening Star*).[4]

After the way they had folded to the smoked fish producers, there was a suspicion that the FDA and the Department of

Agriculture were unwilling to take on the big processed meat manufacturers. The *New York Times* reported a throwaway comment made by the chief toxicologist of the FDA: 'I'm sure the nitrosamines are a carcinogen for humans. I'm sure we're exposed every day to carcinogenic substances. I've even quit smoking, but I like bacon and hot pastrami as well as anybody.'[5]

'WE FEEL QUITE DEFINITELY WE'VE GOT A BIG PROBLEM'

The nitrite curers received powerful support from politicians representing the 'meatpacker states' (such as Illinois, Wisconsin, Texas) and the large industrial farming regions, especially Iowa. The challenge of doing without nitrite was a huge one and had major implications beyond the meat-processing plants. The largest of the factory farms produced between 100,000 and 250,000 hogs per year.[6] Pig meat production was not only enormous in scale, it was highly mechanized: the amount of human labour required to raise a swine (from birth to slaughter) was less than two hours per animal. The 'farmers' bought piglets from specialist operators (the leading pig-breeding firm had a battery of 28,000 sows) and fattened them up as quickly as possible in strictly confined spaces.[7] That good ham and bacon could be achieved using this method was largely due to the nitro-additives; without them the mediocre quality of the meat would be more obvious, particularly in terms of its myoglobin content.[8]

The industry and the cancer specialists went head to head: the meatpackers denounced the 'present day advocates of disaster', and the secretary of agriculture came to the defence of the meat-processors, condemning 'the adverse propaganda which diet and health faddists have heaped upon pork'.[9] He declared: 'The ironic thing is that the food faddists and extremists are fretting about the safest and most nutritious food supply the world has ever known – and to make it even more absurd, food additives and modern processing are the factors helping provide that safety.'[10] In the Chicago region, the epicentre of industrial meat-processing, the trade press came up with spurious statistics to rubbish the warnings. According to *Farmer's Weekly*, a man of average build would need to eat 25,000 pounds of bacon each day to be exposed to any risk of cancer. 'Since 70 percent of the weight of bacon is lost during cooking, the human would have to purchase nearly 82,000 pounds of bacon per day.'[11] *Esquire* drew attention to the public relations campaigns led by the American Meat Institute (AMI), the main arm of the meatpacking lobby: 'The A.M.I. also passes out statements to debunk their critics that have absolutely no basis in reality, such as claiming that "for a human being to run *any* risk from eating bacon he would have to consume 46,000 pounds per day every day of his life".'[12]

Under pressure from the press and consumer groups, the US Department of Agriculture promised in March 1972 that it would set up a panel of experts to look at nitrite and nitrosamines.[13] But when this committee finally assembled, it was evident that it was primarily made up of scientists co-opted by the manufacturers, and at its head the secretary had appointed one of the most passionate supporters of nitrite curing.[14] To one observer, it seemed that 'this panel and USDA are not seeking an answer to the question "how can the hazard of nitrite be minimized", but have re-phrased the question to read "how can the hazard of nitrite be reduced without inconveniencing the meat industry".'[15] A few months later, the *Washington Post* indicated that a congressional committee of enquiry 'charged that federal agencies have known about the potential dangers for years but made "no serious effort to reduce the unnecessary exposure of the public to nitrates and nitrites" while awaiting proof of their potential for causing cancer'.[16] Several industry spokesmen constantly obfuscated the issue by repeating that there was no proof that nitro-meats caused cancer or contributed to it. Nothing was known for certain, they claimed, it was all unbelievable, nothing had been proved, it was impossible, it simply wasn't true ... Until a journalist from the *Washington Post* finally managed to extract an admission in October 1975: 'We feel quite definitely we've got a big problem', said Dr Richard Greenberg, vice-president and chief chemist of the giant meat-processor Swift.[17] The newspaper explained: 'A top official of one of the largest meat-processing companies has acknowledged that cancer-producing agents are formed when bacon is fried, but does not believe that bacon should be withheld from sale until the problem is solved.'[18]

Meanwhile, there was an increasing amount of research on carcinogenic compounds in processed meat.[19] In 1969, the WHO organized a number of meetings on the topic of nitroso compounds, and this led to the International Agency for Research on Cancer dedicating its very first monograph (1972) to nitroso-dimethylamine. The agency subsequently examined other nitrosamines (fourth monograph, 1974) and in late 1975 convened an international conference. The research focused primarily on nitrosamines wherever they occur: those that appear in fried bacon and when nitro-meats are digested; those found in industrial environments (sodium nitrite is used in the vulcanization of rubber and in treating sheet metal); those that crop up in beer if the hops are not properly treated; those – by far the most common – that poison tobacco smoke (when burnt, the nitrate contained in dried tobacco leaves reacts with amines).[20] In 1977, the IARC devoted an entire monograph to the problem of nitroso compounds.[21] There were numerous other scientific conferences, such as the two rounds of a large International Symposium on Nitrite in Meat Products, which were held in Holland in September 1973 and September 1976 and brought together specialists in toxicology, chemistry, microbiology and public health. Not even the US Department of Agriculture could keep turning a blind eye to the scientific evidence.

A BIGGER MONSTER: THE BOTULISM BOGEYMAN

In 1977, the election of Jimmy Carter as US president ushered in a more interventionist American administration. The new assistant secretary of agriculture declared that, 'when faced with problems such as the nitrite one, government can either ignore the health problems or it can help consumers and business adapt to the changes that will be required'.[22] The new head of the FDA, a biologist, noted: 'as science becomes more sophisticated, and as processed foods become more and more a part of our diet, we must expect occasional bad news about substances we thought safe.'[23] In a press release addressed to consumers, he explained why foods that had been found to be carcinogenic in tests on animals must be quickly withdrawn from human consumption: 'Even if it were ethical, it would still be impractical, for example, to expose 20,000 persons to a suspected carcinogen for the required years and under the dietary and other restrictions necessary to identify perhaps the one person in the 20,000 who might get cancer from the substance. And while one cancer in 20,000 sounds rare indeed, it still adds up to more than 10,000 cancers in a population of two hundred million plus.'[24] On the recommendation of cancer specialists who had studied nitroso compounds, the FDA and the US Department of Agriculture sought to drastically reduce the authorized levels of nitrite, and indeed reduce them to zero whenever possible. They recommended that harmless alternatives should be used for the coloration of processed meats. When a preservative was really necessary for protection against bacteria, nitrate and nitrite should be replaced with non-carcinogenic anti-bacterial agents, such as lactates and sorbates. This attempt at reform met with fierce resistance from the industry.

Firstly, the manufacturers claimed that the nitro-additive was essential to the taste of the meat. Scientists working for the AMI claimed that 'The most important organoleptic* reason for using nitrite is its striking effect on flavor.'[25] They pointed to studies indicating that nitrite-cured sausages were tastier: in tests, on a scale from 0 to 9, sausages without nitrite received a score of 4, while nitrite-cured sausages received scores between 5 and 6.2.[26] Consumer groups dismissed the claim by pointing out that the flavour could be adjusted using spices instead. Anita Johnson, a lawyer representing the Environmental Defense Fund, argued that achieving a good taste was 'a trivial reason for exposing consumers to a hazard of cancer'.[27]

Over the course of the 1970s, meat scientists working for the industry ran a variety of tasting tests in order to assess the effect of nitrite curing on the flavour of processed meats. The final report of the expert committee set up by the American Meat Science Association showed no marked preference either way: 'In some tests, panelists have preferred the product containing nitrite. In other studies, tasters had equal preferences for the meat products prepared with or without nitrite.'[28] Given its economic importance, bacon was subjected to particular scrutiny by the biochemists from the US Department of Agriculture. Firstly, they compared commercially produced bacon with bacon cured without use of nitro-additives. They concluded that: 'panelists rated the flavour of the no-nitrite bacon comparably with that of a popular national brand bacon.'[29] Then they prepared batches of bacon with and without nitrite and blind-tested them on a panel of 1,000 people. The results showed that, contrary to what the industry claimed, nitrite was not necessary to create the distinctive taste of bacon. The ministry researchers concluded: 'bacon with flavour acceptable to the average consumer can be made by processing with salt and no nitrite.'[30]

---

* Organoleptic: involving the use of the sense organs.

As the warnings about cancer became ever clearer, one single argument came to the fore. Turning the whole debate on its head, the meat-processing lobby made the audacious claim that nitrate and nitrite were in fact not dangerous, but rather absolutely essential as protection against an even more terrifying danger. Introducing a ban on nitro-additives, they claimed, would lead to botulism-induced carnage. The scientific director of the AMI explained that the botulism toxin was so potent that a glassful of this poison would be enough to wipe out the entire population of the planet.[31] The director of a scientific committee attached to one industry lobby developed the theme: 'if the right conditions existed in our meat-processing industry for the growth of the bacterium, everyone who ate this meat would be killed by botulism.'[32] At a congressional hearing, he made out that botulism was still endemic in certain 'less developed' countries – in particular, he said, in France and Spain.[33]

Another lobbyist claimed that if a consumer bought meats that hadn't undergone any treatment with nitrate or nitrite, he would be 'toying with death'.[34] He even claimed that the manufacturers who used harmless colorants risked turning their meat into 'a pleasant smelling, pleasant tasting and colorful death trap'.[35] Even the most conservative estimates sent a shiver down the spine: one newspaper in Tennessee quoted the warning given out by the lobbyists that 'banning nitrite could save, say, 2,000 lives from cancer each year and lose 10,000 to food poisoning'.[36]

WE'VE BEEN HERE BEFORE: SULPHITE AND 'GERMS'

In July 1972, the trade magazine *Food Manufacture* suggested that 'There has been conflicting opinion concerning the need for nitrite in cured meat products'.[37] It reported that a chemist in Chicago had just published results proving that nitrite was required to prevent botulism: 'According to the American Meat Institute who commissioned research into the subject, the unblemished botulism safety record of canned cured meat products in the USA may be due to the use of nitrite in their processing.'[38] From the outset, this was treated with suspicion, since 70 years earlier manufacturers had used similar arguments to justify the addition of other controversial chemicals, particularly sodium sulphite (or 'sulphite of soda').

Like nitro-additives, sulphite reacts with haemoglobin to create a deep, long-lasting red bloom, which allows sausages and mince to have a longer shelf life without losing their colour.[39] The two methods function according to a similar chemical principle: in the case of nitrite, the colour appears because nitric oxide binds to haemoglobin; in the case of sulphite, the agent is sulphur oxide.[40] In 1904, a chemist at the Preservaline Manufacturing Company explained that 'the most noticeable action of sodium sulphite on meat is the production of a beautiful bright red color'.[41] He pointed out that 'in the presence of nitrites, the meat becomes a ruby red; in the presence of sulphites the meat becomes a scarlet red'.[42] At the same time, Heller & Co. were vaunting the merits of their sulphite-based additive Freeze-Em: 'There are few stomachs that are not whetted to keenness by appearances, and there are few pocketbooks that are not reached through the stomach. [...] The various preparations of Heller give meats such a delicious look as to make them appeal at once to the eye, and it is only necessary for the dealer to use them on his meats in order to attract trade and stimulate the appetite.'[43]

But when the American government wanted to ban use of sulphite in hamburgers and sausages, the additive manufacturers made out that sulphite was indispensable – not for colouring, but to protect the meat 'against germs'. Some manufacturers even funded 'scientific' studies which demonstrated that banning sulphite would cause epidemics, particularly among the most vulnerable: children, the elderly, the sick.[44] And when the health authorities introduced restrictions, the Preservaline Manufacturing

Company took the administration to court, building their case on the argument that sulphite was not a colorant but an anti-bacterial agent.[45]

A chemist and pharmacist called Robert Eccles carved himself out a profitable niche by defending dangerous additives on behalf of manufacturers.[46] Since bactericides allowed products to be sold whose bacteriological condition left something to be desired, Eccles maintained that banning certain additives adjudged to be dangerous would harm less well-off consumers. At a conference entitled 'The Consumer Interest in Food Preservatives', he explained: 'The Pure Food Law has tended to send to the garbage heap great amounts of perfectly wholesome articles, that by coloring, spicing, and putting up in dainty form, were sold at low rates. [...] The butcher, who formerly was permitted to use his perfectly wholesome scrap meat in the manufacture of Hamburg steak, must now either defy the law or lose a large part of the trimmings. This loss, that at first falls on the dealer, soon goes home to the consumer.'[47] Consequently, 'The fight against preservatives, when simmered down to its last analysis, is simply a fight against economy, and a fight in favor of costly methods, that raise the cost of living upon the poor.'[48]

As far as Eccles was concerned, all the purported dangers of food additives were imaginary, based on pure 'superstition'.[49] To the delight of his clients, the rich food processors of Chicago, he stated that it was necessary to treat all food products with powerful germicides in order to protect the population against 'typhoid fever, diarrhoea, dysentery, cholera, cholera morbus, consumption, scarlet fever and diphtheria'.[50] The systematic addition of chemical bactericides to food was in his view justified, because 'To allow these germs to multiply on meats, sausages, milk, puddings, oysters, jellies, ice creams, custards or fish is a terrible thing'.[51] He declared: 'To say that preservatives are injurious to health may or may not be a sufficient reason for objecting to their use. If the injury they cause is less than the injury done by poisonous toxins, why not choose them in preference to the toxins?'[52] It was a trick question; by presenting the issue as a choice between only two options, Eccles omitted to mention that there were other techniques that guaranteed complete safety without the added risks of using sulphite. As the doctors advocating for the Pure Food laws put it, 'foods can be preserved for a proper length of time in wholly unobjectionable ways, namely, by cold storage, desiccation, and sterilization. The only excuse for chemical preservatives is that it is cheaper and more convenient to preserve foods in this way than by any of the other methods.'[53]

Even though sulphite was banned in the end, fraudsters continued to use it in meat products.[54] The public authorities had to warn consumers: 'be not deceived by color, for it sometimes "happens", in violation of pure food regulations, that meat wears an artificial complexion, purchased at the drug store.'[55] Just a few short years before the nitrite affair, a number of sulphite-related scandals made the news, and in the 1950s it was necessary to threaten perpetrators with imprisonment in order to put a stop to use of sulphite.[56] On each occasion, the offenders swore that they weren't interested in the colouring effects of sulphite, but that they felt obliged to use it to protect the consumer 'only because it arrests the spread of bacteria'.[57]

NO REPLACEMENT?

Aware of these relatively recent infractions involving sulphite, one of the American doctors and activists who were seeking a ban on nitrite in the 1970s noted: 'There is little doubt that nitrite *can* prevent the growth of bacteria. The big question is whether, in commercially cured fish and meat, it actually *is needed*.'[58] In the journal *Science*, the cancer specialist William Lijinsky noted: 'Consistency would demand that food manufacturers add nitrite to all products in which botulism hazard exists. One can assume that this is their recommendation for vichyssoise soup and processed mushrooms, large batches of which have been recalled in the past year or so because of the finding of *Clostridium botulinum* contamination in some samples.'[59]

From the very start of their enquiries, US congressmen realized that there were strong reasons to suspect the validity of the arguments put forward by the industry: the use of a chemical antiseptic could no doubt be justified in certain 'at risk' products, where the manufacturer wasn't able to follow proper rules of hygiene, but it didn't appear to be necessary for all the others. In a report entitled 'Need for nitrites in meat to prevent botulism questionable', a congressional committee noted in 1972: 'The subcommittee's investigation, however, did not uncover persuasive evidence that the nitrites are required for this purpose except in special cases, such as canned ham.'[60] One congressional subcommittee was astonished that, in a technologically advanced country such as the USA, manufacturers claimed that it was impossible to devise 'adequate processing, packaging and sanitation practices to protect the public against botulism without the need for questionable chemical preservatives'.[61] The New York congressman Jonathan Bingham pointed out that 'a number of food processors in our country use other safer additives to check botulism'.[62] This was backed up by consumer groups who stressed that 'alternative means of safe processing are available which do not involve risks such as those posed by nitrite'.[63]

Newsweek magazine later confirmed that alternative preservatives did in fact already exist.[64] Manufacturers of chemical products were offering solutions based on sorbate, which eliminated all microbiological risks and ensured a long shelf life.[65] Sorbates had the advantage of not instigating carcinogenic compounds, but according to the meatpackers they were unsatisfactory because they didn't have the same colouring action as nitro-additives; even today, sorbate is often seen as an ideal alternative additive that ensures a long preservation of processed meat products, but the industry tends to reject it, largely because it has no colorant effect.[66]

In the 1970s, a company producing sorbate additives lamented: 'There's no place for a replacement as long as nitrite is available.'[67] A chemical company even offered a 'sorbate-nitrite mixture' which contained a very small quantity of nitrite – just enough to fix the colour.[68] By using this procedure, microbial stability and an extended shelf life would be delivered by the sorbate, while the nitrite would provide the organoleptic advantages. Another manufacturer patented an additive that could 'reproduce the effects obtained with nitrites in cured meat products by means of a composition of authorized food additives',[69] this being a mixture of sorbate and a red colouring agent. But this product didn't enthuse the meat-processors either: at a Senate hearing, one of the principal defenders of nitrite curing explained: 'The limitation of this system is that it can only be used in emulsion-type products. It can be used in frankfurters and bologna, but the red dye would dye bacon completely red, so that both the lean and fat portions of bacon would be colored red.'[70]

Other methods were considered, in particular using gas. A 1971 confidential report by British researchers stated: 'Colour of meat: No substitute for nitrite in the production of traditional red colour in cured meat, had yet been found. It was suggested that carbon monoxide should give this red colour. Dyes cannot be used because they colour both lean meat and fat.'[71] That same year, industrial meat-processors in London suggested replacing sodium nitrite with sulphur dioxide with just a small added dose of nitrite (just enough to ensure a regular colour).[72] In 1972, a confidential report by the FDA made clear that 'in a study of substitutes for colour production in cured meat as alternatives to nitrite, some 120 compounds had been screened, of these some 20 gave colour changes but the colours were either the wrong shade or not stable'.[73]

This shows up the double game the producers were playing: they claimed that nitrite treatment was indispensable to combat bacteria, but they refused to replace nitro-additives with less noxious antiseptics, on the grounds that they didn't colour as well.

CURIOUS TIMING

The press noted that the 'botulism' justification had only really made an appearance very late in the day, exactly when the carcinogenic effect of nitrite-treated meat was starting to become evident. In 1976, the *Washington Post* observed: 'The meat industry claims that sodium nitrite is essential to prevent the formation of the deadly *Clostridium botulinum* spores, which can cause death if ingested. But this claim is recent. The meat industry has fallen back on it since questions about the safety of nitrites have been raised. Previously the industry said it used nitrites primarily to give meats their flavor and characteristic reddish color.'[74] Another journalist noted: 'It was only in the 1960s that the FDA began approving new anti-botulinal uses of nitrite, and it was only after the consumer challenges to color-fixing uses in this decade that the meat industry began claiming that the *main* reason for nitrite use was to prevent botulism.'[75]

Wherever you look – France, Germany, the UK, the USA – this so-called 'indispensable anti-botulinal function' of nitro-additives was never mentioned before the 1970s in the regulatory and technical literature. Nitro-additives had been discussed in countless publications which described their advantages: the immediate appearance of an enticing and lasting colour, the evenness and consistency of coloration, the answer to problems of hygiene in factories, the simplification of production and transportation, the extended shelf life, etc. Indispensable protection against botulism

209

was never brought up. The same goes for the numerous patents relating to nitro-additives, which make no reference to botulism, not even in veiled terms.[76]

What is the reason, for example, why the botulism question was not mentioned in the regulatory texts concerning nitro-additives that were adopted during the first half of the twentieth century? How can it be the case that 'microbiological safety' was the real reason for the use of nitro-additives when the original US regulations (1925) stipulated precisely: 'The function of sodium nitrite in the curing of meats, like that of sodium nitrate and potassium nitrate, is the fixation of the red color [...] Neither the nitrates nor nitrites are of any particular value as preservatives in the quantities used.'[77]

How credible is it that the true purpose of nitro-additives was to combat botulism when in 1948, at a meeting with officials from the US Department of Agriculture, the expert from the FDA declared that their use was questionable as long as their function was 'merely improving the appearance or shelf life'?[78] In the 1965 book *Regulations Governing the Meat Inspection of the United States Department of Agriculture*, which contained all the rules applying to meatpacking factories, the description of nitro-additives was unambiguous, as it indicated: 'purpose: to fix colour'.[79] That same specification can be found in the archives of the British Ministry of Agriculture, Fisheries and Food, in the files for 1968.[80] Decades earlier, the minister and his advisers were openly discussing 'regulations which would allow the use of nitrite for colouring purpose – and that is what it is really wanted for'.[81] For his part, the official in charge of additives at the British Ministry of Health noted in 1942 that nitrite was added to meat products 'in order to preserve their colour and not for the sake of its preservative qualities'.[82] Not a single word about botulism.

Similarly in the French archives: the question of botulism is not even mentioned in the report of the Académie de médecine of June 1964 which gave assent to the use of nitrite,[83] or in the letters that the industrial meat-processors sent in June 1963 to the Ministry of Agriculture to request authorization for the use of sodium nitrite.[84] The *Manuel de l'apprenti charcutier* (Manual for the Apprentice Charcutier), published in 1962 as the industry textbook for young French professionals, stated without any ambiguity that nitro-additives were 'not necessary':[85] nitro treatment was 'optional',[86] only essential if the producer wished to 'give the meat that nice, appetizing pink colour that the charcutier always looks for in his hams and other cured meats'.[87]

In short, before the problem of cancer reared its ugly head, meatpackers were not claiming that nitro-additives were indispensable for microbiological security. The botulism justification seems to have appeared out of nowhere, plucked like a rabbit out of a hat, at the precise moment when nitro-additives were beginning to be suspected of causing cancer.[88] Urged to prove that this was more than merely an opportunistic justification, the processed meat industry has failed to come up with any adequate explanation.

QUESTIONABLE EXPERTS

This strange 'reorientation' of nitro-additives in the 1970s was all the more intriguing given that this notion that they were irreplaceable as protection against botulism was conjured up by scientists widely known to have links to the industry. One of the first authors to insist on this 'anti-botulinal function' of nitrite,[89] Michael Foster, happened to be the chair of a 'Committee of Food Protection' that was heavily criticized for its industry connections.[90] In his technical work during the 1950s, Foster himself had shown that the use of a nitro-additive was not always indispensable for microbiological safety.[91] Nevertheless, he recommended the use of nitro-additives to prolong the life of products and to improve their look. Alongside his academic work, Foster had been a director of one of the largest manufacturers of chemical colorants in Chicago. In particular, the firm produced a colouring agent for hot dogs known as 'Orange B', which ended up being condemned by the FDA.[92] The company also sold synthetic colorants to drinks manufacturers with the reminder that: 'Color is the first overture your product makes to a prospect. Is that color as inviting as it could be? Does your color help make as many sales as it should?'[93]

Described as 'a staunch defender of food additives and a vociferous opponent of consumer groups',[94] Foster headed up a research centre near the base of the giant meat-processing firm Oscar Mayer & Co. in Madison, Wisconsin. Created in 1919, this company had been one of those that had profited most from the legalization of sodium nitrite in 1925. By the systematic use of nitrite curing, Oscar Mayer had become one of the leaders in plastic-wrapped bacon and the largest producer of nitrite-treated hot dogs in the world.[95] When the carcinogenic effects of nitrite curing began to be known, the whole prosperity of the city of Madison seemed to depend on this enormous nitro-processed meat factory, the largest private employer in the region.[96] (Even today, Wisconsin is the US state that has the largest number of processed meat factories.) In the 1970s, Foster's institute was described as 'ensnared in a web of corporate connections'.[97] Over the years, a disproportionate number of articles on the 'necessity of nitrite' has emerged from a handful of Madison laboratories largely funded by the processed meat industry. And that same slant has been evident when it comes to the issue of cancer: since the beginning of the 1970s, a small group of industry-funded researchers in Madison has produced a large proportion of the publications that call into question the carcinogenic effect of nitrite-cured meat.

CHAPTER 14

THE 'NO ALTERNATIVE' LIE

As early as the 1970s, it was clear to American observers that the industrial meat-processors were being economical with the truth when they claimed that it was impossible to cure meat without recourse to nitrate and nitrite. Because even in the USA it was possible to find dozens of small- or medium-sized producers who did not use nitro-additives at all – without incurring a single case of botulism due to their products. Meat scientists had shown that it was possible to make cured meats that were perfectly safe without needing to use carcinogenic additives.[1] Consumer groups suggested that, given a bit of time, the public would adapt to non-nitrited meats and appreciate them just as much as the products with which they were familiar: admittedly they would not have the same colour, they would not last as long, they would need more care and more time to produce, but these drawbacks seemed entirely reasonable since they would remove a cause of cancer.

IS NITRITE-FREE CURED MEAT 'IMPOSSIBLE'?

The most astonishing achievement of the nitro-meat industry was to convince critics that it would give up using carcinogenic additives 'once an alternative method has been discovered'. 'There are no good alternatives for nitrite'[2] was the mantra of the president of the AMI; 'none of the proposed substitutes for nitrites works very well', echoed one of his allies;[3] 'I don't know how long it will take to find a chemical substitute; it could be tomorrow and it could take years',[4] chipped in another chemist in the nitrite camp. In fact, alternative methods had always been available, but the US industry refused to adopt them because none were as profitable as using nitrate or nitrite. When it came to raw ham, for example, only a few American producers knew how to use the traditional European technique: at the end of the 1970s, a report by the American Meat Science Association noted that 'at least 13 federally inspected commercial producers of dry-cured or country hams use no nitrate or nitrite in their products. Most of these are relatively small producers.'[5]

At the same time, an internal report by the US Department of Agriculture (USDA) listed the manufacturers who did not use nitro-additives. For years, the nitro-meat-processors had lobbied the administration to prevent non-nitrite-treated products being sold under their usual name ('bacon', 'frankfurters', 'wieners', etc.). The meatpackers who used nitrite opposed the nitrite-free development because, according to them, it risked making consumers 'unduly concerned'.[6] In their report the USDA experts wrote: 'Presently, there are approximately 34 establishments which market nitrate and/or nitrite-free processed meat products under 168 different product labels. The majority of these are sausage-type products. Since these products cannot be marketed under traditional product names, consumers are, in many cases, unaware of their existence or do not associate their current name with the traditional cured product. This information issue has slowed development of the market.'[7]

The Center for Disease Control in Atlanta confirmed that the processed meats made without nitrate or nitrite had not caused any cases of botulism.[8] Consequently, the *Washington Post* expressed surprise in September 1977 that the large corporations were claiming that it was impossible for them to operate without using carcinogenic additives: 'However, small manufacturers have been producing nitrite-free products for years, using other means to prevent the formation of botulism spores.'[9] Eighteen months later, the *Washington Post* reiterated the point: 'Ever since questions first were raised about the possibility that nitrosamines (a combination of nitrites and amines) might cause cancer, meat-processors have insisted that without the nitrites, which color and flavor processed meats while preserving them, products would be subject to contamination with *Clostridium botulinum* – the botulism toxins often fatal to humans. Despite this position, numerous small, local manufacturers produce nitrite-free meats sold in natural food stores or farmer's markets.'[10]

Consumer groups listed the supermarket chains that stocked them. The public learned that the company James Allen & Sons of San Francisco had been making additive-free sausages since 1971. One meat scientist explained: 'The uncured frank has a greyish white color and must be processed to a higher internal temperature than the normal product. Also, extreme sanitation measures and fast packaging are necessary to maintain quality standards.'[11] These sausages were sold in 26 stores. 'The Berkeley Co-Op, for example, has marketed such a product for some time with great success', said an article in *Science* in 1972.[12] The *New York Times* reported sampling the products of a firm in Philadelphia which offered 'frankfurters that differ from the standard variety only in that they were free of nitrates and nitrites. They looked surprisingly normal, a sort of brownish red in color, and except for a dry, slightly rubbery texture, were almost indistinguishable from a normal garlic and pepper flavored hot dog.'[13] These hot dogs were sold in Pennsylvania, New Jersey and Washington, DC. One group of meat scientists noted in a list of nitrite-free processed meats they had drawn up: 'One type of pre-cooked nitrite-free product which deserves further mention is the "white hot", which is indigenous to certain areas of New York State. This product is basically a non-smoked, nitrite-free hot dog which is marketed in the same manner as traditional hot dogs containing nitrite. Production of white hots is believed to date back more than 100 years. In areas of high popularity, one white hot is supposedly consumed for every 5 cured hot dogs which are eaten.'[14]

The *New York Times* informed its readers of the shops in Manhattan where they could buy nitrite-free sausages as well as pepperoni for pizzas.[15] In another article, the newspaper noted that 'there is a very good nitrite-free bacon, Nodine's, sold by many butchers and small markets'.[16] As for ham, 'nitrite-free ready-to-eat ham is available at Zabar's for $3.99 a pound; baked and glazed it is $4.99. New York state country-style nitrite-free ham is available at $6.50 a pound by the slice from Dean & DeLuca or $5.50 a pound for a whole ham.'[17]

Some distributors offered their meats frozen, as a way of ensuring long shelf life despite the absence of additives. But some processors sold their products in their usual form. The *Washington Post* published a photo of a woman 'frying nitrite-free bacon at breakfast yesterday for reporters'[18] at a press conference organized by a meatpacking firm launching its new product range. Under the headline 'Nitrite-free bacon spices controversy', the newspaper reported: 'In a move that may undercut both industry and government arguments that cured meat products are unsafe if they do not contain sodium nitrite, Gwaltney, a Smithfield (Virginia) meat packer, has begun selling nonfrozen, nitrite-free bacon under the label Williamsburg Old-Fashioned Cure Bacon. The first company to do so on a large-scale commercial basis, Gwaltney products are sold along the East Coast in conventional supermarkets.'[19] The article concluded: 'The company's bacon sales last year were about $20 million. This year Gwaltney is "conservatively" projecting that the nitrite free bacon will gross about $1 million.'[20] In a newsletter, the US Department of Agriculture noted soberly: 'a centuries-old method of preserving food has resurfaced in the search for nitrite alternatives: salt curing.'[21]

NITRITE V. NITRITE-FREE

In several towns, schools switched over to nitrite-free meals; the education authorities were delighted with the result. In Buffalo (New York State), one school official declared: 'We used nitrite-free hotdogs all last year with very good results. The kids all like them. The paler color, according to reports, has not been a detriment at all. We explained to the children what we were doing and they accepted it very well.'[22] Likewise, the *Wall Street Journal* noted that nitrite-free processed meat 'has been around for a while'[23] and seemed to be more and more in favour – to the extent that Safeway supermarkets had started selling it. The newspaper pointed out that its development had been hindered by the inequitable rules imposed by the administration: 'the Agriculture Department requires such unappetizing names as "beef-cooked sausage" for hot dogs and "uncured sausage" for cold cuts.'[24]

A *Washington Post* editorial stressed the cancer risks attached to nitroso compounds and noted: 'Although these health hazards have been known for a long time, the Department of Agriculture diddled and fiddled over the danger and, according to small meat-processors, made it as tough as possible to distribute these kinds of meats without putting nitrites in them.'[25] The magazine *Business Week* reported on a chain of grocery stores that sold nitrite-free hot dogs in its eleven branches. The article closed by depicting an Arkansas-based manufacturer who had been producing nitrite-free meats for ten years and selling them all across America. The director of the company, Warren Clough, declared: 'It's so simple, no one believes us.'[26] He described how 'We've had an uphill battle to get the public to accept a product without the red color, but now the tide is turning because people know the red color means another preservative.'[27] *Esquire* also did a piece on an Iowa-based manufacturer who eschewed nitro-additives, labelling him 'the rebel meat-packer'. This entrepreneur told how ever since he had developed a procedure for making bacon without nitrite, his competitors had never stopped attacking him: 'they're doing their damnedest to destroy us'.[28] They accused him of 'trying to capitalize on the cancer scare'.[29] Repeated, incessant bureaucratic and legal hurdles prevented him from selling his product under the name of 'bacon', and indeed most nitrite-free producers encountered administrative obstacles at every turn. The *Wall Street Journal* described another manufacturer, in Tennessee: he too made nitrite-free bacon, but lawyers acting on behalf of the authorities informed him that he was not allowed to call it 'bacon'. Discouraged, he finally threw in the towel.[30]

In the end, even the US administration recognized that this permanent war against nitrite-free producers was not in the public interest and indeed was a misapplication of food safety laws in favour of the nitrite curers. Assistant Secretary of Agriculture Carol Foreman exclaimed to a Senate committee in 1978: 'I think it is incredible that we have prevented a group of people who want to do business in a fashion that appears to be safe and appears to have some consumer demand from doing business in that fashion. I don't think the Federal regulations were ever intended to be used that way and you don't want them to be used that way.'[31] Commenting on the systematic obstruction faced by innovative nitrite-free processors, a lawyer from an NGO called Public Citizen railed: 'One would think that the Government would encourage such enterprising business. [...] Instead, it is discouraging the sale of nitrite-free products, an action which I am sure was applauded by the processed meat industry. The Department of Agriculture should be acting to eliminate the use of nitrites in meats, which would be applauded by consumers.'[32]

Finally, when the US Department of Agriculture put an end to regulatory restrictions in September 1979, the promoters of nitrite requested that no product made without nitrite could legally be called a 'hot dog' (some of them suggested the term 'cold dog')[33] and they took the government to court over the issue.[34] The specialist newsletter *Food Chemical News* noted: 'The Department of Agriculture has found itself in the ironic position of defending the safety of meat products which do not contain nitrite or nitrate',[35] in particular by producing an affidavit from Dr Ralph Johnston, head of microbiological services at the Department of Agriculture. In the scientific documents and his testimony presented in court, Dr Johnston explained that experiments carried out in the department's own laboratories had shown that meats could be processed without nitro-additives and involve no risk of botulism. According to *Food Chemical News*, 'The affidavit and the brief also noted that the meat industry has had precooked pork sausage links without nitrite on the market for 10 years without a botulism case. "In addition to the precooked sausage links, other nitrate/nitrite-free meats such as roast beef and bratwurst have been produced for years with no occurrences of botulism to my knowledge", Johnston said.'[36]

Nevertheless, nitro-meat producers mounted an ongoing propaganda campaign. Playing on public fears, industry front-groups distributed flyers containing terrifying warnings to consumers. One organization ran a campaign to promote agri-chemicals as the key to happiness and health ('This includes eating processed foods – complete with additives – and using pesticides, herbicides and fungicides to enhance agricultural efficiency').[37] They published brochures against nitrite-free meats which screamed: 'We cannot emphasize the point too strongly: UNCURED MEATS CARRY THE THREAT OF BOTULISM. Nitrites were originally put in meats to prevent botulism poisoning.'[38]

THE PROOF IS IN THE PUDDING

Observers were well aware that many companies were lying when they claimed that it was impossible to make healthy processed meats without the help of nitro-additives – you only needed to speak to older consumers, who could still remember how meat salting was traditionally done on American farms. In 1976, the *New York Times* presented the account of one of their readers: 'The most fascinating and detailed letter came from 74-year old Sybil Ramsing of Laston (Pennsylvania). She describes the way her father and mother did their own butchering, preparing and curing of bacon, ham and sausages on the Indiana farm where she was raised. Her main point was that these meats had no preservatives. "Our meat didn't spoil!," she writes. "It looked brown like cooked, roasted or fried meat."'[39] Mrs Ramsing asked: 'So maybe if we could figure out who started the use of nitrates and nitrites, and when and why, we might be able to figure out how to go back to the previously successful methods ... Could it be that someone makes a profit by selling nitrates and nitrites and the "peepul" have been educated to want the red meat produced by use of profitable chemicals?'[40] The *New York Times* remarked: 'Certainly, she is correct in assuming that people have been conditioned toward red meat and there is no logical reason why they should prefer that color to the normal brown of cooked meat.'[41]

The technical manuals produced for farmers indicated quite clearly that it was possible to make hams using *either* cooking salt alone or a mixture of cooking salt and nitrate. For example, in Alabama, a work published in 1912 by professors in the Department of Animal Industry of the main agricultural college describes in detail the different techniques for making ham, with and without nitro-additives.[42] In another Alabama college, the professor of agricultural science George Washington Carver indicated that a nitro-additive was dispensable unless the farmer wanted to butcher the meat without refrigeration: for 100lbs of meat and 8lbs of salt, Carver recommended '4 ounces of saltpeter in hot weather, and 2 or none in cold weather'.[43]

Later, in the 1950s, many American farmers used nitro-additives, mainly because these chemicals speeded up the curing and hugely simplified the work involved. But many others didn't; they continued to cure with cooking salt and apply the traditional rules of production: hygiene, slow maturing, temperature control. Just a few years before the carcinogenic effect of nitro-meats started to become evident, the US Department of Agriculture conducted an enquiry focusing on meat curing techniques among farmers. The results were published as a detailed set of statistics: in 1951, in addition to the animals they delivered to the processing plants, American farmers slaughtered between 12 and 15 million pigs each year, representing nearly 900,000 tonnes of meat. Some of this was consumed fresh, but the majority was cured.[44] The enquiry broke farmers down into three groups: 41% of them used only salt; 26% used a proprietary mixture (salt + nitrite or salt + nitrate); and 22% a mixture of their own making, which contained nitrate or nitrite in half of all cases. The rest had no particular method or didn't respond.[45] The report showed, then, that in total less than half of American farmers used nitro-additives. The annual production of farmers added up to 700,000 tonnes of processed meat. Yet, with or without nitro-additives, there wasn't *a single case of botulism*: everywhere, even on the ranches, adherence to rules of hygiene and traditional methods of production were enough to ensure microbiological safety.

At an official hearing in March 1971, under oath, the representative of the US Department of Agriculture confirmed that over the course of the previous twenty years his office had not encountered a single case of botulism connected to meats processed without nitrate or nitrite.[46] Even better: a report by a congressional committee included a detailed description of cases of botulism that had occurred in the USA between 1950 and 1963 based on exhaustive data from the US Department of Health.[47] The report covered the whole of American food production:

industrially produced foodstuffs, farm produce and home-made products. Over the fourteen years studied by the Department of Health, a total of 150 'botulinal events' had been recorded. The list included cases of botulism caused by spoiled potatoes, undercooked olives, beans whose canning had been botched, and similar incidents caused by spoiled salmon and tinned fish, lobster, chicken pie, mushrooms, beetroot, spinach, peas, peppers, maize, etc. The congressmen were struck that there was only one mention of a pork product: a pig's trotter in a jar of brine that had been inadequately prepared. In the fourteen years covered by the report, there hadn't been a single case of botulism connected to ham, bacon, sausages, salami or corned beef: not one case in the massive tonnage of processed meat produced over the period in the USA, with or without nitro-additives.[48]

To put it plainly, nitro-additives advocates were lying when they made out that it was not possible to process meats completely safely without using nitrate or nitrite: botulism was their bogeyman (which one dictionary defines as an 'imaginary and terrible creature formerly evoked to frighten children and make them obey').

DISINFECT HAM OR DISINFECT FACTORIES?

In fact, in the 1970s, it wasn't botulism that the meat-processors feared, it was other bacteria, known as 'putrefying bacteria', which thrive in unhygienic factories and can lead to significant financial losses for the meatpackers.* The cancer specialist Paul Newberne remarked: 'They can be a lot less sanitary in their plants, and so they can spend a lot less money cleaning up and using special equipment. Nitrite, you know, is an excellent disinfectant.'[49] During a Senate hearing, one of the main allies of the nitrite lobby

---

* See Chapter 3: The Triumph of Meatpacking.

explained that antiseptics offered a guarantee to sausage makers because some of them might use meat 'which may have erratic microbiological quality'.[50] Noting that nitrite-free meat had an excellent record on microbiological safety, the cancer specialist William Lijinsky posed the question in January 1977: 'The large meatpackers are not enthusiastic about such products and imply that they are not safe. Do they use nitrite to compensate for care-less manufacturing practices?'[51] At a Senate hearing he noted: 'I think they are showing the quite natural conservatism of any busi-ness. They do not want to change, because it will cost them money. I do not blame them for that. However, when we are balancing risks to human health against commercial considerations, I think we have to come down in favor of protecting human health.'[52]

In the late 1960s, the insalubrious state of some meatpack-ing factories was well known. For example, in 1967, an official investigation had shown that around a quarter of American meat production was escaping the scrutiny of federal health inspec-tions. Random spot checks by vets had revealed all sorts of rotten and unauthorized meat being used: the so-called '4-D' meat, from animals arriving at the abattoir already dead, dying, diseased or disabled.[53] During the summer of 1967, US congressmen had heard a long series of testimony from vets, sanitary inspectors and industry professionals.[54] The inspectors had described workplaces full of garbage, rotten wood work tables that were never cleaned, dirty water trickling down from the upper floors, meat polluted with flies and pieces of faecal matter. For example, a sausage fac-tory in Milwaukee was described as follows by inspectors: 'The ceiling in the shipping room was deteriorated, crumbling, and fall-ing on product. Particles of decomposed plaster was observed on the meat. Metal equipment was rusted and corroded. Employees lacked adequate hand washing facilities and were observed handling the rusted equipment and then handling meat products without having removed the rust from their hands. The toilet room for male employees lacked hand washing facilities. These male employees were observed coming directly from the toilet to the work table where they handled meat without washing their hands. Equipment coming in direct contact with meat product was contaminated with dirt and decomposed meat particles.'[55] In such factories, chemical disinfectants compensated for the lack of even the most minimal and elementary standards of hygiene.[56] In nearly every report it was said: 'No control is exercised over excessive use of harmful chemicals such as sodium nitrite and sodium nitrate.'[57] Both these substances were used to 'secure' the products, sometimes to disguise the first signs of putrefaction.[58]

THE BIGGER THEY ARE, THE HARDER THEY RESIST CHANGE

Reports of health inspectors had shown that by subjecting recalcitrant companies to strict controls it was possible to eradicate unhygienic practices.[59] And yet, at the end of the 1970s, the large meat-processors almost unanimously refused to give up using nitro-additives, even in factories equipped with adequate refrigeration. Almost four-fifths of production was controlled by an oligopoly held by a handful of all-powerful corporations (Wilson, Swift, Morrell, Armour, Hormel, Oscar Mayer).[60] For these mammoths, giving up nitrite meant a costly overhaul of their production process and a massive, disruptive change. The *Washington Post* summed up the situation: 'we have a shelf-life problem, a merchandising difficulty. Without nitrites, it would be much more difficult for the great packers to manufacture those enormous varieties of sausage and luncheon meats and ship them every which way to supermarkets that often display them at near room temperature.'[61] Following this analysis, 'the decision to forbid such preservatives would be a threat to large, centralized food manufacturers'.[62] In 1979, in an attempt to explain why the lobby of industrial meat-processors refused to give up using additives that made their products carcinogenic, the biochemist Ross Hall concluded crossly that 'all this maneuvering is designed to avoid disturbing an established mode of processing and distributing'.[63]

Paradoxically, it is the most technologically advanced companies that have fought the most to protect the status quo. They are the ones who presented the botulism problem as an insurmountable obstacle. On several occasions the press questioned representatives of the nitrite lobby: since some meat-processors had managed to do without nitro-additives, why couldn't the others follow suit? At public sessions with expert panels set up by the US Department of Agriculture, doctors and consumer groups put pressure on the producers to turn to less dangerous methods. At one such session in the mid-1970s, the microbiologist Michael Jacobson, the director of an NGO called Center for Science in the Public Interest (CSPI), described a local butcher who 'makes nitrite-free bacon and ham, just like his father used to do. He uses a salt and pepper cure. He spends more time than Oscar Mayer to cure his products, he has a small-scale operation, but people buy his products. His bacon costs $2 a pound, just one penny more than what Briggs bacon costs at the supermarket.'[64] The representatives of the large meatpackers did not respond. At another meeting, Jacobson pointed out that they could simply pasteurize their products, use non-carcinogenic preservatives or even freeze them. But it was a dialogue of the deaf: the official record shows that Richard Greenberg, the scientific director of the Swift company, 'replied that there was no denying that other methods of preservation could provide equal safety against botulism hazard'.[65] But, he said, 'the products would not be frankfurter or luncheon meat – they would be different products'.[66]

A journalist from the *Washington Post* also addressed the members of that panel. She asked why they hadn't examined the techniques employed by the producers who did not use nitro-additives. The response from the head of the committee of experts was positively Kafkaesque: he 'replied that the production of salt-cured products is extremely small in volume and that clearly the large volume products are nitrite and nitrate cured; thus the panel should direct its attention to these products, and develop recommendations on nitrite and nitrate use which will reduce any potential hazard to a minimum.'[67] Instead of examining the methods which allowed meat to be processed without use of sodium nitrite, the experts worked on the principle that it was impossible to ban the additive because it was indispensable as a colorant. Instead of seeking solutions in good faith, they put all their efforts into constructing arguments that could be used to justify nitrite curing.[68] Dr Jacobson refused to fall into this trap. In a declaration recorded in the acts of the Senate, he nicely summed up the situation: 'The goal should be to minimize human exposure to nitrite. Industry, keeping both eyes on its profits, has been trying to pull the wool over the public's eyes by saying that there is no substitute for nitrite.'[69]

CHAPTER 15

CONTESTING THE HEALTH ARGUMENT

In 1977, the US secretary of agriculture and the US secretary of health approached the attorney general for some legal advice: if nitro-additives caused cancer, did the law oblige the health authorities to ban them, even if the producers maintained that they served as protection against botulism? The lawyers from the Justice Department replied that the law allowed food to contain traces of insecticide, fungicide and pesticide but expressly forbade the addition of a toxic or carcinogenic additive if there were other means available to manufacture safer products. The report from the Justice Department pointed out that the risk of botulism was not inherent to the processing of meat; the attorney general stressed that the risk was due to techniques that certain manufacturers chose to use and which they refused to adapt.[1]

SAVING THE PINK

At the end of 1977, the Food and Drug Administration and the US Department of Agriculture gave the industry an ultimatum: they had three months to prove that nitrate and nitrite were harmless, firstly in bacon, then in other processed meats.[2] In the absence of a satisfactory response, these additives should be replaced within 36 months by methods that posed no cancer risk. They could only be authorized under exceptional circumstances, on a case-by-case basis, especially for products that could not be made according to acceptable standards of hygiene. The ban would be introduced in phases: the government proposed to stagger the withdrawal of nitro-additives over three years so that companies had time to adapt their processes and equipment, consumers could be gradually induced to switch to uncoloured products, and the whole of the industry had time to put new methods in place.

Adopting the same tactics as the companies that used nitrite in smoked fish, the meat industry lobby fought the proposal on legal and economic grounds: in a formal statement of April 1978, the AMI repeated almost verbatim the arguments that the salmon producers has just successfully employed to counteract anti-nitrite measures. The AMI explained that, in banning nitrite curing, the government would be adversely affecting competition. They claimed: 'Requiring a process that would be practical only for certain processors or for certain equipment could easily have an anti-competitive effect and might even put small processors out of business.'[3]

In Congress, the representatives from the 'meatpacker states' predicted the ruin of American agriculture: they explained that 65–70% of American pork products were treated with nitrite and claimed that the additive had become so important that without it the whole industry would collapse.[4] The lobbyists declared that reform wasn't an option because 'facilities are not readily adaptable for other uses'.[5] One representative of the meatpackers rewrote history by making out that 'proponents of a ban are asking for removal of an additive we have known for 20 centuries. In its place, to compensate for color and taste deficiencies in cured product, they are saying that other additives could be substituted, a clear case of trading the known for the unknown.'[6] The president of the AMI, Richard Lyng, hammered home the same message: 'Nitrites give meat a color and a flavor that makes cured meats taste like they do taste. It's about as essential to cured meats, I always say, as yeast is to bread.'[7]

In September 1978, one congressman on the side of the meat-packers challenged a government representative: 'Has the administration briefed you, and do you have any idea, about the ripple impact on the economy if all the people throughout the industry that have millions and billions of dollars invested in processing equipment predicated upon the use of this compound, refrigeration equipment, sales equipment, all of the machinery involved, are suddenly told that after 50 years of experience [...] there is a new day and a new technology and all of that investment is going to be wiped out?'[8] On the radio, a scientist criticized the FDA's plan: 'You have a tremendous volume of meat that's been handled in a certain fashion and now all of a sudden the government is going to turn around and say that you can't handle it that way. You have thousands of meat-processing plants. I mean, I just don't understand. What are they talking about? They'll be putting people out of work. It just doesn't make any sense, at all, except in the context that there's a political goal in mind here.'[9]

PROJECTIONS

As president of the AMI, Richard Lyng marshalled his troops for one battle after another throughout the 1970s. Researchers linked to the meat industry fought on multiple fronts to defend nitrite-cured hot dogs and bacon, denying the risk of cancer, exaggerating the risk of botulism, insisting that there had never been a method of meat curing that didn't use nitro-additives, producing surveys that indicated the consumer couldn't eat processed meats unless they were treated with nitrite, because 'the cosmetic features (color, flavor) imparted to meat by nitrite appeal to consumers and have been coded firmly into their buying patterns'.[10] The Council for Agricultural Science and Technology (CAST), an Iowa-based organization connected to the industry, published a report on the consequences of reforming meat-processing.[11] The *New York Times* pointed out that CAST had its headquarters in the heart of 'corn and hog country',[12] a vast swathe of middle America that largely owed its prosperity to the pork sector. The CAST report, drawn up by seven pro-nitrite experts, was then promoted by a wide media campaign; the lobby made great efforts to present CAST as a group of neutral, disinterested academics with no other concern than the public good. CAST argued that consumers would stop buying processed meats because they would be scared of botulism and because the products would be grey. Cue widespread anxiety: Americans simply wouldn't know what to eat!

As if that wasn't bad enough, some industry lobbyists ran macroeconomic models demonstrating that the total disappearance of bacon and its 1.5 billion dollar market would set off a catastrophic chain reaction. If people stopped buying bacon, then there would be no market for pork ... pig farming would grind to a halt ... carcasses of unsold animals would pile up by the roadside. As authors sympathetic to the nitrite lobby put it: 'The financial consequences of a discontinuance of the use of nitrite are enormous and would affect not only the meat-processing industry, but also the farmer and allied supporting industries. The producer of pigs would have no market for certain portions of the carcass; this would, in turn, be reflected in the market for corn.'[13] The animal feed supply chain would collapse. Seed producers might go out of business, the veterinary profession would suffer long-term disruption. No more corn, no more leather – unemployed shoemakers would soon be joining the hordes of bankrupted families who would be forced to abandon their farms. Whole regions would become depopulated. Was the government really determined to bring about such a cataclysm?

In fact, such projections were far from credible, and even the industry's own economists predicted that, in the event of a nitrite ban, production would re-establish itself after a period of adjust-ment.[14] Even in the American Meat Science Association, which had close ties with the AMI, a working group observed: 'consumers prefer the traditional pink color of nitrite-cured products, but they will accept gray nitrite-free products if that's all that is available.'[15] Referring to Norway, which had adopted policies favourable to additive-free meats, the working group noted that the increased consumption of meat products without nitrate or nitrite had had *no impact* on microbiological security (there hadn't been a single case of botulism). As for the economic impact, the group of experts noted: 'The introduction of gray-colored, nitrite-free products into the marketplace initially resulted in a decline in con-sumption of the affected types of products. However, with time, consumption levels returned to normal. Given the choice, consum-ers prefer the red-colored sausages. Companies actively promoting the idea of gray, nitrite-free products found their efforts unsuc-cessful when red-colored products were also available. However, when only the gray products were available, consumers would purchase them.'[16]

The industry knew that a switch to non-nitrite meat was feas-ible. But they also knew that this switch would come at a cost and would depress sales – at least in the short term. Consequently, any argument would do to muddy the waters over the danger of can-cer: several experts claimed that a ban on additives would bring about such a fall in pork production that derivative products from abattoirs would no longer be available for medical uses: 'the drug industry would lose a major source of such medicines as insulin for diabetics and cortisone for arthritis sufferers'.[17] One expert added heparin (used to prevent blood clots) and thyroxin (used to treat thyroid problems) to the list.[18] According to CAST, 'all bacon production would stop and there would be no utilization of the pork bellies; the industry would lose a $1.5 billion market. The additional loss of medicinal byproducts and the effects on ancillary industries would raise the total loss to many billions of dollars.'[19]

A SLICE OF AMERICANA

According to the industry, the government's plan to ban nitro-additives risked perhaps even more serious consequences. Such measures would not simply devastate the health of the nation (botulism epidemic, medicine shortages) or its economy (American agriculture goes bankrupt); to listen to the defenders of nitrite, it was American identity itself that was under attack: its gastronomy, its cultural heritage. The nitrite sausage was presented as a national treasure. The meatpackers' advocates argued that by using the excuse of cancer, the government was trying to kill off the hot dog and all those other pink-coloured meats. Experts from CAST appealed to people's sense of national identity. They talked about the Irish, who first brought over corned beef, the generations of Italians with their salamis, the Jews from central Europe and their pastrami, the quintessentially American delicatessens. Pro-nitrite lobbyists made the point over and over again: corned beef, salami, pastrami, sausages, cooked hams, delicatessens – all would disappear overnight, purely and simply. 'It will be wiped out.'[20] The defenders of nitrite churned out acres of print to assert that all these products could not be made without nitrate or nitrite.[21] Bacon would be doomed. Ham: abolished. Frankfurters: history.

A director of the National Pork Producers Council appeared before senators to argue that there were only two possibilities: pork meat had to be treated with nitrite, otherwise it would have to be sold fresh; there was no other option. He explained that a ban on nitrite would bring about a huge increase in energy usage in North America; if the government wanted to ban nitrite curing, they would first have to decide to build more nuclear power stations.[22] Some lobbyists stressed that the word 'botulism' comes from the Latin *botulus*. They alleged that it shows that the Romans were familiar with the disease and treated their hams with saltpetre accordingly.* Suppressing nitrate and nitrite meant suppressing processed meats.

A senator questioned the secretary of agriculture's representative: 'Tell me, is that going to happen? Are we going to be out of all of those items if we do away with nitrites? [...] Can the argument really be made if we remove nitrites that we lose the idea of bacon and all of these other things as we know them?'[23] The assistant secretary of agriculture replied: 'I think it is a specious argument and I think that it demeans the validity of the other good points that the industry has to make. Why make arguments that aren't supportable, that aren't good, when you do have good arguments to make? I think the fact that we have products out there now that seem to satisfy a large number or a substantial number of people – they ask for them in stores – indicates that it can be done.'[24]

When finally backed into a corner and forced to admit that in fact it was possible to make safe meats without using nitro-additives, the industry argued that such products would not go down well with consumers, because they would have to change their habits. One lobbyist announced the death of polony (as it is known in the UK) or bologna (as it is called in the USA), since it would have to be made without nitrite: 'Such products turn out to be grey-brownish in color, rancid in smell, tasteless, non-sliceable, green in the center and possessing absolutely no holding capacity. Oh yes, they are not toxic and could be considered

---

* In fact, it was traditional salting methods that prevented the occurrence of botulism, and the word itself was not a Roman invention, but rather was coined in the middle of the nineteenth century by German and Belgian doctors (see Chapter 10). As was the scientific tradition at the time, they used the Latin language: for example, the word *bacillum* (1842) comes from the Latin, and *bacteria* (1838) from the Greek.

microbiologically wholesome and unadulterated. But who would eat them? Thus, on some regional products, a ban would have the effect of totally eliminating them from the marketplace ... again denying the consumer the right to another slice of Americana.'[25] By a neat segue, a slice of nitro-meat had become a 'slice of Americana'. In committee, one senator was alarmed at the apparent demise of corned beef, the signature dish of Irish Americans. So the CAST expert then laid on the melodrama: 'As I mentioned, I hate to see the death of a corned beef sandwich on an onion roll, because that is what we are talking about. Corned beef as such would not exist.'[26] In fact, genuine corned beef – without nitrate or nitrite, paler in colour than industrial nitrited corned beef – had been made by artisan producers in Massachusetts since the founding of America. Even today you can still find this grey corned beef in Boston.[27] Much appreciated by foodies, it requires more work to produce because of the need to follow rules of hygiene and its longer cooking time. Incidentally, in most supermarkets in the south of Ireland you can find silverside or topside of beef, without nitrate or nitrite, which has a brownish-grey colour.

CARCINOGENIC MEAT, 'AN ACCEPTABLE RISK'

In the face of government intransigence, lobbyists took up cudgels against the pseudo 'defenders of consumers' who in fact 'infringe on their liberty',[28] 'forcing them to change their shopping patterns, forcing them to pay more than is necessary for a product and force feeding them food of a different color, aroma and taste.'[29] A scientist offered to assume 'the role of an exorcist – hopefully to dispel the fears that many Americans have about the safety of today's American diet'.[30] He explained that one American in 5,000 dies in a car accident and asked: 'Why does the consumer accept the automobile and not the nitrite-treated hot dog or bacon?'[31] In the same vein, one meatpacker pointed out that 50,000 Americans die in car accidents each year, and 8,000 die by drowning, 'But most of us continue to accept the risks of riding in cars and going swimming. We accept the benefits of those activities and the risks that go with them. And so it goes in almost every aspect of life.'[32] He asked why this logic didn't apply in discussions on food additives: 'We will allow all of the benefits of modern food technology but only at the price of absolute safety, with absolutely no risk. Nice goal if you can reach it. But you can't.'[33]

In tandem with this, some organizations attacked 'chemophobia',[34] the 'almost irrational fear of the products of chemistry'.[35] In 1978, Dr Elizabeth Whelan created the American Council on Science and Health (ACSH), an NGO supposedly 'in the service of the public' which, it was later discovered, had been set up with the support of polluting and carcinogenic industries.[36] 'Whelan just makes blanket endorsements of food additives. Her organization is a sham, an industry front', alleged one FDA official.[37] According to ACSH, specific food products should not be singled out; to avoid the effects of deleterious agents, the best thing 'would be to eat a wide variety of foods to minimize the chance that any single carcinogen would be consumed in quantities that would overwhelm the body's natural ability to handle small amounts of hazardous substances with relative safety'.[38] According to Whelan, it was because of an 'ambient toxiphobia' that the government and scientists got it into their heads to attack nitro-additives, and they picked on processed meats as some sort of scapegoat. She mocked the toxicology studies on food additives: 'In terms of food additives currently in use, I know of no reason to ban any',[39] she said. 'We're wasting our money studying rats [...] We cannot keep banning things at the drop of a rat.'[40] In no time at all, the former scientific director of the Swift meatpacking company, Dr Richard Greenberg, became joint editor of the ACSH newsletter.

Elizabeth Whelan portrayed the scientific recommendations on the carcinogenicity of nitro-meats as 'ill-advised', 'unnecessary', 'premature', 'incorrect', 'simply wrong', based on 'misinterpreta-tions' and results that were not 'carved in granite'.[41] In an Ohio newspaper, Whelan explained that the real peril was not cancer but the fear of cancer: the population was 'in the grip of a new and serious disease – cancerphobia'.[42] The article (headlined 'Caution about precautions') concluded with a mollifying comment that you will still hear used today by defenders of nitro-meat: 'The campaign against cancer is too important to have its credibil-ity weakened by warnings and prohibitions that are not soundly based. No step that can check cancer should be neglected, but it will not be eliminated by overstating perils of daily life.'[43] After defending use of sodium nitrite in meat, Whelan and her organiza-tion used similar arguments to press the case for other potentially carcinogenic chemicals, themselves unfortunate victims of what she called the 'cancer scare' or 'nosophobia' (the 'morbid dread of illness').[44]

MASTERFUL EVASION

The propaganda onslaught ultimately bore fruit: as the months went by, the cancer issue became blurred, as if engulfed in a cloud of misdirection – the peril of botulism, the threat of ruin hanging over meat producers, the end of processed meat as we know it. The FDA and the US Department of Agriculture saw their polit-ical position start to crumble under the weight of these attacks. Even though there hadn't been a single case of botulism related to meats produced by American manufacturers who didn't use nitro-additives, the consistent bludgeoning by the nitrite lobby was enough to alter the terms of the debate: by early 1978, they had almost prevailed – in the minds of the public they had turned a cancer crisis into a botulism crisis.

At the end of spring 1978, the government's ultimatum expired. In spite of the propaganda campaign, the industry had not responded to prove the harmlessness of its products, and the administration followed through on its promise: a reform of processed meat production was set in train. On 19 May, *NBC News* announced that there would be a ban on the use of nitro-additives in bacon. Soon, bacon would be brown or grey. The withdrawal plan would have two phases: first, a small reduction in quantities of additive, then a very steep reduction.[45] The industry didn't find the first stage in the programme at all restrictive: it involved only a slight reduction in quantities of sodium nitrite and consisted mainly in adding ascorbate (which served to reduce the occurrence of nitrosamines; for the industry it had the advantage of enhancing the colour and speeding up production even more).[46] This first phase was designed to allow manufacturers the time to prepare the second phase: depending on the product, nitrate and nitrite were to be either eliminated completely or have two-thirds of their dosage replaced by non-carcinogenic preservatives.

But the industry was dead set against the use of any preservative that wasn't also a colorant. The lobbyists launched a flurry of campaigns to have the withdrawal plan repealed. Richard Lyng of the AMI continued to insist that, without sodium nitrite, there would be no processed meats, 'at least not yet'. He requested a delay, because, in his view, 'it would eliminate $12.5 billion worth of food at the very outset, if you did this precipitously'.[47] No stone was left unturned in the bid to win over Washington: political haggling, press campaigns, economic coercion, threats of legal action. Ultimately, the lobby gained the upper hand. Within a few weeks, the withdrawal plan had been rendered more or less toothless: phase two of the reform was firstly made 'conditional' and subsequently dropped from the new regulations.

The press was surprised at how readily the Carter administration had retreated. In an article of June 1978 entitled 'U.S. bacon industry seems to have won nitrite battle with federal regulators', the *Wall Street Journal* announced that 'after lengthy skirmishing, during which the industry claimed that the proposed ruling could mean the end of bacon as we know it, the department modified its proposal. The final regulation, which went into effect last week, permits a level of nitrites high enough that 90% of the nation's bacon already complies.'[48] And the *Chicago Tribune* said: 'As a matter of fact, the products sold in retail stores won't change as a result of the new regulation because it rubber-stamps practices adopted by industry.'[49] The lobbyists were cock-a-hoop: when a government representative suggested that the reduction in nitrite would still take place, one of the editors-in-chief of the *National Provisioner* called the statement 'a mere "publicity gesture" devised to assure the public that the government is "doing something" on the issue'.[50] Another manufacturer crowed: 'This is a clear victory for the industry by virtue of the fact that nitrites aren't being banned as the government led us to believe earlier.'[51] Richard Lyng announced that 'the regulation itself is what we as an industry asked for',[52] but he was already anticipating what lay ahead: 'I don't think we've lost the battle, but we certainly haven't won a war', he declared to the *Wall Street Journal*.[53] And indeed, the fight was set to continue.

DELAY AFTER DELAY

Two months later, the FDA tried a new approach, but they made a mistake. They thought that they would improve their chances by basing their argument on a new study indicating that, in addition to the well-known carcinogenic effects of nitrosamines, sodium nitrite could cause cancer even when nitroso compounds did not occur: in a study conducted at the Massachusetts Institute of Technology, nitrite on its own seemed to have caused lymphomas in rats.[54] The commissioner of the FDA explained: 'This is the first time that nitrite alone has been shown to cause cancer. The mechanism of action clearly is distinct and separate from that of nitrosamines.'[55] But at the commodities exchange in Chicago, the traders shrugged this off as a last-ditch governmental attempt to fight a lost cause. One of them said: 'We've had nitrites in the picture for so long and nothing definitive has yet developed. Traders are saying "Here we go again with the same old thing, the main difference being that another agency is getting involved".'[56]

The traders were right not to be too concerned, as more or less the same scenario played itself out: in August 1978, the FDA put forward a new plan for the withdrawal of nitro-additives;[57] the lobbyists and their allies redoubled their campaigns. In Congress, the allies of the meatpackers tried a number of legislative manoeuvres to 'ban the ban', for example by introducing a proposed legislation entitled 'A bill to prohibit the Secretary of Agriculture from prohibiting the use of nitrites as food preservative on the basis of any carcinogenic effect nitrites may be represented to have until a satisfactory substitute preservative is commercially available'; then another called 'A bill to delay any action which may be taken by the Secretary of Agriculture respecting nitrites used as a food preservative on the basis of any carcinogenic effect nitrites may be represented to have'.[58] Observing the lobbyists at work in Washington, the first secretary at the British embassy noted that these bills 'reflect the strong farm lobby interest now that the Congressional elections are approaching'.[59] Certain of the candidates who supported the meatpackers announced that they had uncovered a genuine 'conspiracy' at the heart of the administration: a 'secret plan to ban nitrite'.[60] One article in the trade press claimed, for example: 'Unfortunately, several government agencies are going ahead with plans to outlaw nitrite use. Currently three plans are under study. They include a phased-in ban of nitrites, spread over a three-year period; a step-by-step phase out of nitrites in certain products; or an immediate, total ban on the use of nitrites in all products. All this is based on one test, in one laboratory, using one particular strain of rats.'[61]

This 'one test' was the latest study – the one which indicated that *as well as* creating nitroso compounds, sodium nitrite could itself be 'directly carcinogenic' – and the lobbyists were often presenting this as the *only* scientific justification put forward by the cancer specialists and the government. A few months later, after many more crises, announcements, promises and threats, new rumours of impending apocalypse, negotiations, then enforced climb-downs, the FDA beat a retreat again: in spring 1979, the administration granted a new delay, postponing the introduction of the ban by a year.[62] The secretary of agriculture gave assurances that processed meats without nitro-additives were 'just around the corner'.[63] One Washington journalist noted that for at least another year 'the cured meat industry will not have to worry about a ban on nitrite'.[64] Government officials announced that 'they can begin getting nitrites out of the nation's food supply next year and rid food of the additive entirely';[65] they even gave a date: nitrate- and nitrite-treated meats could be gone from shops by 1 May 1982.[66] The NGOs criticized what they saw as 'a purely political, nonscientific and concessionary move',[67] to quote Ellen Haas, the director of the Community Nutrition Institute. She predicted that the delay would 'only bring endless special-interest lobbying and procrastination'.[68] In her view, the delay 'could mean hazardous meats for another five years, instead of one and a half'.[69]

'I HAD A HOT DOG FOR LUNCH'

This 'delay' has now lasted four decades. The proposed ban ground to a halt, and the health agencies backtracked when it was confirmed that nitrite was not a *direct* carcinogen. A group of senators approached the government to demand that it 'put to rest all the confusion generated by your previously announced actions regarding nitrite'.[70] At the request of a senator from Iowa, an official report was produced at the start of 1980 which seemed to imply that only one single study had indicted nitro-meats: the latest experiment – the most inconclusive of all – on the direct carcinogenicity of nitrite. The report deliberately ignored the problem of nitroso compounds: as if by magic, the issue of nitrosamines and nitrosamides disappeared in a puff of smoke.[71] A new commissioner was appointed at the FDA, and on 18 August 1980 he held a press conference where he announced that 'it's a situation in which we have to wait until we know more'.[72] Though he seemed to know where he stood: at the end of the press conference he declared: 'I had a hot dog for lunch.'[73]

The ban was definitively shelved when a press release from the FDA and the US Department of Agriculture announced that the plan to abolish nitro-additives had been abandoned.[74] The meatpackers were dancing in the street. They organized conferences, denounced the 'crusade' that had been waged against their businesses, praised the 'pragmatism' that had won the day, and perpetuated the idea that 'the nitrite scare' was provoked by a single study and was built up by a bunch of cranks: activists, fanatics, hippies, vegetarians, ranting know-it-alls on the lookout for the 'carcinogen of the week'.[75] One pro-meatpacker senator brayed: 'I hope that it will, once and for all, reassure American consumers that nitrites are *not directly* linked to cancer.'[76]

Researchers and consumer groups cried foul. They pointed out that there was a welter of conclusive proofs apart from the latest study: the question of the *direct* carcinogenicity of nitrite was not crucial, what really mattered was the carcinogenicity of its metabolites.[77] 'No one has ever thought it was nitrites alone that cause the cancer', argued the NGO Public Interest.[78] Another NGO, Community Nutrition Institute (CNI), called the sudden abandonment of the proposed plan an 'unconscionable' decision and emphasized that nitrite was not directly carcinogenic, but that the risk derived from 'the cancer-causing compounds which occur as a direct result of the use of nitrite'.[79] The magazine *Newsweek* noted that 'the meat industry was quick to hail the government decision'[80] and explained that in fact what troubled scientists was not nitrite itself, but the nitroso compounds, 'which are known to be potent carcinogens'.[81] Similarly, the former commissioner of the FDA pointed out that the fact that nitrite was not directly carcinogenic 'does not challenge other, lower level effects about which we ought to be concerned'.[82]

When the consumer groups denounced the government's capitulation, the new health authority officials said that the file wasn't closed, that 'no one is able to say nitrite is safe',[83] and that everything would be done to 'continue to try to reduce the use of nitrite in food'.[84] After announcing that cured meats now had a 'clean bill of health',[85] one US Department of Agriculture official quickly backtracked and gave the assurance that science 'does not really give nitrites a clean bill of health'[86] and that he was releasing $2 million of funding for 'research on a safer substitute'.[87] The new commissioner of the FDA stated that 'this is not the end of the nitrite issue',[88] that 'nitrites are not home-free by any means'[89] and that 'they will eventually be phased out of the food supply'.[90] He guaranteed that the file would be reopened with renewed vigour. But in fact, it was all over: the storm had abated, the ban on sodium nitrite had been avoided, the reform of processed meat was dead and buried. In the *Wall Street Journal* the AMI was jubilant, declaring that the consumer was 'the real beneficiary'.[91]

CHAPTER 16

AFTER THE WAR

The election of Ronald Reagan in November 1980 put paid to any hopes of reform: ardent defenders of nitro-meats were given top positions in the American administration. Richard Lyng, the president of the AMI, was initially appointed to the special post of vice secretary at the US Department of Agriculture (USDA);[1] in 1986, he became the secretary of agriculture. In other words, the head of the nitro-meat lobby took charge of the body that was responsible for overseeing the industry. The journalist Eric Schlosser described that period: 'The USDA became largely indistinguishable from the industries it was meant to police. President Reagan's first secretary of agriculture was in the hog business. His second was the president of the American Meat Institute (formerly known as the American Meat Packers Association). And his choice to run the USDA's Food Marketing and Inspection Service was a vice president of the National Cattlemen's Association.'[2] Lyng would remain in his post until the end of Reagan's presidency in 1989: in those eight years, the US Department of Agriculture was transformed into a global champion of nitro-meats.

THE AMI IN CONTROL

The government promised to undertake 'more complete' research into nitro-additives and commissioned a new report. The journal *Science* recorded that 'privately, FDA officials see the study as a device for getting both Congress and consumer activist groups off the backs of both FDA and USDA'.[3] The cancer issue was put on the back burner. From time to time, journalists would express surprise that sodium nitrite was still authorized.[4] When scientists published findings that drew an explicit link between processed meats and cancer, their work was nipped in the bud or else subjected to a blitz of disinformation.[5] The meat lobby threatened researchers with financial reprisals, sometimes publicly.[6] By way of positive PR, the industry set up some reassuringly named organizations – such as the 'Nitrite Safety Council', officially tasked with assessing the presence of nitrosamines in commercially available processed meats.[7] Unsurprisingly, their results were very encouraging ...[8]

In early 1981, the magazine *The Nation* noted that the nitrite problem had simply 'disappeared' and denounced the political sleight of hand that the nitrite lobby had pulled off. In an article entitled 'The nitrite fiasco', the journalist Ralph Moss wrote that 'the rehabilitation of nitrites may represent the dawn of a new era in which science abdicates its primary responsibility to protect the health of the public in favor of deregulation'.[9] His prediction was correct: the historian Robert Proctor has described this catastrophic (in terms of cancer prevention) period that began in November 1980. In a chapter called 'The Reagan Effect' in his book *Cancer Wars*, published in 1995, he recounts how, in a carbon copy of the appointment of the head of the AMI as agriculture secretary, the government bodies in charge of public health were systematically handed over to ultra-liberal, pro-business reformers with a mandate to loosen constraints and eliminate regulations.[10] Proctor describes how the agencies in charge of oversight fell into the hands of people who were their fierce opponents – often lawyers or legal advisers of companies that had been prosecuted – and the health authorities were systematically 'cleansed' to suit business interests. Scientific work on carcinogenic foods was denounced as deluded pseudo-science carried out by 'chemo-phobes', 'enemies of the pleasure of eating' and 'health ayatollahs'. As Proctor put it: 'At the Food and Drug Administration, rules for food additives were revised to allow carcinogenic contaminants in foods and cosmetics, so long as they were introduced *unintentionally* as a side effect of the manufacturing process.'[11]

The new report was eventually published in 1982 under the title *The Health Effects of Nitrate, Nitrite and N-Nitroso Compounds*.[12] The US Department of Agriculture, now under the thumb of the AMI, put the most positive spin possible on the report's findings and gave the impression that the scientists had dismissed the case entirely. As if there were suddenly no more concerns, and everything could go back to the status quo ante.[13] The report stated that nitro-meats presented certain risks ... but that public taste also had to be taken into account: consumers liked their meat pink.[14] The report recognized that there were alternative ways to cure meat without the risk of botulism ... but did not consider these methods entirely satisfactory since they didn't provide coloration.[15]

The press indicated that the report's conclusion was that the risks of nitro-meats were not fully understood, but in any case 'neither nitrates nor nitrites directly caused cancer in animals'.[16] One senator from Iowa, a staunch defender of the meatpackers, claimed that this was proof that the health authorities should never have tried to ban nitro-additives; according to him, this showed 'just how governmental agencies should not go about making a decision'.[17] Again, the newspapers expressed surprise and reminded their readers that the *direct* carcinogenicity of nitrite was not the main issue; the real problem was not nitrite itself, but the carcinogenicity of nitrite-treated meat due to nitroso compounds.[18] For the *New York Times*, 'the report has by no means cleared the air for health-conscious consumers';[19] 'cured meats are as potentially dangerous as ever'.[20] The press were especially taken aback at 'the report's downplaying of the relative dangers of nitrites in cured meat',[21] for example the estimate that 'only six to 138 cases of cancer could be prevented annually if nitrites were removed from cured meats'.[22] The journalist of the *New York Times* noted: 'Even the report, while suggesting that nitrites in cured meat are not a major risk, said an alternative to them should be found.'[23]

By concentrating on the idea that nitrite was not directly carcinogenic, the US Department of Agriculture tried to nullify the perception of risk. The official press release was a masterpiece of 'confusion manufacturing', full of ambiguity and equivocation. Instead of dealing with the problem of nitro-meat and cancer, the press release recommended that studies should be undertaken to find ways of reducing nitrate levels ... in vegetables.[24] The *Washington Post* noted that, according to the information released to the public, 'most persons need not make any major changes in what they eat and certainly should not stop eating healthful fruits and vegetables'.[25] The *New York Times* said that the report was trying to pull off a remarkable tour de force: 'Its conclusions appeared to change public perceptions about what is, and isn't, safe to eat: meats cured with sodium nitrite are not so hazardous, it seemed to say, and vegetables took the role of villains.'[26]

'BY NO MEANS THE WORST'

The basic line in all this waffle was that nitro-meats can indeed generate carcinogenic nitroso compounds ... but they are by no means the worst! According to the USDA press release, there are other much more dangerous sources. It claimed that the average consumer would receive a daily dose of nitrosamine of 'about 0.17μg [microgram] from cured meat products, especially bacon. In contrast, a daily pack of American filter cigarettes provides 17μg.'[27] There was no reason, then, to *stigmatize* nitrite-treated meats, since they were responsible for 'only a small proportion of the total exposure to nitrosamines'.[28] A newspaper in Arizona tried this reassuring approach in an article on processed meat and cancer entitled 'Smokers risk more exposure to nitrosamine'.[29] Was the public really comforted by the idea that the harmfulness of a slice of ham was best measured by comparison with a packet of cigarettes?[30]

Echoing the language of the press release, the *Los Angeles Times* claimed that the danger of nitrited meat was insignificant because there were other sources of nitroso compounds: you could be exposed to nitrosamines by handling agricultural chemicals or if you worked in a factory producing rubber or military rocket fuel. (This seemed to take its inspiration from a CAST report indicating that nitroso compounds were equally abundant in 'tobacco smoke, herbicides and such industrial processes as the production of hydrazine fuel for rockets'.[31]) You could also be exposed to dangerous nitroso compounds by handling dimethylamine sulphate (a chemical used in tanneries as a hide depilatory agent), and even the plastic in the interiors of new cars could sometimes release nitrosamines (though the *Los Angeles Times* failed to mention that not many people actually eat their dashboards). In short, nitro-meats were merely 'a minor contributor to a very big problem'.[32] Under the headline 'A report on nitrites finds cured meats are relatively safe', the *International Herald Tribune* said: 'By far the greatest exposure to nitrosamines occurs among people who work in such industries as rocket fuel, leather-tanning and rubber manufacture [...] Outside of occupational exposures, the committee reported, cigarette smoke is the main source of nitrosamines. A person who smokes one pack of filtered cigarettes a day would inhale 17 micrograms of nitrosamines.'[33] In response to this line of argument, the former commissioner of the FDA had previously noted: 'there is absolutely no denying that if we could get people to stop smoking, we would have a larger impact on cancer rates in this country than we would have by any single regulatory action of the intervention type known to me now and probably larger than any that are not known to me now. But to say as a consequence of that conclusion that we should therefore ignore other exposures of less significance seems to me to go in exactly the wrong direction.'[34]

In 1982, the American National Academy of Sciences published a new report entitled *Diet, Nutrition, and Cancer*, produced by the most renowned cancer specialists in the USA, gathered under the auspices of the Commission of Life Science. The document explained: 'The public is now asking about the causes of cancers that are not associated with smoking. What are these causes, and how can these cancers be avoided?'[35] The report explained that nitro-additives are not directly carcinogenic, but that nitrite 'can interact with specific components of diets consumed by humans and animals or with endogenous metabolites to produce N-nitroso compounds that induce cancer'.[36] The text cautiously pointed out that 'We are in an interim stage of knowledge similar to that for cigarettes 20 years ago. Therefore, in the judgment of the committee, it is now the time to offer some interim guidelines on diet and cancer.'[37] The AMI responded by presenting elements of the published report in the press but distorting their meaning. Dr Alvin Lazen, the executive director of the Commission of Life Science of the National Academy of Sciences, wrote a firm riposte in the *Washington Post*. Entitled 'Diet, cancer and the American Meat Institute', his editorial accused the AMI of seeking to 'confuse the issue' and deliberately offering mischaracterizations and misinterpretations of the National Academy of Sciences report. Dr Lazen accused the AMI of manipulating the results of official enquiries to make out that nitrited meat posed no risk, for example by stating 'that the recent diet and cancer report reaffirmed a previous report's finding that "neither sodium nitrate nor sodium nitrite are carcinogenic nor mutagenic in mammals"'.[38] Dr Lazen explained that the 'AMI misused that finding. The report on nitrite actually found neither sodium nitrate nor sodium nitrite to cause cancer *directly* in animals. It also explained that both substances are converted to nitrosamines, which are known to cause cancer in animals and are suspected to cause cancer in humans as well. It is this indirect activity leading to nitrosamine formation that is the cause of concern.'[39]

SORCERY

In 1978, at the time when the health authorities were trying to force the meatpackers to renounce nitro-additives, the assistant secretary of agriculture, Carol Foreman, had warned that the meat-processing industry would kick back hard to protect the status quo. She warned one conference audience: 'You're going to hear a lot about how government's going to ban bacon and about how farmers are going to go out of business and about how packers are going to go out of business, and about how much bacon you'd have to eat in order to reproduce the laboratory tests on rats, and how prices are going to go up, and how bacon is everybody's basic human right, and finally, how government is messing around in your business unnecessarily once again.'[40] She was right: the meat-packers' organizations and their PR companies tried all this and more. But what really saved the day for nitro-meats was success-ful litigation. A salient characteristic of the 'nitrite war' was the protracted legal wrangle. In a 1978 interview a radio journalist noted that the problem of nitrite and cancer seemed no longer to be a matter for scientists but was being fought over by lawyers. A scientist funded by the AMI replied: 'Well, I think that that's where the final battle always takes place. I mean, there is a set of laws which governs the legislation of food additives, and the interpretation of those laws usually falls into the hands of the lawyers. And, I think it's how they ultimately will interpret the meaning of some of those words, like "induce," that I think will influence the outcome.'[41]

The legal battle focused on defining a 'carcinogenic agent'. To reiterate: it is not nitrite or nitrate that cause cancer, but nitroso compounds. American law stipulates that no food additive 'shall be deemed to be safe if it is found to induce cancer when ingested by man or animal, or if it is found, after tests which are appropriate for the evaluation of the safety of food additives, to induce cancer in man or animal'.[42] The pro-nitrite Republican congressman Bud Brown (Ohio) pointed out that the law 'does not speak to what happens to food additives which are safe in and of themselves after they are in the food supply and interact chemically with other substances (either additive or naturally occurring) during cooking or in the process of ingestion or digestion'.[43] Crucially, according to Brown, this technicality seemed to offer an opportunity for legal action: 'presumably, court interpretation of the clause might provide answers in these hazy areas'.[44] The lawyers of the meat lobby made hay with the fact that, since nitrate and nitrite were not *in themselves* carcinogenic, they could not be banned under anti-cancer arrangements: they managed to prove that American law banned additives that *were* carcinogenic agents but had nothing to say on additives that *gave rise to* carcinogenic agents. Or to put it another way, taken literally, American law forbade 'carcinogenic additives' but had nothing to say about additives that made meat carcinogenic.[45] As a journalist explained, 'since nitrosamines are not the original components of the preservatives, they are not subject to these provisions'.[46]

British observers who followed the legal developments noted that the American law on carcinogenic additives 'would not require an immediate ban on the products affected, as it deals with additives and nobody is adding nitrosamines to food'.[47] The director of the Bureau of Foods at the FDA sought to clarify this point: 'The law would forbid the use of nitrite as a food additive if the use of nitrite produced cancer. There is no evidence that I am aware of that the use of nitrite does. When nitrite and secondary amines are simultaneously administered, the carcinogen is generated in the stomach, and *the nitrite alone* has not been so incriminated.'[48]

These might seem like very contrived legal arguments, but they were effective in blocking regulatory measures by the health agencies.[49] Denouncing the 'great nitrite scandal', the biochemist Ross Hall wrote in 1979: 'It seems USDA views its jurisdiction as ending the instant the bacon is chewed. The nitrosamines formed in the digestive system from the nitrites/nitrates added to bacon fail to fit any bureaucratic niche, and hence escape regulation. In effect, we have legal carcinogens and illegal ones. We wonder if your body tissues can tell the difference?'[50] Even more surprisingly, the industry representatives claimed that the administration was not allowed to examine the impact of nitrate and nitrite on health on the basis that these additives had been previously authorized or 'prior-sanctioned': following regulatory arrangements adopted in the late 1950s,[51] nitrate and nitrite were classified in a category of additives which, according to the industry counsels, had been 'affirmatively approved' in the 1920s and could not be unilaterally re-evaluated.[52] Throughout the 1970s, then, there were constant legal battles over the hypothetical 'prior-sanctioned' status of nitro-additives and the existence – or not – of a special administrative approval which would protect nitro-additives against toxicological re-evaluation by the public health authorities.[53]

NITRITE FALSELY ACCUSED

In 1995, Robert Proctor blamed the American Meat Institute for trying to discredit the research on colorectal cancer in the same way that the Tobacco Institute was trying to paper over the risks of lung cancer, or the Asbestos Information Association was warning the public against 'fear' of asbestos (or 'asbestopho-bia').[54] Today, the nitro-meat lobby tries hard to sell the idea of a 'temporary aberration' in public opinion that happened 'a long time ago'. For example, in 2013, the AMI produced a twenty-page brochure entitled *Processed Meats: Convenience, Nutrition, Taste – American Traditions and Iconic Foods*. Illustrated with sumptuous pictures of crispy bacon, slices of ham and different types of sausages, it presented itself as 'a consumer guide' on pro-cessed meats ('What are they?' 'Are they safe and nutritious?'). In the section that touched on health matters, the brochure said: 'In the 1970s, a single study that was later discounted cast a dark cloud over nitrite, alleging that its use in cured meats could cause cancer.'[55] It went on to explain that, in fact, toxicologists had subsequently carefully tested sodium nitrite and had 'concluded that nitrite was safe at the levels used and did not belong on the national list of carcinogens'.[56] In another publication, the nitro-meat lobby asserted: 'After an intense review of the risk, the issues were resolved in the early 1980s.'[57] The AMI also published a flyer entitled 'Sodium nitrite: the facts'. The second page opened with the question: 'Years ago, I heard some people say that nitrite causes cancer. Is sodium nitrite safe?'[58] The answer was that the top medical authorities were now satisfied that nitro-curing pre-sented no risk; one scientist was quoted as saying: 'The idea it's bad for you has not played out.'[59]

Although processed meats are now listed in group 1 of the International Agency for Research on Cancer classification, the American Meat Institute has remained bold and defiant. It has argued that claims about the carcinogenicity of nitrite-cured meats are absurd, no more than a fad. When the WHO officially classified processed meats as definitely carcinogenic, the head of the AMI's research foundation set her sights on the 22 experts of the IARC: 'They tortured the data to ensure a specific outcome.'[60] Following this, the pro-nitrite lobby campaigned to discredit the

IARC, its mission and its researchers[61] and reduce its source of public funding.[62] When, nine years before the IARC, the World Cancer Research Fund published its conclusions on the carcinogenicity of processed meats, the head of the AMI Foundation proclaimed: 'If someone today said the world was flat, we'd laugh because that's such an uninformed and disproved hypothesis. We need to put some of our notions about meat and cancer, nitrite risks and other issues into that same mythological category.'[63] In a leaflet for the general public, one branch of the American lobby has said of the cancer risk linked to nitrite in processed meat that it was 'one of five health scares you can ignore'.[64] Likewise, the veteran nitro-additive producer Aula Werk in Austria, holder of one of the first patents on nitrite curing, had no hesitation in declaring on its website that sodium nitrite was 'now rehabilitated'. Cannily, the company positioned itself as the victim – along with consumers of nitro-meat – of a kind of holier-than-thou propaganda campaign by an 'anti-meat' lobby: 'Increasingly, they try to make us feel guilty because we enjoy the taste of meat. There are so many of these people in the media, trying to lecture us. They go so far as to say that nitrited curing salt may be partly responsible for certain types of cancer. But it has now been rehabilitated.'[65]

In short, manufacturers of nitro-meat and nitro-additives tend to present themselves as the innocent victims of bullying scaremongers. According to them, the carcinogenicity of nitro-meats is nothing but a hoax, fake news peddled by a handful of cancer McCarthyites, an enormous collective error committed decades ago, a mass delusion which real scientists have awoken from – but which continues to exert a tenacious hold on the popular imagination: nitro-additives have been falsely accused, the carcinogenic effect of nitro-meats is a mirage, an urban legend, an ancient myth. 'We actively ignore what does not fit into our cultural paradigm',[66] stated one food industry toxicologist in January 2019, as one of his colleagues mounted a defence of nitro-additives: 'You have to consume a lot for them to induce cancer, but at the amount we consume, that's not possible at all.'[67] According to him, 'you would have to eat maybe a container of bacon a day. Who does that? No one.'[68]

As a way of distracting attention away from the fact that the metabolites of nitrite are carcinogenic, many industry advocates have kept plugging the message that nitrite itself is not carcinogenic. For example, one defence of nitro-meats reads: 'In 2003, the FAO/WHO Expert Committee, based on its review of the scientific literature, concluded that the findings "do not provide evidence that nitrite is carcinogenic to humans"',[69] when in reality the committee, composed of 47 doctors assembled by the WHO and the FAO (the UN organization for food and agriculture), had issued a recommendation designed to limit the consumption of processed meat because of cancer.[70]

One of the most successful counterattacks from the meat lobby plays on the fact that vegetables contain nitrate. Several authors linked to AMI express puzzlement about this apparent paradox: since vegetables clearly help protect against cancer, how can it possibly be the case that foods treated with nitrite are carcinogenic? For example, one article says: 'It is hard to believe that the ingestion of nitrite from cured meats or nitrate from fruits and vegetables could have any potential adverse toxic outcomes',[71] before concluding: 'Consumers should avoid getting caught up in fads to the point where they might ignore sound science and common sense.'[72] Green vegetables may well contain nitrate, but they do not contain nitroso compounds.

Along the same lines, some authors exploit the fact that saliva contains permanent, albeit minute traces of nitrite. One of their favourite tropes is: 'if nitrite caused cancer, people would be advised to avoid swallowing';[73] or else 'if the government agencies had proceeded with a nitrite ban, would a move against saliva have followed?';[74] or 'a position that would require saliva to be designated as "hazardous to health" is clearly in no one's best interest'.[75] The approach is always the same: talk about nitrate and nitrite (known not to be directly carcinogenic) instead of nitroso compounds. Nitro-meat lobbyists pulled this stroke way back in the 1980s – and it was dismissed at the time by the National Academy of Science[76] – but the stratagem is still employed by industry websites today.[77] It seems to be designed to make non-specialists feel that the question of nitrited meats and cancer is strange and complicated, too confusing and difficult to ever understand.

Today, just like 40 years ago, a small group of researchers centred around the city of Madison are in the vanguard of such attempts to 'bring clarity', to use the expression of the *National Provisioner*.[78] According to that newspaper, the task is now to 'effectively educate a broad community of public health scientists, nutritionists, and the general public about the fundamental role of nitrite in biology in order to address their fears and concerns'.[79] The strategists of nitrite come out with ever new arguments to frame the problem of carcinogenic meat as a mere 'controversy' in which there are just two camps: those who 'think' that nitrite-cured meats are carcinogenic and those who are sure that they aren't. Oddly enough, the latter camp consists mostly of manufacturers – and of some of the scientists who work, or have worked, for them.

CHAPTER 17

ALICE IN NITROLAND

'Now, here, you see, it takes all the running you can do, to keep in the same place', says the Red Queen to Alice in *Through the Looking-Glass*.[1] It could be the motto of the land of pink ham: the nitro-meat lobby misses no opportunity to proclaim loudly that it is really doing its very best to stop using nitro-additives – and, when it comes to cancer, that it is necessary to 'continue to do research'. Will we ever know how many deaths could have been avoided but for these constant delaying tactics?

It seems that the fantasies of Wonderland have been used as a model by the peddlers of pink ham in more ways than one. For when it comes to defending profits, no tale is too outlandish. In the face of mounting scientific evidence, some nitrite curers have had to abandon their defensive positions and adopt ever more audacious strategies, even to the extent of counterattacking with what you might call 'pre-emptive vindication'. Their message is stunningly simple: contrary to what you might believe, nitro-meats are in fact *good for you*.

THE SAUSAGE DOCTORS

Some defenders of nitrited meats claim that sodium nitrite is good for the health because the molecule can sometimes be used as an antidote for cyanide poisoning. They forget to mention that

intravenous infusion of sodium nitrite is an extremely dangerous medical procedure used only *in extremis*: it is itself a form
of controlled poisoning (by methaemoglobinaemia). The nitrite
counteracts the cyanide because the two substances work on the
blood in opposing ways.[2]

More than a century ago, similar arguments were already
being dismantled by doctors who opposed the use of new chemical preservatives. In 1905, Dr Harvey Wiley put it thus: 'No one
will deny that the materials used for the chemical preservation of
food are often useful for specific purposes as medicines. Boric acid,
salicylic acid, benzoic acid, sulphurous acid, and formaldehyde,
all have undoubted uses. This, however, is no excuse for placing
them in foods. It is not necessary to medicate the food of a thousand citizens in order to reach one who may need it. Drugs should
be reserved for dispensation by physicians and pharmacists and
not by the food purveyor.'[3] Following the same line of thought,
a famous British toxicologist wrote a stinging article headlined
'On the use of food preservatives' which denounced the use of
pharmaceutical substances as colorants and antiseptics in food: 'I
object to be physicked indiscriminately by persons not qualified
to administer medicine whilst I am in health; I object still more
when I am ill. [...] It is no consolation to me to know that the
physic is not immediately fatal, or not even violently dangerous.
The practice is utterly unjustifiable except from the point of view
of a dealer, who wants to make an extra profit, who wants to palm
off a stale or ill-prepared article upon the public.'[4]

Today, some lobbyists promote sodium nitrite as a new
panacea: in a factsheet entitled 'Emerging research suggests nitrites
may have health benefits', Pork Checkoff, an industrial organization in Iowa, said that 'scientists at the National Institutes of
Health have concluded that nitrite is a potential new treatment for
organ transplantation, heart attacks, sickle cell disease, brain aneurysms, leg vascular problems and even pulmonary hypertension,

an illness that suffocates babies'.[5] This would be laughable if the matter weren't so serious. Let's look at this a bit more closely.

Because of its effect on the blood and on the walls of the blood vessels, sodium nitrite can be used in specific cases of cardiac arrest or to treat pulmonary hypertension, brain aneurysms (an abnormal dilation of the wall of an intracranial artery) and sickle cell disease (a genetic disorder of haemoglobin). There are many substances locked away in pharmacy cabinets that have potential medical applications, but you would never dream of using them as food additives. Morphine sulphate (or sulphate salt of morphine) is a reliable painkiller, but would that be a good reason to allow it to be used as a preservative in cheese or a colorant in meat? Has anyone thought to improve the colour of orange juice or extend its shelf life by adding a dose of arsenic or digitalin, on the basis that these substances have been part of the Western pharmacopoeia for centuries? But this is how it is with sodium nitrite: today some lobbyists try to fool the public by convincing them that nitro-meats pose no risk because sodium nitrite can be used in an operating theatre to dilate coronary arteries or pulmonary veins. It's a rather baroque piece of skulduggery: sodium nitrite was indeed employed in cardiology well before being used to cure hams.* But it is ironic to see industrial meat-processors reviving the pharmaceutical history of sodium nitrite. Hitherto, the fact that, prior to being used in bacon, sodium nitrite had a brief early career in British hospitals in the 1880s was something they seemed desperate to hide!

'THE NEW HEALTH FOOD'?

An AMI factsheet, downloadable from the internet, says: 'Nitrite can prevent injury from a heart attack, control blood pressure,

---

* See Chapter 7: Wonder Product of the Twentieth Century: Sodium Nitrite.

promote wound healing, help treat sickle cell anemia and many other health conditions and may even prevent disease progression.'[6] The document concludes: 'Consumers should consume – and enjoy – nitrite-containing cured meats with confidence.'[7] For us today, the 'therapeutic argument' may seem absurd, but that doesn't mean it isn't effective, at least in sowing doubt in the minds of the public. Thus, an article published in 2005 promised that 'Hot dog preservative could be new medication',[8] and posed the question: 'Could the salt that preserves hot dogs also preserve your health?' The article explained that scientists were conducting experiments on nitrite with the aim of developing 'a cheap but potent treatment for sickle cell anemia, heart attacks, brain aneurysms, even an illness that suffocates babies'.[9] Apparently, this helped in 'repairing the reputation of this often maligned meat preservative'. The article concluded: 'It's a surprising revival for a substance once suspected of spurring cancer.'[10]

This line of reasoning is put forward on www.knowyour nitrites.com, the website set up by a Canadian organization called the Canadian Meat Council, or CMC. (This organization's former name – the Meat Packers Council of Canada – reveals its true nature. Like its US counterpart, it has nitrite at the heart of its operation.) The CMC's argument is worth quoting at some length: 'While this is surprising to many people who for years thought they should avoid dietary nitrite, numerous studies have shown that nitrite is an essential part of human health. Research has shown that nitrite: regulates blood pressure; prevents injury from heart attack; prevents brain damage following a stroke; prevents pre-eclampsia in pregnant women;* promotes wound healing; promotes successful organ transplantation; treats sickle cell anemia; prevents stomach ulcers. All of these conditions have been shown to be affected positively by dietary nitrite interventions.'[11]

---

* Pre-eclampsia is a serious increase in blood pressure during pregnancy.

Another page on the CMC website adds even more benefits: apparently, studies have shown that nitrite 'improves exercise performance', 'treats sleep apnea', 'prevents atherosclerosis', 'lowers blood pressure', 'reduces damage in acute stroke', 'treats erectile dysfunction', 'treats major depression', 'decreases ocular pressure in glaucoma', etc.[12]

The CMC asserts that: 'Once considered to be a harmful food additive, nitrite is now being viewed increasingly by members of the scientific community as not only a beneficial molecule, but one with vital medicinal properties.'[13] 'The emerging health benefits of nitrite represent a profound change in paradigm from the debate that occurred during the previous five decades.'[14] To the extent that: 'In the context of these dramatic new and evolving discoveries, the historical classification of nitrite as a "cure" is acquiring new meaning.'[15]

According to the CMC, 'thorough and ground breaking research has further dispelled the linkage between nitrite and cancer'.[16] At the risk of repeating ourselves, let us make the point once more: lawyers may indeed argue that this sentence is strictly speaking correct, if interpreted literally, as sodium nitrite itself is not directly carcinogenic. But the statement is deceptive in that it disguises the crucial point: it is not sodium nitrite that is carcinogenic, but its metabolites. It is not the additive that is carcinogenic, it is the meat that has been treated with this additive. The CMC's text is a bit like the tobacco lobby encouraging the public to smoke by saying: 'Contrary to what some once believed, scientists now know that nicotine is not carcinogenic.' This statement would technically be true from a legal point of view (the carcinogenic agent in cigarettes is not nicotine itself), but false in its implication.

Just like their Canadian colleagues, many American nitro-meat-processors have been keen to propagate disinformation about the risks of their products. Claiming that 'much confusion and even mythology surrounds nitrite',[17] the American Meat Institute has been running a 'media myth crusher' campaign professedly designed to 'improve accuracy when writing about the use of sodium nitrite in cured meats'.[18] The information was picked up by a number of websites often unaware of the fallaciousness of the lobby's arguments. Under the heading 'The nitrate and nitrite myth: another reason not to fear bacon', one American nutrition blog described how 'the belief' that nitro-additives were harmful 'has been entrenched in popular consciousness and media'.[19] It described how the 'fears' had been swept aside and replaced with positive findings in favour of nitrate and nitrite, and stated that 'recent research suggests that nitrates and nitrites may not only be harmless, they may be beneficial, especially for immunity and heart health'.[20] The blog went on to pose the question: 'Bacon: the new health food?'[21] In 2019, the industry-friendly Institute of Food Technologists published an article that, without asserting anything categorically, posed a series of questions: 'are nitrates and nitrites misunderstood?'; 'what if all of the warnings and links to poor health outcomes were based on inconclusive science?'; 'what if nitrates and nitrites are actually good for human health, and the avoidance of them may be causing another health crisis?'[22] Or, in short: 'Nitrates and nitrites have had a reputation for being toxic to human health for a number of decades, but their association with poor health outcomes may be unfounded.'[23] Other publications constructed false dichotomies (when in fact the two options are not incompatible). For example, one article entitled 'Nitrites and nitrates in the human diet: carcinogens *or* beneficial hypotensive agents?'[24] explained that it was once believed (in the past) that the presence of nitro substances in food represented a health risk. The introduction to the article stated that 'nitrates have been inaccurately linked to the development of cancer in the modern west and, as recent evidence suggests, are actually beneficial to health, particularly in reducing blood pressure'.[25]

'FROM MENACE TO MARVEL'

In France, the arguments of the North American lobby were first picked up by agri-chemical groups that were campaigning for a revision of rules against water pollution, which limited the use of nitrogen fertilizer – ammonium nitrate, calcium nitrate, potassium nitrate, sodium nitrate – and restricted the spreading of animal manure as fertilizer. By some strange irony, the pro-nitrate activists are often themselves linked to the pork industry, because European nitrate regulations limit the expansion of industrial pig farms, which produce large amounts of manure. Hence the surprising collusion between the two ends of the production chain: certain industrial pig farmers maintain that *nitrate* is harmless so that they can more easily dispose of their slurry; certain industrial meat-processors allege that *nitrite* is harmless so that they can carry on putting it in their hams. In March 2013, the magazine *La France agricole* (Agricultural France) thus concocted an 'exceptional plan of action' on the theme of 'Nitrate and health: a surprising counter-inquiry', claiming to report on a 'spectacular reversal' of science. Known for its close links to the agro-industrial complex, *La France agricole* announced that: 'Experiments carried out mainly in the Anglo-Saxon countries show that nitrate in food offers more advantages than drawbacks when it comes to health, contrary to the dogma that has dominated for at least fifty years.'[26] The magazine warned: 'This inquiry goes against the grain and will be met with incredulity and even suspicion.'[27] The suspicion was undoubtedly well deserved: most of these 'Anglo-Saxon experiments' were the work of researchers linked to the AMI.

A prominent figure in this 'new paradigm' was Nathan S. Bryan,[28] an entrepreneur and researcher from Texas. In many publications he has championed the merits of nitrite and some of its derivatives. On flyers produced by the AMI entitled 'Sodium nitrite: the facts', Bryan is quoted talking about the 'old myths about an alleged link to cancer'.[29] In an article entitled 'Cured meats found to hold health benefits', an American website explained that Nathan Bryan 'is concerned by what he calls the misinformation being distributed by some groups regarding the "health risks" of nitrites and nitrates'.[30] According to Bryan, 'There is absolutely no scientific basis to single out cured and processed meats as a culprit in certain cancers'.[31] In smaller doses, nitrite appeared rather to offer protection against cancer, and would only pose a danger to those who were already very ill: 'Taken together, our data indicate that nitrite can inhibit cancer cell progression at low doses and early stage but may promote cancer cell progression at higher doses in stage 4 colon cancer.'[32] In other words, sodium nitrite was actually beneficial, and its harmful effects only kicked in when the person already had generalized cancer (at 'stage 4', the metastases have already invaded the organism).

At a conference in 2010 Bryan presented a paper entitled 'Dietary nitrite and nitrate: from menace to marvel'.[33] One of his articles for the journal *Meat Science* claimed that 'foods or diets *enriched* with nitrite can have profound positive health benefits'[34] and he complained that 'despite decades of rigorous research on its safety and efficacy as a curing agent, it is still regarded by many as a toxic undesirable food additive'.[35] The French magazine *Agrodistribution* even published an interview with Bryan under the title 'Nitrites act like vitamins'.[36]

Documents of the American Meat Institute show that Nathan Bryan has been working with the AMI to defend its interests[37] and contest the scientific research critical of nitrited meats.[38] In particular, Dr Bryan was part of an 'IARC challenge committee' set up by the AMI to bring about a change of classification of nitro-additives as carcinogenic agents so as to 'preempt any regulatory action'.[39] Bryan also appears in an AMI-produced video interview entitled 'Myth: nitrite in cured meat is linked to diseases like cancer'.[40] In the video, he points out that, thanks to technical progress, scientists have discovered that the human body itself produces nitrite, and that it turns out that this molecule actually offers interesting benefits from the cardiovascular point of view (in fact the endogenous generation of minute amounts of nitrite has been known about for more than a century).[41] In the video, an AMI employee exclaims: 'Well, this is very different from what most consumers are hearing about, it's a little bit of a "man bites dog" story: you're saying that nitrite doesn't cause cancer, and it actually has health benefits!'[42] Bryan's advice to consumers is 'to not get caught up in the fads and the scare tactics on the use of nitrite in foods'.[43] He declares: 'in fact recent data suggests that nitrite can actually inhibit cancer cell growth in certain animal models.'[44]

This video is one of a series that the AMI has put out on its 'Meat myth crushers' YouTube channel. Another title on the channel is 'Nitrite sources and benefits'. On it, the AMI's head of communication interviews a professor from Madison. Question: 'So, what's your message to consumers who have concerns about nitrite?' The professor replies that obviously 'nitrites from cured meats are a very safe compound to consume'![45] What he neglects to mention is that the carcinogenic agent is not the nitro-additive itself but the nitroso compounds to which the nitrite can give rise. In another video on the channel, the past president of the AMI Foundation himself appears in a white lab coat (as if clothes maketh a scientist). The video is called 'Nitrite-cured meats, are they safe? Ask the meat-science guy', and the answer given to that question is: 'Consumers shouldn't fear nitrite! The medical community doesn't any more.'[46] Again, this neglects to point out that it is nitrite-treated meats that are carcinogenic, while concentrating on the message that nitrite itself is not a direct carcinogen, which is of course true.

These online videos are a troubling example of the tactics of the nitro-meat industry. For who else are these messages aimed at other than worried members of the public looking for clear information?

SMOKESCREENS

In the 1940s, with the scientific case against cigarettes mounting, the tobacco industry published proofs of the positive therapeutic effects of nicotine in the treatment of Raynaud syndrome (a disorder of the blood circulation).[47] Later, the tobacco companies funded researchers to identify other benefits of nicotine. One cardiologist demonstrated that it could help with angiogenesis, that is, the growth of new blood vessels. And following other similar discoveries in the field of neurology, the cigarette lobby started celebrating tobacco as the 'healing weed'. Thus, one industry-friendly journal claimed: 'The news has been conveniently ignored by tobacco's adversaries, but recent studies indicate that, aside from its well-known risks, the golden leaf may have some health benefits. More and more research is showing that certain tobacco compounds offer protection against medical disorders, such as brain diseases like Parkinson's disease and Tourette's syndrome.'[48]

Historians have identified the techniques of 'uncertainty manufacturing' employed by pro-tobacco 'scientists' to 'create a climate of broad public confusion surrounding the topic'.[49] Thanks to documents revealed in major tobacco-related court cases, we now have a much better insight into the strategies that were employed. The cancer specialist Siddhartha Mukherjee summed up these manipulations in a few words: when the extent of the machinations began to be revealed, 'the depth of deception made even the industry's own attorneys cringe in horror. Cover-ups were covered up with nonsensical statistics; lies concealed within other lies.'[50] For those the historian Robert Proctor labelled the 'tobacco charlatans' or 'denialist scholars', the mission was simple:[51] convince people that there was not yet enough proof to claim that cigarettes were carcinogenic. Proctor wrote of the Tobacco Institute: 'The industry's vast PR machinery churned out its message – "not yet proven" – while never actually funding research that might have yielded such proof. Hundreds of press releases urged the need for "more research", with the claim sometimes even made that it was dangerous to jump to conclusions, given that the case was not yet closed. And that, of course, is how the industry wanted the health "question" kept: forever open.'[52]

The cigarette firms used many different methods, including trying to convince people that cigarettes were only carcinogenic 'when consumed to excess', and even that moderate consumption was actually beneficial ('a pack a day keeps lung cancer away', said one tobacco industry publication in the early days of their defence campaign).[53] Later, the strategy was to fool the public over the real risks posed by 'light' cigarettes or to conceal the carcinogenic effects of nitrosamines present in tobacco smoke[54] (nitrosamines are present not only in the smoke inhaled by the smoker but *especially* in the 'sidestream smoke' or 'second-hand smoke', that is, the smoke inhaled by people nearby).

We know today that the cigarette companies spent millions of dollars to conceal the full seriousness of passive smoking from the public. They ran fake studies so that they could tar passive smoking as a 'myth',[55] even manipulating members of the IARC in order to prevent the reality becoming known through the WHO agency.[56] One of the doctors who closely studied the strategies of influence deployed by the tobacco industry explained why, in the 1990s, nitrosamines became such a preoccupation of the cigarette companies and their lawyers: that smokers risked cancer was nothing new, unfortunately, 'but they were obsessed by the evidence that tobacco poisons released into the air could affect the health of non-smokers'.[57] They had to neutralize the fear of passive smoking and nitrosamines: 'An army of public relations experts, front organisations, and corrupted consultants served the lawyers, not the truth – the companies, not the public.'[58] And today, documents from the tobacco court cases reveal how, sometimes, the nitro-meat lobby made common cause with the fake news merchants of the tobacco industry to hide the risks of nitroso compounds from the public.

'CURED MEAT STRATEGY'

Among the documents brought to light during litigation between US states and tobacco industry organizations, there is a series of faxes which show how some lobbyists pooled resources to defend both carcinogenic meats and passive smoking. In a letter of February 1996, a lobbyist called Jim Tozzi[59] described his ongoing meetings with some of the officials of the government departments in Washington who might take measures against nitrited meats because of the cancer risks: he reports on the contacts he formed, the phone calls he made, the strokes he pulled as he wove his web of intrigue.[60] Recently released confidential emails and memos show him asking for extra funds and describing his plans to 'forestall any future regulatory action' by the government authorities.[61] He knew his subject well: for years, he had been putting the brake on the health agencies by offering 'evidence' that the carcinogenicity of passive smoking was 'not proved'.[62]

Hidden in this trove of secret documents, there are also details of operations that nitro-meat companies ran to snare researchers who were studying infantile cerebral tumours linked to the consumption of nitrite-cured sausages. In a confidential document of 1996, Tozzi offered a six-page description of his 'tobacco and cured meats' strategy. First, he reviewed recent scientific findings that confirmed that it was dangerous to expose young children to tobacco smoke and that embryos were at risk – not only if the mother smoked during pregnancy but also if she inhaled the smoke of others.

Tozzi flagged up to his clients that these types of results posed a serious problem for them because there was a risk that the administration would use the findings 'in a mistaken effort to protect children'[63] because 'activists are always looking for impacts on children because they can use them to attempt to compel immediate regulatory action'.[64] So he outlined a response 'to counter the potential damage which could occur':[65] his company would find a scientist for hire who would have the task of disseminating well-prepared and thought-through counter-arguments 'throughout the government as well as with outside groups and the press'[66] in order to protect the interests of the tobacco manufacturers.

In a second section entitled 'Cured meat strategy', Tozzi then described how he proposed to deal with a cancer specialist who was a particular thorn in the side of the nitro-meat industry: Dr Susan Preston-Martin, a professor at the School of Medicine of the University of Southern California. Since the early 1980s, Dr Preston-Martin had been analysing the relationship between the occurrence of cancer and the consumption of nitrited meats. She was particularly interested in the abnormally high frequency of cerebral tumours in children whose mothers had eaten nitrited meats during pregnancy.[67] (In a Powerpoint presentation of the AMI Foundation, the results published in 1994 by Dr Preston-Martin are presented in somewhat laconic terms: 'Maternal hot dog consumption during pregnancy related to childhood brain cancers. Carried in popular media. Summer hot dog sales dropped 8%.'[68]) When, two years later, the same medical research team was preparing to publish new results which confirmed the impact of nitroso compounds on the human foetus, arch tactician Jim Tozzi warned his clients: 'Dr Preston-Martin is a very well known and respected epidemiologist; she is widely published and her work appears in important and well respected medical journals. Her reputation as a researcher is "rock-solid" according to several of our contacts.'[69] The problem was: 'Dr Preston-Martin's work has become increasingly specific in pinpointing the consumption of nitrite from cured meats as a cause of tumors (and, possibly, leukemias).'[70] In another document, the lobbyist stressed that 'The forthcoming February publication of two articles by Dr Susan Preston-Martin in the journal *Cancer Epidemiology, Biomarkers and Prevention* as well as the anticipated release in the last half of 1996 of the research results from her large-scale study on the sources of cancer causation, could have a significant impact on the tobacco and meat businesses.'[71]

Given the serious implications of these results, manufacturers should have brought in measures to protect consumers. Instead, they tried to obstruct the work of Dr Preston-Martin: they investigated her funding sources and tried to prevent her findings having an impact on regulations. A fax from Jim Tozzi described their secret strategy: the lobbyists proposed to approach her without revealing their true motives (contacting her under the pretence of asking for advice on a technical question). Tozzi specified that, in order not to give the game away, it would be necessary to bury discussion of the nitrosamines of cured meats among other themes that would serve as camouflage.[72] Tozzi hoped that it would permit his team 'to work with [Dr Preston-Martin] as she completes the large epidemiological study she currently has underway. This will afford us the opportunity to help shape the conclusions of that study and get early information on what those conclusions are likely to be.'[73] In the end, the nitro-meat industry was able to launch a finely tuned media campaign to neutralize the impact of Dr Preston-Martin's results regarding hot dog consumption during pregnancy and the risk of brain tumour in the offspring.

THE DOUBT-MONGERS

In the USA there is now a whole cottage industry specializing in 'product defence'. Exponent, Inc., Weinberg Group, ChemRisk are just a few of the highly paid firms that combine public relations with scientific spin-doctoring. Mathias Girel, a French historian of science, has written about how the public is gradually becoming aware that 'alongside research that increases our knowledge there is another type of research that sets out to undermine existing knowledge and has no other purpose than to sow doubt, a science with the sole purpose of offering reassurance and hindering the implementation of regulations'.[74] The epidemiologist David Michaels has described how 'it has become standard operating procedure for corporations to attempt to manufacture scientific uncertainty when they're faced with allegations that their products or activities cause harm'.[75] Michaels spells it out: 'It is not an exaggeration to say that in the product defence model, the investigator starts with an answer, then figures the best way to support it.'[76] 'The work has one overriding motivation: advocacy for the sponsor's position in civil court, the court of public opinion, and the regulatory arena.'[77] Their aim is: delay, delay, delay, perpetually postponing the day of reckoning for the industry.

In 1994, the AMI commissioned a report to contest the results of three studies that had just been published. One of the joint signatories of the report was AMI consultant David Klurfeld[78] – an ardent ideologue of the 'no proof' approach. The report provided a list of arguments that might be used to question the epidemiological studies' methodology and results. It stressed that 'Statistical correlations do not prove causality' – a line that might be considered the mantra of the 'technicians of doubt' – and went on to claim: 'these findings merely suggest possible relationships that should be studied further.'[79] Of course, little effort was subsequently made to ensure that these 'further studies' did actually take place! The studies demonstrated that the more children ate sausages treated with sodium nitrite, the higher the incidence of cancer (this affected mostly children from poor black families). The report suggested an 'alternative' interpretation which gives some insight into the contorted thought processes of some AMI consultants: 'even before cancer patients know they have the disease, there is a change in metabolism that occurs to increase the utilization of fat and calories. The body's natural response to this might be to unconsciously choose higher fat and higher calorie foods. This would mean that studies of childhood diet and cancer could actually be measuring cancer's effect on the diet, rather than the other way around.'[80] In other words: if children who eat sausages made with sodium nitrite have more cancers than other children, it does not mean that the sausages themselves cause the cancer. Rather, it might be because the cancer makes them want to eat sausages!

Nitro-meat literature contains a seemingly endless catalogue of such sophistry: some commercial scientists have added the defence of nitrited meats to an already long list of dubious causes.[81] Is this a wise approach for the industry to follow? Won't the fact that their pink ham is being defended in such a way actually alarm consumers? When they come to make a choice between a natural, pale-grey ham sandwich and a nitrited ham sandwich, won't shoppers be rather troubled by the idea that, in order to counter the findings of epidemiologists and cancer specialists, the processed meat lobby has had to seek the help of these strategists of lost causes?

CHAPTER 18

BLUFF, SUBTERFUGE, HOGWASH: THE ART OF NITRO-POKER

What could the nitro-meat lobby do in 2007, when the World Cancer Research Fund recommended that processed meats should not be eaten at all?[1] What to do in 2015 when the International Agency for Research on Cancer confirmed 30 years of epidemiological results by classifying processed meats in 'group 1: carcinogenic to humans'? Once you are on board the train, it's quite hard to get off. As it hurtles onwards, you have to keep upping the ante: as the science advances, the strategies have to become ever more sophisticated. You have to come up with new ploys, new reasons to keep on saying: 'we just don't have anywhere near enough information to make recommendations that we would feel quite certain would really result in a dramatic change in cancer'[2] (1981), or that 'there is virtually no scientific rationale' for concluding that nitrited meats are carcinogenic[3] (1998), or that nitrate and nitrite have 'only theoretical long-term risk that remains unproven'[4] (2011), or that the classification by the IARC 'defied common sense'[5] and 'simply cannot be applied to people's health because it considers just one piece of the health puzzle'[6] (2015), or that 'cancer is a very complex disease with many potential causes. Linking cancer to one food or food group is irresponsible and reflects poor scientific judgement'[7] (2018).

Keeping this going is a risky strategy, since the more it escalates, the more it risks overstepping the bounds of legality.

CONTAMINATION

In London, Paris, Brussels, Maastricht, Oslo, Valencia, the tacticians of the European nitro lobby are hard at work. Pseudo-scientific conferences, fake data, PR campaigns designed to sow doubt and confuse the public, to buy a bit more time, to fool the health authorities: behind the lovely pink hams and tasty-looking sausages, the world of nitro-meat is a shadowy one, populated by paid consultants posing as 'medical nutritionists', by professional jugglers of facts and statistics, by unscrupulous lobbyists masquerading as researchers – a world full of pseudo-science, fake news and unconscionable trade-offs between health and profit.

In the last few years, the falsifications of the nitrite lobby have gained remarkable traction in the scientific literature. For example, consider an article in the journal *Trends in Food Science & Technology* from 2015 that sets out to offer a definitive review of 'alternatives to nitrite in processed meat'.[8] Written by four Asian academics (from Korea and Sri Lanka), it aims at presenting a complete evaluation of current knowledge in order to find ways in which processed meat products might be made less noxious. The authors acknowledge that epidemiological studies indicate that nitrited meat is carcinogenic. Unfortunately, their bibliography is influenced by around fifteen articles produced by researchers in Madison and by authors close to the AMI. These works are enough to distort the arguments, since at the end of their piece the authors say: 'in contrast, some studies have elucidated the beneficial effect of nitrite on human health.'[9] As a result, the four Asian academics are unable to draw a definitive conclusion regarding the desirability of nitrite suppression, and consequently they leave the question open. For the 'doubt manufacturers' of the nitro-meat lobby, it's job done. But what a waste and what a vile trick: these four academics were artfully caught in the trap, and so were the organizations that funded their work, the civil servants who read their report, the students and other researchers who accessed it, even perhaps some pork butchers who looked to it for alternatives to using nitro-additives. In the end, how many victims, how many future cases of cancer could have been avoided in Sri Lanka and Korea and everywhere else this article was consulted? To protect the interests of a few American and Western factories, how many people on the other side of the world will have to suffer?

Not that you need to look so far afield to see how the doubt-mongering has distorted scientific knowledge as much as it has corrupted food itself. Consider, for example, a 2014 study carried out by three researchers at a Scandinavian university. Funded by the Danish food council Fødevarestyrelsen, it set out to evaluate the necessity of nitro-additives in order to bring Denmark into congruence with European regulations.[10] But the Danish rapporteurs were unaware that the nitro-meat industry was using the same tactics that asbestos, pesticide and tobacco firms had used. Throughout their report, the authors put their trust in publications which they considered authentic, unbiased scientific articles, without realizing that they had in fact been written by scientists who were exceedingly close to the AMI. For instance, the Danish authors referred to several articles from the 1970s without understanding that the works in question were produced by defenders of the nitro-meat lobby in the specific context of the 'nitrite war'. What the Danish researchers took as a genuine evaluation of the 'minimum level of nitrite necessary to make sausages safe' was nothing more than findings produced under the aegis of manufacturers of nitrite-cured sausages in order to justify use of the chemical.[11] Likewise, what the Danish authors thought was a reference work on meat-curing traditions was in fact a revisionist 'history' written by two prominent American meatpackers working on behalf of the AMI which attempted to prove that 'cured meats' and 'nitro-cured meats' have been synonymous since the dawn of civilization.[12]

It was all par for the course: recent articles by AMI-supported authors insinuated their way into the bibliography used by the Danish authors, influencing their field of reference to such an extent that they felt compelled to write: 'For the sake of completeness it should be added that it has recently been suggested that small amounts of nitrite may even be beneficial to health.'[13] Unfortunately, the Danish report was not a one-off. For the strategy of the AMI has been given a significant boost in Europe: one of the latest expert reports delivered to the European Commission draws upon the work of the AMI and adopts its approach so as to justify the continued use of carcinogenic additives by European manufacturers.[14]

NEWSPEAK

The term 'cured meats' used to designate meats that had been cured using salt; but in the USA, the legal definition of curing has been revised so that it now refers to treatment using salt *with added nitrate or nitrite*.[15] Meat scientists have noted this: 'meat curing, historically defined as the addition of *salt* (sodium chloride) to meat, is now referred to as the intentional addition of *nitrite and salt* to meat.'[16] In Canada, when meat curers work without nitro-additives – and even when they produce hams whose superior quality is recognized – they receive no encouragement; on the contrary, because they don't follow the rules imposed by the industry, artisan producers are sometimes bullied, threatened with legal action and have their products confiscated.[17] As a consequence of this hostile regulatory and commercial environment, there is only a single company in Canada – McLean Meats in Vancouver – that produces nitrite-free processed meats on an industrial scale.

Everywhere, the nitrite curers have succeeded in imposing definitions, recipes and rules that are favourable to them. Unfortunately, Europe hasn't escaped this chemical hijack: here too the nitro-curers managed to rig things to the point that even the meaning of words changed. The European food authority EFSA uncritically adopted the industry's newspeak. In defiance of tradition, they redefined the process of curing to bring it into line with the North American concept by making nitro-additives an integral feature: in a scientific notice on sodium nitrite the EFSA announced: 'By definition, "cured meat products" contain curing salts, usually salt (sodium chloride) and either nitrites or nitrates.'[18]

Even if there wasn't the problem of cancer, this verbal land-grab would be a scandal, in that it totally violates the European tradition of meat curing. Let us repeat for the umpteenth time: in Parma, the location of EFSA's headquarters, the meat curers banned nitro-additives in their famous hams and reverted to the traditional methods. It should be noted too that, from its very first edition, the *Code of Practice* of French charcutiers explicitly stated that in Parma no nitro-additives were used in the curing of ham: curing 'is carried out using only commercial cooking salt'.[19] So how can it be that the EFSA betrayed European consumers and, to placate the nitro lobby, decreed that henceforth 'curing' was synonymous with 'nitrite curing'? Imagine if the word 'bread' was redefined just at the behest of some lobby group? If, in order to suit a few frozen food manufacturers, the word 'fish' came to mean 'fish finger'? Or if 'beef' was modified to mean 'hormone-treated beef'? It's as shameful as it is absurd.

An example of the relentless advance of chemical curing: currently the type of Spanish ham known as *pata negra* is undergoing changes. Traditionally, this product was obtained through a slow maturation, without any recourse to nitro-additives, as we have

seen. It represented the peak of the Iberian meat-curing tradition. To simplify production and increase volumes and margins,
European and Spanish regulations have recently allowed the use
of nitro-additives. Likewise in Corsica, where the true Corsican
dry sausage (*u salsicciu*), authentic Corsican ham (*prisuttu*), the
lonzu (cured loin) and the *coppa* are still to be found on farms
in the centre of the island. Elsewhere, all these products have
been replaced by ersatz versions made with meat imported from
the continent, produced with the help of nitro-additives and then
sold to summer visitors who believe they are buying 'authentic
Corsican food'. The same process has more or less killed off
the traditional French *saucisson* (dry-cured sausage), now ubiquitously replaced with rapidly produced versions that are very
'real-looking'. For let us recall once again: if you take the trouble
to avoid the 'manuals' and 'treatises' published by the food additive manufacturers,[20] you can find written confirmation by the
most reputable technical experts that to produce real *saucisson*
'only salt and pepper are indispensable'.[21] How can we tolerate the
gastronomic and cultural treasure that is our cured meats being
replaced by carcinogenic copies?

THE NITRO-ORGANIC SCANDAL

Probably the most revolting phenomenon of all is the inroads that
nitro-additives have made into the organic industry. Historically,
the organic sector totally forbade the use of nitro-additives. 'When
you know what you're doing, you can make cured meats of superior quality without any need for the commonly used additives
– phosphate, nitrited curing salt and flavour enhancers. And it is
tasty!'[22] exclaimed Hermann Jakob, a German professor of meat-
processing with an expertise in organic production. 'Nitrited curing
salt is taboo at Demeter',[23] said a German manufacturer about the
specifications of an organic certification mark.[24] 'Demeter does

not allow any additives. Its sausages are distinguished by their paler colour', explained the professional magazine *Bio Press*.[25] The Demeter organization clarified: 'At Demeter, nitrited curing salt is banned. In its place, the curer uses his knowledge, his care and his skill. In conventional transformation, most additives serve to accelerate production and standardize the process in order to obtain a constant result despite variations in the raw material.'[26] Similarly, use of nitrate and nitrite is forbidden by the rules laid down by the organization Nature et Progrès (Nature and Progress, an initiator of the International Federation of Organic Agricultural Movements, IFOAM).[27]

But the growth in organic foods has sharpened appetites. As early as 1999, the *Sunday Times* was questioning the obscure circumstances under which the Soil Association had authorized the use of sodium nitrite in British organic production, allegedly at the request of Sainsbury's and other supermarket chains.[28] In Belgium, Nature et Progrès was also sounding the alarm: 'In the past, thanks to the action of Nature et Progrès, the Belgian organic sector has banned nitrited curing salt in charcuterie. Under the influence of large retailers, some organic actors are currently considering reopening this question, rather too easily forgetting that these additives are bad for human health.'[29] To restore some order in the organic sector, the European Commission decided in 2006 that 'sodium nitrite and potassium nitrate shall be re-examined before 31 December 2007, with a view to limiting or withdrawing the use of these additives'.[30] Microbiologists provided evidence that nitro treatment was not necessary, concluding that 'any effect of nitrite on product safety and stability may be compensated for by modification of formulations and processes'.[31] They explained: 'the epidemiological data do not support the common view that the addition of nitrite is essential for protection against botulism or other food intoxications.'[32] A group of experts was summoned to Brussels, and they recommended introducing 'a general and efficient education programme in alternative processing methods and hygiene to organic meat manufacturers'.[33] The experts' official report was unequivocal: 'the group recommends that nitrate and nitrite within a reasonable time scale should be eliminated from organic meat products.'[34]

In 2008, a European Commission regulation recapitulated: 'the additives sodium nitrite and potassium nitrate were to be re-examined before 31 December 2007, with a view to limiting or withdrawing their use. A panel of independent experts has in its conclusion of 5 July 2007 recommended eliminating sodium nitrite and potassium nitrate in organic meat products within a reasonable time scale.'[35] But because the industry fought it tooth and nail, this measure was never implemented. British manufacturers in particular claimed that a ban on nitro-additives would 'ruin an organic bacon industry in the UK that is booming'[36] and lead to the 'collapse of the market for organic bacon'.[37] The scientific results showed that it was perfectly possible to manufacture organic bacon without recourse to nitro-additives,[38] and a number of Danish and German studies proved that nitro-additives could be gradually phased out without compromising health safety.[39] But the British meat companies made out that there was no way of assuring that nitrite-free bacon would not pose a bacterial risk,[40] 'nor does it seem likely that many shoppers would be willing to accept a change to a rather unappetising looking, greyish product'.[41] In 2008, the industry magazine *Food Manufacture* reported that British producers 'are up in arms about European Commission proposals to ban the use of nitrites in organic food by the end of 2010, which they fear could kill off the sector in the UK'.[42] Under pressure from lobbyists, the European Commission agreed to offer the industry a short stay of execution. Then they offered another one. In November 2010, *Food Manufacture* noted that the industry was 'seeking to indefinitely delay an EU regulation that would ban their use from 2011, arguing there are no suitable alternatives'.[43] Finally, the European Commission gave way, and the plan to ban nitro-additives was never enacted.

Lionel Rostain, who heads one of the main French producers of organic meat products, said in an interview: 'I was consulted when specifications were being drawn up for processed meats and personally I was against the use of this additive, because you can do without it. Basically, it is used purely to enhance the visual aspect, especially to turn ham pink, but provides nothing in the way of preservation. The problem is: if we stop using nitrited curing salt, will the consumer buy grey ham? I myself have experimented in doing without nitrite and yet have managed to find ways to preserve the pink colour. It is once the package is opened that the product starts to turn grey and the customer might believe that it is unfit for consumption, even though that is not the case! The solution would be for all organic producers to give up using this additive and provide the consumer with clear information to set their minds at rest.'[44] So Rostain marketed his nitrite-free organic ham by including a note on the label: 'We favour quality over pleasing appearance. That is why our nitrite-free cooked ham is paler, and changes colour once opened. This is a normal reaction and is due to the absence of nitrited curing salt; it has no effect on the taste and nutritional value of the ham.'[45] But by allowing nitrite treatment, the European regulators had created an irresistible market force. Once they clapped eyes on pink-coloured 'organic' ham, customers progressively gave up on nitrite-free ham. Orders dropped, and finally the Rostain company was forced to cease production of its original untinted ham.

NOW YOU SEE IT, NOW YOU DON'T

In the USA in the 1970s, a few producers attempted – before being rumbled – to pass off their nitro-meats as nitrite-free, simply by indicating that the product was made without additives.

'Our inspectors clearly failed to check the mixing of the pickle',[46] complained the assistant secretary of agriculture. 'It's inexcusable that the processor could make a label that said no nitrite and then put the additive in',[47] lamented another official of the US Department of Agriculture in the *Chicago Tribune*.

In the twenty-first century the tricks are a great deal more sophisticated. 'The curing code has recently been cracked', announced the *New York Times* in 2006: 'Instead of relying on sodium nitrates or the more common sodium nitrites for color, texture and shelf life, hot dog makers have found a magic solution of celery juice, lactic acid and sea salt that rescues the organic dog from its tough brown reputation and rockets it to pink juiciness.'[48] One producer of this miracle hot dog was quick to exclaim: 'If we can introduce a whole new group of people to healthy food through a hot dog, oh my god, the lateral impact is enormous.'[49]

The only problem is that the recipe for these hot dogs *does* contain nitrite. Instead of using cooking salt with a bit of sodium nitrite mixed in, additives manufacturers have developed a clever technique: they use a vegetable extract with a high nitrate content, to which they add a special bacterial culture (for example, staphylococci). In presence of nitrate, these micro-organisms generate nitrite. A patent for a 'mixture for reddening meat products'[50] granted to some German chemists explains that the bacteria 'are able to provide the required nitrite from the nitrate of the vegetable products. Nitrite is the basis for the formation of nitric oxide that is required for the reddening.'[51] Meat-processors only need to mix this nitrite-generating compound with salt water to achieve exactly the same effect as if they had used the usual nitrited curing salt.

The point of this process is to use nitrite without having to write 'sodium nitrite' on the label, as more and more consumers are learning to avoid it. This technique has no positive impact on the cancer problem, but it is highly advantageous when it comes to marketing: rather than eliminate nitro-ingredients from their recipes, the manufacturers thus manage to remove those damn nouns from their labels. The words 'potassium nitrate', 'sodium nitrate' and 'sodium nitrite' have been magicked away. In the jargon of the agri-food business, this is described as a 'clean label' or 'clear label'.

One additives manufacturer explains: 'As even highly processed meat products such as sausages are now subject to clean and clear label trends, manufacturers need to find ways to keep the color appeal of these products while losing the artificial additives.'[52] The Irish company Kerry Foods obtained a patent which describes 'a process for preparing the curing agent comprising contacting a plant material with an organism capable of converting nitrate to nitrite'.[53] One Scandinavian food additives manufacturer vaunted the merits of its specially selected bacterial culture: it 'provides the desired colour much faster than the market standard'[54] because 'the enzyme complex reduces the nitrate into nitrite, which provides the appealing reddish colour'.[55] Describing its special powder for making nitrited pink ham 'without added nitrites', the French firm Solina explained: 'The culture transforms natural nitrate obtained from the vegetables into nitrite which interacts with the haemoglobin in the meat to form nitrosomyoglobin, which is responsible for the pink colour of the hams.'[56]

The industry website meatpoultry.com has described the advantages of this so-called plant-based nitrite by citing the director of one of the main US suppliers: 'It provides curing and shelf-stability in an easy-to-use powder format.'[57] He emphasized that the active ingredient was 'naturally occurring' nitrate/nitrite, 'yet is labeled simply as celery powder'.[58] (Consumers not versed in biochemistry or meat science might read the label and suppose it is referring to celery salt, a common cooking ingredient; they might find this reassuring.) Another executive explained that the powder, sold under the label VegStable®, 'acts as a curing agent' and allows meatpackers to replace the nitro-additives 'so they can market their products as natural and capitalize on the double digit growth rate in this segment'.[59] As one food processing researcher put it, 'Development of meat products by replacing the synthetic nitrite with nitrite from a natural source might provide a new growth to the meat-processing industry.'[60] He went on to say: 'We showed that fermented spinach extract contained pre-converted nitrite which could substitute synthetic nitrite to maintain the color development of cured meats.'[61] Another study which set out 'to improve the health image of processed meat products'[62] noted that nitrite obtained from fermenting chard powder gave equally good results and concluded that the procedure was interesting 'because synthetic nitrite has a negative health image while the nitrite from a natural source such as vegetables receives the perception as natural and healthy'.[63] One staunch defender of nitrite treatment said: 'They are the exact same molecules ... just from a different source.'[64]

What do a few more pennies matter if you can buy a ham without carcinogenic additives? In the belief that they are looking after their health or that of their children, consumers are willing to pay 15% to 20% more for their bacon and ham. In fact, they are often simply paying more for the same dangerous substance.

STEALTH NITRITE

In an article in a professional journal entitled 'Nitrite free: Where does the truth end?', one meat-processing expert wrote in 2007: 'By adding nitrite "indirectly" it does not need to be declared on the label of the product and the term "preservative" therefore can be avoided on the label. In most countries the term "preservative" in a meat product is not well liked.'[65] He noted that 'this material also has the potential to form nitrosamines. As a result, the health benefit as portrayed to the consumer that "no nitrite is added" is not exactly the entire truth. The described process of obtaining nitrite out of nitrate is just another way of introducing nitrite into meat, with all its desired technological functions, but without having to declare nitrite on the label of the product.'[66] In a professional journal, pro-nitrite scientist Nathan Bryan, whom we met earlier, denounced this process as a 'public deception'.[67] And yet a growing number of additive manufacturers who formerly sold sodium nitrite today sell this 'furtive' nitrite that escapes the requirement of labelling. Certain additive manufacturers have even thought of developing genetically modified bacteria capable of producing nitric oxide without the input of nitrate at all.[68]

In the USA, there has been a string of unsuccessful court cases against manufacturers.[69] In Germany, after a long legal battle in the organic branch, it was established that it was illegal to use the pretence of 'plant-based' nitrite to make consumers believe that their organic sausages were free of nitro-ingredients.[70] In France, manufacturers accused each other of betraying the trust of the public.[71] The French Fraud Prevention Office (DGCCRF) adjudged that this technique 'basically boils down to adding the same substance in the end, while replacing mention of this additive in the ingredients with that of a simple vegetable broth'.[72]

Recently, some European countries have clamped down on these tricks. The Republic of Ireland, which was one of the countries most affected, has adopted a very unambiguous position. In its instructions to manufacturers, the Food Safety Authority of Ireland (FSAI) now indicates: 'your label must not mislead the consumer to the extent that they buy your product in preference to similar products'.[73] To the question 'Can I use ingredients that are also sources of nitrite/nitrate to replace sodium or potassium nitrite/nitrate?', the FSAI replied: 'You cannot use ingredients that are also a source of nitrite/nitrate if used for the intended technological purpose of preservation or colouring in the final food.'[74] The European Commission published several official statements, firstly on 'spinach extract containing high levels of nitrate used in sausages'[75] (2006) then on 'the use of fermented vegetable broth, enriched with nitrite'[76] (2010). In September 2018, the European authorities noted: 'The Commission has been made aware by Member States of industry practices which consist in adding plant extracts to food primarily for food additive functions while being erroneously claimed not to be used as food additives.'[77] An official committee of European experts met in Brussels and decided unanimously not to allow use of this stratagem. Nevertheless, the practice remains endemic, and all around the world increasing numbers of consumers are being taken in by it.

These techniques are reminiscent of similar strategies used at the start of the twentieth century. In 1905, Heller & Co. was selling its additive 'Freezine' to a credulous public and telling them that 'Freezine is a liquid gas, *made from vegetable materials*'.[78] It was in fact formaldehyde, which was produced from the oxidation of methanol, at the time obtained from the distillation of wood. In government commissions of enquiry, senators and doctors rejected this as a malicious attempt to whitewash a notoriously noxious substance.[79] Other companies used a preservative called 'Zero' so that they could claim on their labels that they used 'zero preservative'![80]

CONCLUSION

The nitro-meat-processors protect their interests by deliberately ignoring alternative techniques and making out that it is not possible to ensure safety without recourse to nitro-additives. This is not exactly a new ruse. In France, nearly 200 years ago, suppliers to the navy managed to fool the royal authorities by playing on another fear. At the time, sailors often suffered from vitamin C deficiency, due to a lack of fresh fruit and vegetables. Their teeth fell out, their gums bled: symptoms of the dreaded scurvy, which killed explorers and decimated ships' crews. Looking into curing methods, the chief physician of the French navy was surprised to discover that certain manufacturers of 'barrelled beef' used a paste containing a colorant ('dyer's madder') and alum (aluminium sulphate). This bizarre, nauseating compound served to speed up production and increase yields. But in order to have it approved, the shrewd suppliers made out that this mixture was endowed with prophylactic properties that made it indispensable. They claimed that it was an 'anti-scurvy brine' which made salted meat safer to eat. In 1829, the chief physician of the navy publicly denounced this subterfuge: 'The very title of this amalgam indicates the specious pretext on which it has been introduced into cured meats destined for the navy [...] I had seen for myself, on board vessels, recently cured meats coated with a substance the origin of which I was unable to divine: it was the gum and madder of the so-called anti-scurvy brine.'[1]

THE THREE BIG LIES

Today, the lying is more complex, and a great deal more ambitious in scope. Forty years after the 'nitrite war' of the 1970s, the meatpackers of America and Europe are trying to keep their show on the road: leave them in peace to carry on using nitrite, 'to protect the consumer', is their line. Their obsession is that nitro-products should not be 'stigmatized'. Their strategy is to maintain the idea that it is necessary to treat processed meats with nitro-additives – that there is no choice, that there never has been a choice. At all costs make sure the public never really grasps that processed meats have been *made* dangerous for no useful purpose.

In order to try and instil confidence in their pink products, producers are deceiving the consumer. Perhaps even more seriously, they are deceiving themselves. By denying the risk, they have mortgaged their future: by protecting the short-term profits of an industry forever in thrall to perpetual growth, industrial meat-processing has failed to develop a process for making its products safer. In this, European artisans and farmers are perhaps the first victims of the manoeuvres of the lobby and the large interests it represents. Unfortunately, the cancer will not go away. For as long as they are injected with additives that make them carcinogenic, processed meats will present a problem for public health. The lobby's strategies and inexhaustible supply of dirty tricks will not avoid that reality.

But the cheats don't realize that the world has moved on. Honest meat curers are becoming aware of the fact that they have been manipulated by industrial companies willing to do whatever it takes not to have to change their methods, for fear of losing their market share. The public will discover that a campaign of falsification has been operating on three fronts: firstly, to deny the risks of cancer and confuse the issue; secondly, to disguise as much as possible the colouring and accelerating functions of nitro-additives; thirdly, to construct an excuse, the botulism scare, and exaggerate this so that people accept the risk of cancer as the price to be paid and see no alternative. On all three scores, these meat-processors are lying. Perhaps that is the most surprising aspect of the nitro-meat fraud: there isn't just one enormous lie, there are three. Nitro-meat promoters deceive the public on cancer; they deceive the public on the risk of botulism; they deceive the public by hiding the real reasons why their clients use nitro-additives. All three lies prop up this house of cards.

In the end, who decides? Will the consumer go on buying chemically tinted and accelerated meats? In the age of social media, it will become more and more difficult to hide the fact that nitro-treatments make meat carcinogenic for no good purpose; it will become more and more costly to maintain the 'inevitable botulism' fiction; it will require ever more audacious means to make consumers accept that, in order to have their proteins pink, it is worth running the risk of cancer. Consumers can already see in their supermarkets that a return to processed meat without nitrate or nitrite is perfectly possible, even on an industrial scale: if some manufacturers have done it, why can't others? Who will go on buying nitro-ham and nitro-bacon when some manufacturers are offering untreated ham and bacon? Which parent will carry on feeding dangerous sausages and bacon to their children once they find out that they can buy sausages and bacon without the cosmetic treatment?

Can't European industrial meat-processors see the risk in remaining hand-in-glove with the American lobby? How do they not see that pressing on regardless is no longer an option? That the deceit is no longer viable? That they should give up on this mad strategy of 'move on, nothing to see here'?

ENDGAME

In the nineteenth century, when saltpetre curing became widespread, no one could have foreseen that nitro-additives would react with meat proteins and generate carcinogenic metabolites. The scandal of nitro-meats is not the initial fault in the product;

the scandal is in the fact that the industrial meat-processors did not take the necessary radical measures to correct this fault, even though they have known for decades that nitro-additives make processed meat more dangerous than it would otherwise be. The scandal is all the past, present and future cases of cancer. The scandal is in the modern lobby's propaganda, in its efforts to muzzle cancer specialists, in the deceitful reassuring messages that the meat companies put out in the press and on their websites. The scandal is in the constant growth in nitrited meat production – instead of the healthier meats which could have been developed in their place. In short, the scandal is in this long rap sheet of dirty tricks and lies.

But one day, inevitably, the wind will change. The historians Naomi Oreskes and Erik Conway have shown how the tobacco companies were condemned not only because they manufactured dangerous products, but mainly because they 'knew the dangers of smoking as early as 1953 and conspired to suppress this knowledge. They conspired to fight the facts, and to merchandise doubt.'[2] Dr David Kessler, a former FDA commissioner, has described how he managed to gather together sufficient evidence to take on the cigarette lobby in court: the manufacturers were certainly prosecuted because they sold a dangerous product, but more than that *because they lied*. Quoting Horace, Kessler wrote: 'The guilty have a head start, and retribution is always slow of foot, but it catches up.'[3]

What will happen when the lawyers come to subject the arguments of the meat-processors to close scrutiny and ask why some of them can happily do without nitro-additives but not others? Will the lawyers start collecting the medical files of consumers of nitro-meat? Will they claim compensation for cases of cancer that could have been avoided? According to the law, a product is defective if it 'doesn't offer the level of safety which might legitimately be expected'.[4] Once the notions of defect and responsibility come into play, interesting legal prospects open up. Since it can be proved that the meat companies can manufacture less dangerous products if they so wish, who is responsible for these cancer cases?

TRUE COLOURS

It is also a matter of ethical responsibility: today, the European processed meat industry is trying to open up new export markets. And there are huge markets out there to conquer. China is the new El Dorado: the Chinese eat very little processed meat, but they love pork. 'Why miss out on a country that represents 50% of world pork production and consumption?' asked the president of a major French meat-processor.[5] So meat companies are conducting broad campaigns to educate foreign populations to appreciate the taste of processed meats and are adapting Western products to the habits of local consumers.[6] It is all about teaching Koreans and Mexicans to like our sausages, all about converting Southeast Asia and Africa to the joys of ham, salami and bacon. But the poorer countries do not always enjoy all the benefits of colonoscopy, therapy or surgery: a diagnosis of cancer of the colon means a death sentence much more often than it does in Europe.

Is it right to go around the world selling carcinogenic food? Is it moral? On dedicated websites targeted at Asian audiences, European meatpackers show appetizing pictures of 'sausages to cook', 'sausage kebabs' and 'ham macaroni'.[7] Images of medieval pigs roaming the woods evoke a venerable tradition of quality and authenticity. But in order to be truly authentic – assuming there is no change in the methods of production – shouldn't their brochures include a warning that these products are classified in the highest category of carcinogenic substances in the typology of the IARC? Shouldn't they say that the WHO regularly warns about the growing export of carcinogenic food to populations that were formerly largely exempt from diet-induced colorectal cancer?[8] Shouldn't they say that in the wake of North America, Europe and Australia, incidences of cancer are exploding in developing countries as they start to adopt a Western lifestyle and food habits?[9] Since we are introducing a virgin population to York ham and Frankfurt sausages and Ardennes pâté, why not take the opportunity to start from scratch with an honest product: processed meats which have a natural colour?

In May 2015, on the fiftieth anniversary of the IARC, its director said: 'During our short history we have seen cancer shift from being a problem affecting mainly rich countries to becoming a truly global problem. Two thirds of cancer deaths already occur in the less developed countries. This is a problem set to worsen: in the next 20 years the annual number of new cases is estimated to rise from around 15 million to 24 million – nine million additional new cancer patients every year – with the greatest increases occurring in the less developed countries.'[10]

Epidemiologists predict that in 2030 the global number for colorectal cancer will reach 2.2 million new cases per year, resulting in 1.1 million deaths.[11] Why contribute to this catastrophe, when specialists in bowel cancer constantly recommend that we introduce a genuine prevention policy relating to food?[12]

Does Europe have to cling to the shirt-tails of the American processed meat industry? Can't it instead point the way and take up the challenge to produce meats that will not poison anyone?

NOTES

Prologue

1 Roger Sohier and A.G.B. Sutherland, *The Origin of the International Agency for Research on Cancer*, IARC technical report no. 6, IARC, Lyon, 2015.

2 Véronique Bouvard et al., 'Carcinogenicity of consumption of red and processed meat', *The Lancet Oncology*, 16, 16, December 2015.

3 World Cancer Research Fund/American Institute for Cancer Research, *Food, Nutrition, Physical Activity and the Prevention of Cancer: A Global Perspective*, WCRF-AICR, Washington, 2007, p. 382. The IARC gives the following definition: 'Processed meat refers to meat that has been transformed through salting, curing, fermentation, smoking or other processes to enhance flavour or improve preservation.' (Bouvard et al., 'Carcinogenicity of consumption of red and processed meat', art. cit.)

4 Apart from alcohol, and before processed meats were classified as 'carcinogenic to humans', there was only a single food product classified in group 1: *Chinese-style salted fish*, a cause of cancer of the nasopharynx. See IARC, 'List of classifications by cancer sites with sufficient or limited evidence in humans, volumes 1 to 124', IARC, October 2019.

5 Mandy Oaklander and Heather Jones, 'The science behind how bacon causes cancer', *Time*, 26 October 2015.

6 *Financial Times*, 28 October 2015, p. 1.

7 Tim Hayward, 'You can either savour the bacon or relish the hysteria', *Financial Times*, 28 October 2015, p. 9.

8 'WHO setzt Wurst auf die Krebsliste' [WHO puts sausage on the cancer list], *Die Welt*, 27 October 2015, p. 1.

9 *Taz die Tageszeitung*, 27 October 2015, p. 1.

10 'La charcuterie est cancérogène, la viande rouge "probablement" aussi selon l'OMS' [Charcuterie is carcinogenic, and 'probably' red meat too, according to the WHO', *Le Figaro*, 26 October 2015; 'OMS: la charcuterie cancérogène, la viande rouge aussi "probablement"', *Le Parisien*, 26 October 2015.

11 Chris Smyth, 'Processed meats blamed for thousands of cancer deaths a year', *The Times*, 27 October 2015, p. 1.

12 IARC/WHO, 'Population fact sheet: Europe', May 2019, available at gco.iarc.fr.

13 IARC/WHO, 'Population fact sheet: World', May 2019, available at gco.iarc.fr.

[14] Freddie Bray et al., 'Global cancer statistics 2018: GLOBOCAN estimates of incidence and mortality worldwide for 36 cancers in 185 countries', *CA: A Cancer Journal for Clinicians*, 68, 6, September 2018.

[15] Ibid.

[16] IARC, press release no. 240, 'IARC monographs evaluate consumption of red meat and processed meat', 26 October 2015.

[17] Kathryn Bradbury et al., 'Diet and colorectal cancer in UK Biobank: a prospective study', *International Journal of Epidemiology*, April 2019. According to this study, each 25g/day increment in processed meat intake is associated with a 19% increase of colorectal cancer. The risk being on a log scale, each 50g/day increment in processed meat intake is associated with a 42% higher risk. Each 75g/day increment in processed meat intake is associated with a 69% higher risk.

[18] Ibid.

[19] The French National Cancer Institute says: 'The category of processed meats includes all meats preserved by smoking, drying, salting or addition of preservatives (including mince, if it is chemically preserved, corned beef …) […] They include those that are eaten on their own (including ham) and those included in other dishes, sandwiches, savoury tarts …' (Institut national du cancer, *Nutrition et prévention des cancers: des connaissances scientifiques aux recommandations* [Nutrition and Prevention of Cancers: From Scientific Knowledge to Recommendations], INCa, Paris, 2009, pp. 24–5.)

[20] Aisha Gani, 'UK shoppers give pork the chop after processed meats linked to cancer', *The Guardian*, 23 November 2015.

[21] Kurt Straif quoted in Nuño Dominguez, 'Que el público decida en quién confiar, la industria o nosotros' [The public has a choice: believe us or believe the industry], *El Pais*, 28 October 2015, p. 27.

[22] For example, see this statement (November 2016) on the website set up by French meat-processing companies (www.info-nitrites.fr): 'Cancer is an illness involving multiple factors both food-related and non-food-related, and the essentially international research by the IARC does not take account of specific local factors (nature of the products, actual consumption …) […] Another important point to be taken into account: data are relevant only to populations consuming over 50g of processed meat a day. In France the daily consumption of processed meat is around 36g per day.' This statement is false: on the contrary, the IARC experts were clear that they were not suggesting a dose below which the consumption of processed meat involved no risk of cancer. The benchmark 50g is not the point where risk comes into play but is a threshold of more elevated risk.

[23] Ian Johnson, 'The cancer risk related to meat and meat products', *British Medical Bulletin*, 121, 1, January 2017.

[24] Ibid.

[25] Farhad Islami et al., 'Proportion and number of cancer cases and deaths attributable to potentially modifiable risk factors in the United States', *CA: A Cancer Journal for Clinicians*, 68, 1, January–February 2018, p. 39.

26 Ibid., pp. 38–9.
27 David Kim et al., 'Cost effectiveness of nutrition policies on processed meat: implications for cancer burden in the U.S.', *American Journal of Preventive Medicine*, 57, 5, November 2019, p. 6.
28 'Viande/cancer: Le Foll met en garde contre la "panique"' [Meat/cancer: Le Foll warns against 'panic'], europe1.fr, 26 October 2015.
29 *Frankfurter Allgemeine Zeitung*, 28 October 2015, p. 17.
30 'Carni lavorate e cancro, Lorenzin: da OMS allarmismo ingiustificato' [Processed meats and cancer, Lorenzin: unjustified scaremongering from the WHO], *Corriere della sera*, 29 October 2015.
31 *Sydney Morning Herald*, 27 October 2015.
32 Hugh Lockhart-Mummery, 'Cancer of the rectum', in Basil Morson (ed.) *Diseases of the Colon, Rectum and Anus*, William Heinemann Medical Books Ltd, London, 1969.
33 Teresa Norat et al., 'Meat consumption and colorectal cancer risk: dose-response meta-analysis of epidemiological studies', *International Journal of Cancer*, 98, 2002.
34 Joint WHO/FAO expert consultation, *Diet, Nutrition and the Prevention of Chronic Disease*, OMS, Geneva, 2003, p. 101.
35 Recommendation no. 5 in World Cancer Research Fund/American Institute for Cancer Research, *Food, Nutrition, Physical Activity and the Prevention of Cancer: A Global Perspective*, op. cit., p. 382.
36 Lawrence Kushi et al., 'American Cancer Society guidelines on nutrition and physical activity for cancer prevention', *CA: A Cancer Journal for Clinicians*, 62, 1, 2012, p. 39.
37 Conseil supérieur de la santé, 'Viande rouge, charcuterie à base de viande rouge et prévention du cancer colorectal' [Red meat, red processed meat and prevention of colorectal cancer], *Brève* (Notice no. 8858), Brussels, 2013, p. 2.
38 Ibid.
39 Denis Corpet, 'Red meat and colon cancer: Should we become vegetarians, or can we make meat safer?', *Meat Science*, 89, 2011.
40 Susan Preston-Martin et al., 'Maternal consumption of cured meats and vitamins in relation to pediatric brain tumors', *Cancer Epidemiology, Biomarkers & Prevention*, 5, 1999, p. 600.
41 William Lijinsky, 'N-Nitroso compounds in the diet', *Mutation Research*, 443, 1–2, 1999.
42 Nadia Bastide et al., 'A central role for heme iron in colon carcinogenesis associated with red meat intake', *Cancer Research*, 75, 5, March 2015; Diane de La Pomélie et al., 'Mechanisms and kinetics of heme iron nitrosylation in an *in vitro* gastro-intestinal model', *Food Chemistry*, 239, January 2018.
43 William Grady, 'Molecular biology of colon cancer', in Leonard Saltz (ed.), *Colorectal Cancer: Evidence-Based Chemotherapy Strategies*, Humana Press, Totowa, 2007, p. 1.

44 Denis Corpet, letter addressed to the EU commissioner for health and food safety and the UK Secretary of State for Health and Social Care, 7 February 2019.

45 Ibid.

46 Denis Corpet, 'Red meat and colon cancer: Should we become vegetarians, or can we make meat safer?', art. cit.; Raphaelle Santarelli et al., 'Meat-processing and colon carcinogenesis: cooked, nitrite-treated, and oxidized high-heme cured meat promotes mucin-depleted foci in rats', *Cancer Prevention Research*, no. 3, 2010.

47 Denis Corpet, letter addressed to the EU commissioner for health and food safety and the UK Secretary of State for Health and Social Care (letter cited).

48 Interview with the CEO of Herta France in a film directed by Sandrine Rigaud, co-authored by Guillaume Coudray, *Cash Investigation*: 'Industrie agroalimentaire: business contre santé' [Agro-food industry: business versus health], France 2, September 2016 (available on www.youtube.com).

49 Friedrich-Karl Lücke, 'Nitrit und die Haltbarkeit und Sicherheit erhitzter Fleischerzeugnisse' [Nitrite and the shelf life and safety of heated meat products], *Mitteilungsblatt der Fleischforschung Kulmbach* [Kulmbach Bulletin of Meat Research], 47, 181, 2008.

50 See, for example, EFSA, 'Opinion of the scientific panel on biological hazards on a request from the Commission related to the effects of nitrites/nitrates on the microbiological safety of meat products', *EFSA Journal*, 14, 2003 (p. 24, 'Germany' and 'Italy'); Food Chain Evaluation Consortium, *Study on the Monitoring of the Implementation of Directive 2006/52/EC as Regards the Use of Nitrites by Industry in Different Categories of Meat Products*, European Commission/Civic Consulting, Brussels, 2016, p. 6.

51 Daniel Demeyer et al., 'The World Cancer Research Fund report 2007: a challenge for the meat-processing industry', *Meat Science*, 80, 2008, p. 956.

52 Didier Majou and Souad Christieans, 'Mechanisms of the bactericidal effects of nitrate and nitrite in cured meats', *Meat Science*, 145, 2018.

53 Jambon Biocoop (Ensemble label), Bio direct/SBV viande, Le Petit Brevelay.

54 'Les nitrites au cœur des charcuteries' [Nitrites at the heart of processed meats], *Infos Viandes et Charcuteries* [Meat and Processed Meat Info], November 2013, Ifip-Institut du porc, p. 1.

55 See, for example, 'Rillettes du Mans, de Tours et rillons' [Rillettes of Le Mans and Tours and rillons], in Louis-François Dronne, *Charcuterie ancienne et moderne. Traité historique et pratique* [Ancient and Modern Charcuterie: Historical and Practical Guide], E. Lacroix, Paris, 1869, pp. 177–8 or 'Rillons' and 'Rillettes', in Marc Berthoud, *Charcuterie pratique* [Practical Charcuterie], Hetzel, Paris, 1884, pp. 201–02.

56 Centre technique de la salaison, de la charcuterie et des conserves de viandes [Technical Centre for Cured, Processed and Preserved Meats],

Code des usages en charcuterie et conserves de viandes [Code of Practice for Charcuterie and Preserved Meats], Paris, 1969, section V, p. 29.

57 Jean-Claude Frentz and Michel Poulain, *Livre du compagnon charcutier-traiteur* [Pork Butcher and Delicatessen Companion Book], Éditions LT Jacques Lanore/Jérôme Vilette, Les Lilas, 2001, p. 260 (our emphasis).

58 Answer E-3840/2010 (23 June 2010); Commission Regulation No. 1129/2011 of 11 November 2011, *Official Journal of the European Union*, L 295/1.

59 Fabrice Pierre et al., 'Calcium and α-tocopherol suppress cured meat promotion of chemically-induced colon carcinogenesis in rats and reduce associated biomarkers in human volunteers', *American Journal of Clinical Nutrition*, 98, 5, November 2013.

60 See, for example, the 10th reunion of the Expert Panel on Nitrites and Nitrosamines (March 1977), in Expert Panel on Nitrites and Nitrosamines, *Final Report on Nitrites and Nitrosamines. Report to the Secretary of Agriculture*, USDA, Washington, February 1978, pp. 78–9; also William Mergens et al., 'Stability of tocopherol in bacon', *Food Technology*, November 1978, pp. 40–4.

61 J. Ian Gray et al., 'Inhibition of N-nitrosamines in bacon', *Food Technology*, 6, 6, 1982; Walter Wilkens et al., 'Alpha-tocopherol', *Meat Processing*, September 1982; 'Bacon breakthrough', *Meat Industry*, September 1982.

62 'Nitrites and nitrosamines advisory panel challenged', *Congressional Record – Senate*, 19 December 1975, p. 42089.

63 Ibid.

64 See, for example, Expert Panel on Nitrites and Nitrosamines, *Final Report on Nitrites and Nitrosamines*, op. cit., p. 14.

65 Cancer Research UK, 'Bowel cancer statistics' and 'Bowel cancer survival statistics', Cancer Research UK (data for England and Wales, 2010–2011).

66 See Raymond Oliphant et al., 'The changing association between socioeconomic circumstances and the incidence of colorectal cancer: a population-based study', *British Journal of Cancer*, 104, May 2011; Emily Tweed et al., 'Socio-economic inequalities in the incidence of four common cancers: a population-based registry study', *Public Health*, 154, January 2018.

67 Denis Burkitt, 'Epidemiology of cancer of the colon and rectum', *Cancer*, 28, 1, July 1971, pp. 5–6.

68 'Modern diet may play role in cancer of the bowel', *JAMA*, 215, 5, 1 February 1971, p. 717.

69 Ibid.

70 Robert Proctor, *Cancer Wars. How Politics Shapes What We Know and Don't Know About Cancer*, Basic Books, New York, 1995, p. 261.

Chapter 1: Miracle Additives

1 Nadia Bastide et al., 'A central role for heme iron in colon carcinogenesis associated with red meat intake', art. cit.; Nadia Bastide, Fabrice Pierre,

Denis Corpet, 'Heme iron from meat and risk of colorectal cancer: a meta-analysis and a review of the mechanisms involved', *Cancer Prevention Research*, 4, 2, 2011.

2 See, for example, Bernard Moinier and Olivier Weller, *Le Sel dans l'Antiquité* [Salt in Antiquity], Les Belles Lettres, Paris, 2015, pp. 167–70.

3 Strabo, IV, 3, 2, quoted in Marie-Yvane Daire, *Le Sel des Gaulois* [Salt of the Gauls], Errance, Paris, 2003, p. 116.

4 Jun-ichi Wakamatsu et al., 'Direct demonstration of the presence of zinc in the acetone-extractable red pigment from Parma ham', *Meat Science*, 76, 2007, pp. 385–7.

5 Christina Adamsen et al., 'Changes in Zn-porphyrin and proteinous pigments in Italian dry-cured ham during processing and maturation', *Meat Science*, 74, 2006, p. 379.

6 See, for example, Confédération de la charcuterie de France et de l'Union Française, *Manuel de l'apprenti charcutier* [Manual for the Apprentice Charcutier], Paris, 1962, p. 157.

7 'Étiquetage d'un jambon de Bayonne' [Labelling of a Bayonne Ham], Letter from the Fraud Prevention Office, 22 April 1969, French National Archives, 830510/9-3Cons9.

8 René Pallu, *La Charcuterie en France, Tome III: techniques nouvelles ou inédites, salaison au sel nitrité* [French Charcuterie, Volume 3: New or Unpublished Techniques, Salting with Nitrited Curing Salt], Éditions Pallu, Paris, 1965, pp. 82–3.

9 Ibid., our emphasis.

10 Ibid.

11 Ibid., p. 145.

12 Christina Adamsen et al., 'Zn-porphyrin formation in cured meat products: effect of added salt and nitrite', *Meat Science*, 72, 2006, p. 676; Ralph Hoagland, 'The action of saltpeter upon the color of meat', *Report of the Bureau of Animal Industry* (year 1908), USDA, Washington, 1910, p. 301.

13 See Pallu, *La Charcuterie en France, Tome III*, op. cit., p. 193.

14 Adamsen et al., 'Zn-porphyrin formation in cured meat products: effect of added salt and nitrite', art. cit., p. 674.

15 Alberto Martin et al., 'Characterisation of microbial deep spoilage in Iberian dry-cured ham', *Meat Science*, 78, 2008.

16 M. Henry et al., 'Opportunité de l'addition de nitrates alcalins ou de nitrites alcalins ou des deux sels simultanément aux salaisons de viandes', [Opportunity of addition of alkali metal nitrates or alkali metal nitrites or both salts simultaneously in the curing of meats], *Revue de pathologie générale et comparée*, 655, 1954, pp. 1280, 1287.

17 'Nitrite safety reviewed by microbiologist', *Processed Food*, December 1978.

18 See, for example, in June 2014, EU Regulation 601/2014, which authorized the use of nitrite in a further dozen traditional foodstuffs.

[19] Karl-Otto Honikel, 'The use and control of nitrate and nitrite for the processing of meat products', *Meat Science*, 78, 2008. Similarly, see 'The enigma of hams cured without nitrite', in Harold McGee, *On Food and Cooking, the Science and Lore of the Kitchen*, Scribner, New York, 2004, p. 174.

[20] Jens Moller et al., 'Color', in Fidel Toldra (ed.), *Handbook of Fermented Meat and Poultry*, 2nd edn, Wiley-Blackwell, Hoboken, 2015, p. 196.

[21] Hidetoshi Morita et al., 'Red pigment of Parma ham and bacterial influence on its formation', *Journal of Food Science*, 61, 5, 1996, pp. 1021–3.

[22] René Pallu, *La Charcuterie en France, Tome I: généralités, charcuterie crue* [French Charcuterie, Volume 1: Generalities, Raw Charcuterie], Éditions Pallu, Paris, 1971 (2nd edn), p. 336.

[23] As is the case with certain traditional Chinese hams, where the salt is contaminated by nitrate. See Ryoichi Sakata, 'Prospects for new technology of meat-processing in Japan', *Meat Science*, 86, 2010, pp. 244–5.

[24] Moller et al., 'Color', art. cit., p. 195.

[25] Christina Adamsen et al., 'Thermal and photochemical degradation of myoglobin pigments in relation to colour stability of sliced dry-cured Parma ham and sliced dry-cured ham produced with nitrite salt', *European Food Research and Technology*, 218, 2004, p. 405.

[26] See Daniel Demeyer et al., 'Fermentation', in Michael Dikeman et Carrick Devine (eds), *Encyclopedia of Meat Science*, vol. 2, 2nd edn, Academic Press, London, 2014, p. 7.

[27] 'How to give a bright, red color to Bologna and Frankfort sausage without artificial coloring', *Secrets of Meat Curing and Sausage Making*, Heller, Chicago, 2nd edn, 1911, p. 260.

[28] Fred Wilder, *The Modern Packing House*, Nickerson & Collins, Chicago, 1905, p. 311.

[29] Robert Hinman and Robert Harris, *The Story of Meat*, Swift and Co., Chicago, 2nd edn, 1942, p. 129.

[30] Robert Ganz (ed.), *Directory and Hand-Book of the Meat and Provision Trades and their Allied Industries for the United States and Canada*, National Provisioner Pub. Co., Chicago, 1895, p. 354.

[31] *Prague Powder*®, brochure, Griffith laboratories, Chicago, 1951, p. 15.

[32] See W. Berry, *Coloring Matters for Foodstuff and Methods for Their Detection*, USDA, Washington, 1906, p. 35.

[33] Ronald Pegg and Fereidoon Shahidi, *Nitrite Curing of Meat, the N-nitrosamine Problem and Nitrite Alternatives*, Food and Nutrition Press, Trumbull, 2000, p. 23.

[34] Food Manufacturers Federation, Inc., 'Memorandum on the curing of meats with salt, saltpetre and/or nitrites', March 1939.

[35] British Food Manufacturers' Research Association, 'The Curing of Meat', *Food Research Reports*, no. 6, January 1929, p. 1.

[36] Ibid., p. 14.

37 Gordon Wickham Monier-Williams, 'The use of nitrates and nitrites in the curing of meat', October 1929, p. 3.

38 Clifford Hollenbeck, 'Curing of meat' (assigned to Merck & Co.), US Patent 2739899, March 1956.

39 Charles Breizy, 'Chemical compound and method of producing the same', US Patent 1976831, October 1934.

40 *Prague Powder*®, brochure, op. cit., p. 5.

41 *Curing Pork Country Style*, leaflet 273, USDA, Washington, 1953.

42 Advertising insert 'Nitral' (by the company La Bovida), in Pallu, *La Charcuterie en France, Tome III*, op. cit., facing p. 93.

43 Advert 'Les fameuses spécialités Bovida' [The famous Bovida specialities], in Laszlo Hennel, *Le Charcutier et la Loi* [The Pork Butcher and the Law], Éditions La Bovida, Paris, 1947, p. 171.

44 Hennel, *Le Charcutier et la Loi*, op. cit., p. 59.

45 Ibid.

46 *Tarif Berty*, 1969, Paris, pp. 21-3.

47 Advertising insert 'Colorants Klotz S.A.' [Klotz S.A. colouring agents], in Pallu, *La Charcuterie en France, Tome III*, op. cit., facing p. 192.

48 Marie-Jeanne Diot, *Premier inventaire analytique des additifs utilisés en charcuterie et salaison* [First Analytical Inventory of Additives Used in Meat-processing and Curing], Pharmacy dissertation, Clermont-Ferrand, 1978.

49 'Herta® met-il du nitrite pour donner une couleur rose à son jambon?' [Does Herta® add nitrite to give its ham a pink colour?', available at www.herta.fr, subheading 'Herta et les nitrites' [Herta and nitrites] (accessed November 2019).

50 See, for example, Joseph Sebranek and James Bacus, 'Cured meat products without direct addition of nitrate or nitrite: what are the issues?', *Meat Science*, 77, 2007, p. 145.

51 Winford Lewis, 'Producing stable color in meats', US Patent 2147261 of 23 May 1936.

52 Ibid.

53 Jeffrey Sindelar and Terry Houser, 'Alternative curing systems', in Rodrigo Tarté (ed.), *Ingredients in Meat Products: Properties, Functionality and Applications*, Springer Science + Business, 2009, p. 380.

54 Hugh Gardner, 'Sowbelly blues, the links between bacon and cancer', *Esquire*, November 1976, p. 144.

55 Robert Bogda, 'Most pork-belly futures prices increase, indicating ban on nitrite isn't seen soon', *Wall Street Journal*, 15 August 1978, p. 38.

56 Henry et al., 'Opportunité de l'addition de nitrates alcalins ou de nitrites alcalins ou des deux sels simultanément aux salaisons de viandes', art. cit., p. 1280.

Chapter 2: The Truth Behind the Golden Legend

1 Pallu, *La Charcuterie en France, Tome III*, op. cit., p. 113. For the discussion on authentic Bayonne ham, see p. 82.

2   Quoted by Françoise Desportes, 'Les métiers de l'alimentation' [The food trades], in Jean-Louis Flandrin and Massimo Montanari (eds), *Histoire de l'alimentation* [History of Food], Fayard, Paris, 1996, p. 442.

3   Stéphane Malandain and Inès Peyret, *Éloge du saucisson. De Confucius à Bocuse, un trésor de l'humanité* [In Praise of Dry Sausage. From Confucius to Bocuse, a Human Treasure], Éditions du Dauphin, Paris, 2014, pp. 87–8.

4   On Arles dry sausage, see Pallu, *La Charcuterie en France, Tome III*, op. cit., p. 129.

5   See, for example, the old recipe for Tuscan mortadella reproduced in Alberto Capatti and Massimo Montanari, *La Cuisine italienne, histoire d'une culture* [Italian Cooking, History of a Culture], Seuil, Paris, 2002, pp. 380–1.

6   See, for example, Konrad Jurisch, *Salpeter und sein Ersatz* [Saltpetre and its Substitute], Hirzel, Leipzig, 1908, pp. 1–4; Andrew Sandison, 'The use of natron in mummification in ancient Egypt', *Journal of Near Eastern Studies*, 22, 4, October 1963 (see in particular p. 261 for references to saltpetre); Alfred Lucas, *Ancient Egyptian Materials and Industries*, Arnold, London, 2nd edn, 1934, pp. 245–6. For a comparison of mummification/food preservation: Ambrose Abel, *The Preservation of Food*, Case & Lockwood, Hartford, 1857, pp. 6–10.

7   Mark Gladwin et al., 'Meeting report: The emerging biology of the nitrite anion', *Nature Chemical Biology*, 1, 6, 2005, p. 313.

8   Testimony of Stephen Krut (American Association of Meat Processors), Subcommittee on Agricultural Research and General Legislation, *Food Safety and Quality – Nitrites*, US Government Printing Office (US GPO), Washington, 1978, pp. 106–10.

9   Frederick Ray, 'Meat curing', *Oklahoma Cooperative Extension Fact Sheet*, n.d., p. 1. *Idem* in Christopher Kevil et al., 'Inorganic nitrite therapy, historical perspective and future directions', *Free Radical Biology & Medicine*, 51, 2011, p. 577.

10 Sonia Pittioni, *Nitrosohemoglobin synthesis*, dissertation, University of Toronto, 1998, p. 1.

11 Amanda Gipe McKeith, 'Alternative curing', *Factsheet PIG (Pork Information Gateway)*, June 2014, p. 1.

12 Jean Pantaleon, 'Chimie et technologie des viandes' [Chemistry and technology of meats], *Revue technique de l'industrie alimentaire* [Technical Review of the Food Industry], 43–7, May–October 1957 (reissued by Hoffmann-Laroche laboratories), p. 6.

13 Jeanine Louis-Sylvestre et al., 'Les charcuteries' [Processed meats], *Cahiers de diététique et de nutrition* [Notebooks on Diet and Nutrition], 45, 2010.

14 Jean-Luc Martin, 'Impact du salage et de la cuisson sur la couleur des jambons et lardons' [Impact of salting and cooking on the colour of hams and diced bacon], *TechniPorc*, 33, 5, 2010, p. 26.

15 info-nitrites.fr/nitrites-et-nitrates, accessed August 2019.

16 info-nitrites.fr/categorie/questions-reponses, accessed August 2019.

17 Randy Huffman and Nathan Bryan, 'Nitrite and nitrate in the meat industry', in Nathan Bryan (ed.), *Food, Nutrition, and the Nitric-Oxide Pathway*, Destech, Lancaster, 2010, p. 79.

18 Frank Frost, 'Sausage and meat preservation in Antiquity', *Greek, Roman and Byzantine Studies*, 40, 1999, p. 245.

19 Lucius Columella, *De re rustica* [On Rural Economy], Book XII. On the other hand, Columella explicitly cites another product (natron) in recipes for preserved vegetables, which suggests that he wouldn't have forgotten if he had used it in curing meat too.

20 See the chapters 'How to Use White Salt' and 'Salting of Hams' in Cato, *De agricultura*. For a good synthesis, see Maria Dembińska, 'Methods of Meat and Fish Preservation in the Light of Archeological and Historical Sources', in Astri Riddervold and Andreas Ropeid (eds), *Food Conservation Ethnological Studies*, Prospect Books, London, 1988, p. 16.

21 See the recipes published by the *Revue de la conserve* [Food Preservation Review], special edition 'Charcuteries et salaisons' [Charcuterie and cured meat], November 1962. On the archaeology of salting techniques, see, for example, Sabine Deschler-Erb, 'Viandes salées et fumées chez les Celtes et les Romains de l'Arc jurassien' [Salted and smoked meats of the Celts and the Romans of the Jura region], in *Actes des premières journées archéologiques frontalières de l'Arc jurassien* [Acts of the First Days of Border Archaeology in the Jura Region], Besançon, 2005.

22 Pallu, *La Charcuterie en France, Tome III*, op. cit., p. 23. Harold McGee, *On Food and Cooking, the Science and Lore of the Kitchen*, op. cit., p. 173. This hypothesis was picked up by various authors advocating the use of nitro-additives, for example: Evan Binkerd and Olaf Kolari, 'The history and use of nitrate and nitrite in the curing of meat', *Food and Cosmetics Toxicology*, 13, 1975; 'Nitrates, Nitrites, and Salt (Notice of proposed rulemaking)', *Federal Register*, Washington, 11 November 1975, p. 52614.

23 Frank Gerrard, 'Introduction', in R.E. Davies, *Pigs and Bacon Curing. A Practical Manual on the Various Methods of Feeding Pigs, Curing Hams and Bacon, and Utilising the By-Products of Swine*, 4th edn, The Technical Press, London, 1950.

24 'At first, only salt was added, but now we add sugar and sodium nitrate or nitrite to fix the colour, make it stable under heat, and spices.' Jacques Rivière, 'Contribution à l'étude de la transformation des nitrates en nitrites dans les salaisons' [Contribution to the study of the transformation of nitrates into nitrites in curing], *Annales de l'Institut National Agronomique* [Annals of the National Agronomic Institute], 51, 35, 1948.

25 Ibid., p. 8.

26 On the history of salt, see in particular the work of Robert Multhauf, *Neptune's Gift: A History of Common Salt*, Johns Hopkins University Press, Baltimore, 1978.

27 Anthony Bridbury, *England and the Salt Trade in the Later Middle Ages*, Clarendon Press, London, 1955, pp. 6–7.

28 See for example Sebranek and Bacus, 'Cured meat products without direct addition of nitrate or nitrite: what are the issues?', art. cit., p. 141.

29 Raphael Koller, *Salz, Rauch und Fleisch* [Salt, Smoke and Meat], Das Bergland-Buch, Salzburg, 1941.

30 Robert Curtis, *Ancient Food Technology*, Brill, Leiden, 2001.

31 Salima Ikram, *Choice Cuts: Meat Production in Ancient Egypt*, Peeters, Louvain, 1995.

32 British Food Manufacturers' Research Association, 'The Curing of Meat', *Food Research Reports*, art. cit., p. 13.

33 Louis Truffert and Henri Cheftel, 'À propos de l'emploi de nitrite de sodium, et de ortho-meta et pyrophosphate de sodium dans certaines conserves de viandes' [Concerning the use of sodium nitrite and sodium ortho-, meta- and pyrophosphate in some preserved meats], *Rapport au Conseil supérieur d'hygiène public de France* [Report to the French High Council for Public Hygiene], Ministère de la Santé [Ministry of Health], November 1952, p. 1.

34 A. Böhm, 'Untersuchungen an Pökelhilfsstoffen' [Research on Curing Agents], *Die Fleischwirtschaft* [Meat Business], July 1955, no. 7.

35 Manuscript reproduced in Constance Hieatt and Sharon Butler, *Curye on Inglysch. English Culinary Manuscripts of the Fourteenth Century*, Oxford University Press, Oxford, 1985, p. 73.

36 Robert Lovell (1661), William Thraster (1669) and William Salmon (1693), quoted by David Cressy, *Saltpeter: The Mother of Gunpowder*, Oxford University Press, Oxford, 2013, pp. 30–2.

37 Thomas Chaloner, *A Shorte Discourse of the Most Rare and Excellent Vertue of Nitre*, London, 1584. Commentary by Anthony Butler and Martin Feelisch, 'Therapeutic uses of nitrite and nitrate: from the past to the future', *Circulation*, 117, 2008.

38 Bee Wilson, *Swindled: The Dark History of Food Fraud*, Princeton University Press, Princeton, 2008. See pp. 15–20.

39 J. Rivière, 'Contribution à l'étude de la transformation des nitrates en nitrites dans les salaisons', art. cit., p. 9.

40 Truffert and Cheftel, 'À propos de l'emploi de nitrite de sodium ...', art. cit., pp. 1, 11.

41 Jennifer Stead, 'Necessities and luxuries: food preservation from the Elizabethan to the Georgian era', in Anne Wilson (ed.), *Waste Not, Want Not: Food Preservation from Early Times to the Present Day*, Edinburgh University Press, Edinburgh, 1991, p. 68.

42 Mark Kurlansky, *Salt: A World History*, Jonathan Cape, London, 2002, pp. 293–4.

43 Maria Dembińska (revised and adapted by William Woys Weaver), *Food and Drink in Medieval Poland: Rediscovering a Cuisine of the Past*, University of Pennsylvania Press, Philadelphia, 1999, pp. 85–6 and 205.

44 Klaus Lauer, 'The history of nitrite in human nutrition: a contribution of German cookery books', *Journal of Clinical Epidemiology*, 44, 3, 1991.

45 Robert Boyle, *Some Considerations Touching the Usefulness of Experimental Naturall Philosophy*, part II, 2nd edn, Hall, Oxford, 1664, p. 99.

46 William Clarke, *The Natural History of Nitre*, Nathaniel Brook, London, 1670, p. 92.

47 John Collins, *Salt and Fishery: A Discourse Thereof*, Godbid & Playford, London, 1682, pp. 9–12, 69–70, 121–6, 135–6.

48 See, for example, the description of 'clod salt' by Thomas Rastel (1678), quoted by M. Kurlansky, *Salt: A World History*, op. cit., pp. 294–5.

49 Collins, *Salt and Fishery*, op. cit., p. 9.

50 Ibid., p. 124.

51 Ibid., p. 125.

52 Lauer, 'The history of nitrite in human nutrition …', art. cit., p. 263.

53 William Salmon, *The Family Dictionary or Household Companion*, Rhodes, London, 1710, p. 31 (quoted by Harold McGee, *On Food and Cooking, the Science and Lore of the Kitchen*, op. cit., p. 173). William Salmon also recommends saltpetre in pickled pork (p. 378) and for tongues (pp. 334–7, 378). Moreover, he cites several alchemical formulae that include saltpetre (see, for example, his recipe for the 'universal solvent', *alkahest*, p. 184).

54 Émile Zola, *Le Ventre de Paris* [*The Belly of Paris*] (1873), quoted by Théodore Bourrier, *Le Porc et les Produits de la charcuterie, hygiène, inspection, réglementation* [Pork and Charcuterie Products: hygiene, inspection, regulation], Asselin et Houzeau, Paris, 1888, p. 294.

55 Recipe of Le Cointe (1790) in Madeleine Ferrières, *Histoires de cuisines et trésors des fourneaux* [Cooking Stories and Treasures of the Stove], Larousse, Paris, 2008.

56 Wilson, *Swindled*, op. cit., pp. 77–84.

57 *Douglas's Encyclopaedia. Book of Reference for Bacon Curers, Bacon Factory Managers, etc.*, Douglas, London, 1893, p. 48.

58 According to those in the additives industry, 'since 1792, preserving salts for pickles and minced meats were sold in France' (Letter of the Syndicat National des Fabricants de Sels Conservateurs et Produits Similaires pour Charcuterie et Salaison [National Union of Manufacturers of Preserving Salt and Similar Products for Meat-Processing and Curing] to the Ministry of Agriculture, 22 July 1969, French National Archives, 880495/P590).

59 René Pallu, *Réglementation et usages en charcuterie* [Regulation and Practices in Charcuterie], Éditions Pallu, Paris, 1960, p. 41.

60 Christian Martfeld, 'Mémoire sur les procédés employés en Irlande pour saler les viandes' [Memoir on the procedures used in Ireland to cure meats], *Annales de l'industrie nationale et étrangère* [Annals of the National and Foreign Industry], 9, 39, 1823, p. 250. See also T.C. Bruun-Neergaard, *Traité sur la salaison des viandes et du beurre en Irlande et manière de fumer le boeuf à Hambourg* [Treatise on the Curing of Meats and Butter in Ireland and Method of Smoking Beef in Hamburg], Imprimerie Royale, Paris, 1821, p. 35. And in ibid., p. 72 (on the hams of

Hamburg): 'To preserve as long as possible the colour of the meat after curing it, sprinkle it with a certain amount of nitre.'

61 *Farmer's Register*, mentioned by Wolfgang Arneth, 'Chemische Grundlagen der Umrötung' [Chemical basis of Reddening], *Fleischwirtschaft* [Meat Business], 78, 8, 1998.

62 Élisabeth Celnart, *Manuel du charcutier, ou l'art de préparer et de conserver les différentes parties du cochon, d'après les plus nouveaux procédés* [The Charcutier's Manual, or the Art of Preparing and Preserving the Different Parts of the Pig Following the Most Up-to-Date Procedures], Roret, Paris, 1827, p. 115. The same instructions would appear in later editions of the manual under the signature of M. Lebrun and W. Maigne, *Nouveau manuel complet du charcutier, du boucher et de l'équarisseur* [New Complete Manual for the Charcutier, the Butcher and the Renderer], Roret, Paris, 1869, p. 126.

63 Entry 'Salaison' [Curing], in Jean-Baptiste Glaire and Joseph-Alexis Walsch, *Encyclopédie catholique* [Catholic Encyclopedia], vol. 17, Desbarres, Paris, 1848, p. 81.

64 James Robinson, *The Whole Art of Curing, Pickling, and Smoking Meat and Fish*, Longman & Co., London, 1847, pp. 42–55; *Dictionnaire général de la cuisine française ancienne et moderne ainsi que de l'office et de la pharmacie domestique* [General Dictionary of Old and Modern French Cookery, Housekeeping and Home Pharmacy], Plon, Paris, 1853, p. 347.

65 Glaire and Walsch, *Encyclopédie catholique*, op. cit., p. 82.

66 *Dictionnaire général de la cuisine française*, op. cit., p. 241.

67 M. Piérard, 'Sur la préparation du bœuf fumé, d'après les procédés suivis à Hambourg' [On the preparation of smoked beef, according to procedures followed in Hamburg], *Annales des arts et manufactures* [Annals of Arts and Manufactures], 2, 1, 1818, p. 221.

68 See Daniel Baugh, *British Naval Administration in the Age of Walpole*, Princeton University Press, Princeton, 1965, p. 423.

69 William Moore, *Remarks on the Subject of Packing and Re-packing Beef and Pork*, Mahum Mower, Montreal, 1820, p. 6.

70 Martfeld, 'Mémoire sur les procédés employés en Irlande pour saler les viandes', art. cit., p. 249. See also Bruun-Neergaard, *Traité sur la salaison des viandes et du beurre en Irlande*, op. cit., p. 35.

71 Pierre François Keraudren, 'De la nourriture des équipages et de l'amélioration des salaisons dans la marine française' [On food for crews and the improvement of cured meats in the French navy], *Annales maritimes et coloniales* [Maritime and Colonial Annals], Part 2, Imprimerie royale, Paris, 1829, p. 366.

72 John Nott, *The Cook's and Confectioner's Dictionary*, Rivington, London, 1723; Hannah Glasse, *The Art of Cookery Made Plain and Easy*, Wangford, London, 1774 (first edn, 1747).

73 Josef von Kurzböck, *Spectacle de la nature et des arts* [The Spectacle of Nature and the Arts], vol. 5, Kurzböck, Vienna, 1777, articles 'Les saucisses' [Sausages] et 'Chaircuitier' [Charcuterie/Cooked meats].

74 Erik Viborg, 'Mémoire sur le porc' [Memoir on pork] (1805) in Erik Viborg and M. Young, *Mémoires sur l'éducation, les maladies, l'engrais et l'emploi du porc* [Memoirs on Education, Illnesses, Feeding and Use of Pork], 2nd expanded edn, Huzard, Paris, 1835, pp. 144–6.

75 Ibid., pp. 136 and 142–3.

76 *Dictionnaire général de la cuisine française ancienne et moderne*, op. cit., pp. 82–5, 141–2, 165, 356 (our emphasis).

77 Jules Breteuil, *Le Cuisinier européen* [The European Cook], Garnier, Paris, 1863, pp. 242–51, 290.

78 Bourrier, *Le Porc et les produits de la charcuterie*, op. cit., p. 372.

79 Théodore Bourrier, *Les Industries des abattoirs* [Abattoir Industries], Baillières, Paris, 1897, pp. 236–7, 241.

80 Madeleine Ferrières, *Histoire des peurs alimentaires* [History of Food Fears], Seuil, Paris, 2002, p. 413.

81 Marc Berthoud, *Charcuterie pratique* [Practical Charcuterie], Hetzel, Paris, 1884, pp. 325–49.

82 Lebrun and Maigne, *Nouveau manuel*, op. cit., p. 172.

83 Dronne, *Charcuterie ancienne et moderne*, op. cit., pp. 112, 107, 156.

84 Robert Henderson, *Treatise on the Breeding of Swine and Curing of Bacon*, A. Allardice, Leith, 1811.

85 Loudon Douglas, *Manual of the Pork Trade: A Practical Guide for Bacon Curers, Pork Butchers, Sausage and Pie Makers*, Upcott Gill, London, 1893.

86 See the recipes in William Youatt, *The Pig*, Routledge, London, 1860, pp. 107–15.

87 Alphonse Gobin, *Précis pratique de l'élevage du porc* [Practical Guide to the Raising of Pork], Audot, Paris, 1882, pp. 209, 211–12.

88 Extract from *The Grocer* (4 June 1898), reproduced in Committee on Manufactures, *Adulteration of Food Products*, US Senate, 56th Congress, 1st session, Report 516, 1900, p. 271.

89 Albert Fulton, *Home Pork Making*, Orange Judd Company, Chicago, 1906, pp. 44, 48.

90 *Iron County Register* (Iron County, Missouri), 3 February 1881 (our emphasis).

91 Borel (pseudonym of Charles-Yves Cousin d'Avallon), *Le Cuisinier moderne mis à la portée de tout le monde* [The Modern Cook Made Accessible to Everyone], Corbet, Paris, 1836, pp. 98–109.

92 Sanson procedure described by Alphonse Chevallier and Alphonse Chevallier Jr, 'Recherches chronologiques sur les moyens appliqués à la conservation des substances alimentaires' [Chronological research into means applied to the preservation of foodstuffs], *Annales d'hygiène publique et de médecine légale* [Annals of Public Hygiene and Legal Medecine], 2nd series, 9, 1858, pp. 81–2.

93 Jules Gouffé, *Le Livre des conserves, ou recettes pour préparer et conserver* [The Book of Preserves, or Recipes for Preparing and Preserving], Hachette, Paris, 1869, pp. 14, 20–21.

94 Adolphe Fosset, *Encyclopédie domestique, recueil de procédés et de recettes* [Domestic Encyclopedia: Collected Procedures and Recipes], Salmon, Paris, 1830, p. 6; Glaire and Walsch, *Encyclopédie*, op. cit., p. 81.

95 Article 'Curing', in Nicolas François de Neufchâteau, *Dictionnaire d'agriculture pratique* [Dictionary of Practical Agriculture], Aucher-Eloy, Paris, 1828, p. 562.

96 Binkerd and Kolari, 'The history and use of nitrate and nitrite in the curing of meat', art. cit., p. 656.

97 See *Douglas's Encyclopedia. A Book of Reference for Bacon*, op. cit., pp. 332–3; Binkerd and Kolari stated that the fusion of saltpetre into saltprunelle caused nitrite to appear (E. Binkerd and O. Kolari, 'The history and use of nitrate and nitrite in the curing of meat', art. cit., p. 656).

98 Nicolas Lemery, *Pharmacopée universelle, contenant toutes les compositions de pharmacie qui sont en usage dans la médecine* [Universal Pharmacopoeia, Containing All Pharmaceutical Compositions Used in Medicine], 5th edn, D'Houry, Paris, 1761, p. 36; Louis Vitet, *Pharmacopée de Lyon ou exposition méthodique des médicaments simples et composés* [Phamacopoeia of Lyon or Methodical Demonstration of Simple and Compound Medicines], Perisse, Lyon, 1778, pp. 119–20.

99 Christophle Glaser, *Traité de la chymie* [Treatise on Chemistry], Paris, 1663, p. 205.

100 Nott, *The Cook's and Confectioner's Dictionary*, op. cit., p. BE74; Glasse, *The Art of Cookery*, op. cit., p. 252.

101 Ann Cook, *Professed Cookery*, the author, London, 1760, p. 63.

102 *New System of Domestic Cookery*, by 'A Lady', Murray, London, 1807, pp. 41–2, 67–8; Charlotte Bury, *The Lady's Own Cookery Book*, Colburn, London, 1844, pp. 113–14.

103 Robinson, *The Whole Art of Curing*, op. cit., pp. 45, 49, 67.

104 J.R., *The Art and Mystery of Curing, Preserving, and Potting All Kinds of Meats, Game, and Fish*, Chapman and Hall, London, 1864.

105 Recipe for treating beef in *Canadian Grocer* (Toronto), 1899, p. 487.

106 Recipe for 'Westphalian ham' in Loudon Douglas, *Douglas's Receipt Book: A Complete Set of Recipes for Bacon Curers, Pork and Meat Purveyors, Sausage Makers, and Provision Merchants*, William Douglas & Sons, London, 1896, p. 20. A few years earlier, the same author also gave a recipe for a brine to inject into bacon and recommended saltprunelle: Douglas, *Manual of the Pork Trade*, op. cit., p. 87.

107 Jack Brooks et al., 'The function of nitrate, nitrite and bacteria in the curing of bacon and hams', His Majesty's Stationery Office, London, 1940.

Chapter 3: The Triumph of Meatpacking

1 'Pork on the farm. Killing, curing and canning', *Farmer's Bulletin*, 1186, USDA, 1949, p. 21.

2 Ganz (ed.), *Directory and Hand-Book of the Meat and Provision Trades*, op. cit., p. 352.

3   James Duff, *The Manufacture of Sausage*, The National Provisioner, Chicago, 1899, p. 26.

4   Advert for *Preservaline* in ibid., p. II.

5   Adolf Juckenack and Rudolf Sendtner, 'Über das Färben und die Zusammensetzung der Rohwurstwaaren des Handels mit Berücksichtigung der Färbung des Hackfleisches' [On the colouring and composition of raw sausage products on the market taking into account the colouring of minced meat], *Zeitschrift für Untersuchung der Nahrungs- und Genussmittel* [Journal of Research into Foodstuffs and Comestibles], February 1899, Book 2, p. 1 (our emphasis).

6   F. Schwartz, *Jahresbericht des Chemischen Untersuchungsamtes Hannover* [Annual Report of the Chemical Inspectorate Hanover], 1902, no. 14, quoted in *Adulteration of Food, Report. Returns and Statistics of the Inland Revenue of the Dominion of Canada*, Dawson, 1907, Part III, p. 20.

7   Committee on Ways and Means, *Adulterated Foods Exported to the United States*, 54th Congress, Washington, 1896, pp. 130–1.

8   Charles Stevenson, *The Preservation of Fishery Products*, US Commission of Fish and Fisheries, Washington, 1899, p. 561.

9   In France, see, for example, the uses of bleach (for meat) and formaldehyde 'Lactine' (for milk), in Paul Brouardel, *Cours de médecine légale de la faculté de Paris. Les empoisonnements criminels et accidentels* [Course in Legal Medicine at Paris University. Cases of Criminal and Accidental Poisoning], Baillière, Paris, 1902, p. 286.

10 *Adulteration of Food, Report*, op. cit., p. 18.

11 J. Fränkel, 'Untersuchung von Farbstoffen, welche zum Färben von Wurst, Fleisch und Konserven dienen' [Investigation of colorants used to colour sausages, meat and preserved food], review in *Zeitschrift für Untersuchung der Nahrungs- und Genussmittel*, October 1902.

12 Adolf Günther, 'Chemische Untersuchung eines neuen im Handel befindlichen Dauerwurstsalzes Borolin und eines Dauerwurstgewürzes' [Chemical investigation of a new preservative sausage salt Borolin and a preservative sausage seasoning on the market], review in *Zeitschrift für Untersuchung der Nahrungs- und Genussmittel*, September 1903, p. 802.

13 Louis Kickton, 'Über die Wirkung einiger sogenannter Konservierungsmittel auf Hackfleisch' [On the effects of certain so-called preservatives on minced meat], *Zeitschrift für Untersuchung der Nahrungs- und Genussmittel*, May 1907, p. 534.

14 Ibid., p. 541.

15 Bernhard Fischer, 'Falsifications observées dans les différents pays' [Falsifications observed in different countries], *Revue internationale des falsifications* [International Review of Falsifications], 10th year, July–August 1897, p. 109.

16 Ibid.

17 Bernhard Fischer et al., 'Färbung der Wurst' [Colouring of sausage], *Jahresbericht des chemischen Untersuchungsamtes Breslau* [Annual Report of the Chemical Inspectorate Breslau], 1899/1900, p. 14.

18 For example Eduard Polenske, 'Chemische Untersuchung verschiedener, im Händel vorkommender Konservirungsmittel für Fleisch und Fleischwaaren' [Chemical investigation of various preservatives available on the market for meat and meat products], *Arbeiten aus dem Kaiserlichen Gesundheitsamte* [Papers of the Imperial Health Ministry], 1889, devoted to the chemical analysis of eleven new preservatives and preservative-colorants for meat (powders and liquids) with different formulae.

19 M. Haefelin, 'Recherche de l'acide borique dans la viande et les saucissons' [A search for boric acid in meat and dry sausage], *Revue internationale des falsifications*, 10th year, Book 4, 1897, p. 166.

20 From a vast bibliography, one might cite: Eduard Polenske, 'Über den Borsäuregehalt von frischen und geräucherten Schweineschinken nach längerer Aufbewahrung in Boraxpulver oder pulverisirter Borsäure' [On the boric acid content of fresh and smoked pork ham after prolonged storage in borax powder or powdered boric acid], *Arbeiten aus dem Kaiserlichen Gesundheitsamte*, 1902; Rudolf Abel, 'Zum Kampfe gegen die Konservierung von Nahrungsmitteln durch Antiseptica' [On the fight against preservation of foodstuffs using antiseptics], *Hygienische Rundschau* [Hygienic Review], 1901.

21 'Preservatives in meat foods', *The Lancet*, 11 July 1908, p. 101.

22 P.F. Keraudren, 'De la nourriture des équipages et de l'amélioration des salaisons dans la marine française', art. cit., p. 371. In the eighteenth century, France had tried to build up salt beef production following the Irish model. After a successful early experiment in Nantes, the project failed to take off. See Léon Vignols, 'L'importation en France au XVIIIème siècle du boeuf salé d'Irlande, ses emplois, les tentatives pour s'en passer' [The import into France in the eighteenth century of salt beef from Ireland, its uses, the attempts to live without it], *Revue Historique* [Historical Review], vol. 159, fasc. 1, 1928.

23 Through the Cattle Acts of 1663 and 1667. See Daniel Baugh, *British Naval Administration in the Age of Walpole*, op. cit., p. 416.

24 See Bertie Mandelblatt, 'A transatlantic commodity: Irish salt beef in the French Atlantic world', *History Workshop Journal*, 2007, 10, p. 20; Vignols, 'L'importation en France au XVIIIème siècle du boeuf salé d'Irlande, ses emplois, les tentatives pour s'en passer', art. cit., p. 94; Baugh, *British Naval Administration in the Age of Walpole*, op. cit., p. 416.

25 Admiralty Records (1740) quoted in Baugh, *British Naval Administration in the Age of Walpole*, op. cit., p. 427.

26 See Máirtín Mac Con Iomaire, 'The pig in Irish cuisine past and present', in Harlan Walker (ed.), *The Fat of the Land: Proceedings of the Oxford Symposium on Food and Cooking*, Footwork, Bristol, 2003; Ruth Guiry, *Pigtown: A History of Limerick's Bacon Industry*, Limerick City and County Council, Limerick, 2016, pp. 15–16.

27 This is how they are described in the French version of the first patent for nitrited curing salt (patent Doran 'Procédé pour la conservation des

viandes' [Procedure for the preservation of meats], patent no. 486077 of 1917).

28 Arthur Cushman, 'The packing plant and its equipment', in Institute of American Meat Packers, *The Packing Industry, a Series of Lectures*, University of Chicago/Institute of American Meat Packers, Chicago, 1924, pp. 100, 103; Hinman and Harris, *The Story of Meat*, op. cit., pp. 15–16.

29 Margaret Walsh, *The Rise of the Midwestern Meat Packing Industry*, University Press of Kentucky, Lexington, 1982, pp. 16–17, 26.

30 'Pork packing in Cincinnati', *Harpers's Weekly*, 1868.

31 'Police boucherie et charcuterie' [Police report: butcher and pork-butcher], Mauriac, 1826 (Archives départementales du Cantal, series 105M1).

32 L.-F. Grognier, 'Du régime des porcs, à Maurs, département du Cantal' [Feeding pork in Maurs, department of Cantal], in Viborg and Young, *Mémoires sur l'éducation, les maladies, l'engrais et l'emploi du porc*, op. cit., pp. 258–9.

33 Jérôme Martin, 'Ethnographie du phénomène "salaison" autour de Saint-Symphorien sur Coise' [Ethnography of the 'curing' phenomenon around Saint-Symphorien sur Coise], *L'Araire* [The Plough], 127, 2001.

34 Mac Con Iomaire, 'The pig in Irish cuisine past and present', art. cit.

35 R.E. Davies, *Pigs and Bacon Curing. A Practical Manual on the Various Methods of Feeding Pigs, Curing Hams and Bacon, and Utilising the By-Products of Swine*, 1923. Quoted from the 4th edn (The Technical Press, London, 1950), p. 51. Similarly, see Bruce Walker, 'Meat preservation in Scotland', *Journal of the Royal Society for the Promotion of Health*, 1, 1981, p. 22.

36 Paul Geib, 'Everything but the squeal: the Milwaukee stockyards and meat-packing industry', *Wisconsin Magazine of History*, 78, 1, 1994, p. 6.

37 Hinman and Harris, *The Story of Meat*, op. cit., p. 42.

38 Milwaukee Chamber of Commerce Annual Reports 1860 and 1862, quoted by Geib, 'Everything but the squeal', art. cit., p. 8.

39 Roger Horowitz, *Putting Meat on the American Table: Taste, Technology, Transformation*, Johns Hopkins University Press, Baltimore, 2006, p. 50.

40 Ibid., p. 48.

41 'The making of ice', *The National Provisioner*, 5 July 1902, p. 14.

42 Cushman, 'The packing plant and its equipment', art. cit.

43 Ibid., p. 114.

44 Jonathan Ogden Armour, 'The packers and the people', *The National Provisioner*, 3 March 1906, p. 37.

45 Geib, 'Everything but the squeal', art. cit., pp. 3–22.

46 Wilder, *The Modern Packing House*, op. cit., p. 248.

47 Walsh, *The Rise of the Midwestern Meat Packing Industry*, op. cit., p. 85.

48 Milwaukee Chamber of Commerce, quoted by Geib, 'Everything but the squeal', art. cit., p. 14.

49 E. Binkerd and O. Kolari, 'The history and use of nitrate and nitrite in the curing of meat', art. cit., p. 656. Previously, the technique involving slicing and use of a studded brush was practised in Ireland (using salt)

and is described in Bruun-Neergaard, *Traité sur la salaison des viandes et du beurre en Irlande*, op. cit., p. 33–34.

50   Bourrier, *Le Porc et les produits de la charcuterie*, op. cit., p. 379.

51   Viborg, 'Mémoire sur le porc', art. cit., pp. 138–9.

52   Richard Perren, *Taste, Trade and Technology: The Development of the International Meat Industry since 1840*, Ashgate, Aldershot, 2006, p. 39.

53   Binkerd and Kolari, 'The history and use of nitrate and nitrite in the curing of meat', art. cit., p. 656. See also M.W. Maigne, *Nouveau Manuel Complet de l'Alimentation, Seconde Partie: Conserves alimentaires* [New Complete Manual of Food, Second Part: Preserved Food], Roret, Paris, 1892, p. 92: 'Antiseptiques par injection' [Antiseptics by injection].

54   See Jean-Nicolas Gannal, *Histoire des embaumements et de la préparation des pièces d'anatomie normale, d'anatomie pathologique et d'histoire naturelle, suivie de procédés nouveaux*, Ferra, Paris, 1838. The book appeared in several translations, including: J.-N. Gannal, *History of Embalming and of Preparations in Anatomy, Pathology and Natural History, Including an Account of a New Process for Embalming*, Judah Dobson, Philadelphia, 1840.

55   Chevallier and Chevallier Jr, 'Recherches chronologiques sur les moyens appliqués à la conservation des substances alimentaires', art. cit., p. 92.

56   Ambrose Abel and Elizur Goodrich Smith, *The Preservation of Food*, Case & Lockwood, Hartford, 1857, pp. 38–40. See also Maigne, *Nouveau Manuel Complet de l'Alimentation*, op. cit., pp. 105–8.

57   John Alberger, 'Processes for preserving flesh', Patent No. 194569, 1877.

58   Douglas, *Manual of the Pork Trade*, op. cit., pp. 82–4.

59   C.L. Sterling, 'Device for salting meat', Letter patent 341357, 1886. Several different versions of these 'injector sticks' were proposed, for injecting either a salt/saltpetre mixture or other compounds, especially salt/saltpetre/boric acid. (W. Flinn, 'Device for salting meat', Patent 609799, 1898.)

60   'Pork packing in Cincinnati', art. cit.

61   M. Walsh, *The Rise of the Midwestern Meat Packing Industry*, op. cit., p. 85.

62   Mérice, *Journal d'agriculture pratique* [Journal of Practical Agriculture], 1880, quoted by Bourrier, *Le Porc et les produits de la charcuterie*, op. cit., p. 386.

63   James Sinclair, *Report on the Hog-Raising and Pork-Packing Industry in the United States*, Department of Agriculture, Victoria, 1895, p. 16.

64   James Young, *Pure Food: Securing the Federal Food and Drugs Act of 1906*, Princeton University Press, 1989, p. 130; and Jimmy Skaggs, *Prime Cut: Livestock Raising and Meatpacking in the United States, 1607–1983*, Texas A & M University Press, College Station, 1986, p. 109.

65   Daniel Boorstin, *Histoire des Américains* [The Americans], Robert Laffont, Paris, 1991, p. 1208.

66   Sinclair, *Report on the Hog-Raising and Pork-Packing Industry in the United States*, op. cit., p. 14.

[67] Hoagland, 'The action of saltpeter upon the color of meat', art. cit., p. 301.

[68] Charles McBryde, *A Bacteriological Study of Ham Souring*, USDA, Washington, 1911, pp. 33–45.

[69] Ibid., p. 33.

[70] Jesse White, 'Cured and smoked meats', *The Quartermaster Corps Subsistence School Bulletin*, 29, series X, 1926, p. 25.

[71] William Richardson, 'The occurrence of nitrates in vegetable foods, cured meats and elsewhere', *Journal of the American Chemical Society*, 29, 1907. Quoted by J. Ian Gray and A.M. Pearson, 'Cured meat flavor', *Advances in Food Research*, 29, 1984, pp. 13–14.

[72] Michael French and Jim Phillips, *Cheated Not Poisoned?*, Manchester University Press, 2000, p. 86, referring to *The Lancet*, 14 January 1905, pp. 120–3.

Chapter 4: Rapid Curing Takes Over the World

[1] For a good introduction to the emergence of the global economy in meat products, see, for example, Peter Koolmees, 'From stable to table: The development of the meat industry in the Netherlands, 1850–1990', in Yves Segers, Jean Bieleman and Erik Buyst (eds), *Exploring the Food Chain: Food Production and Food Processing in Western Europe, 1850–1990*, Brepols, Turnhout, 2009.

[2] Horowitz, *Putting Meat on the American Table*, op. cit., p. 49.

[3] George Holmes, *Meat Supply and Surplus with Consideration of Consumption and Exports*, USDA, 1907, pp. 52, 42.

[4] Máirtín Mac Con Iomaire and Pádraic Óg Gallagher, 'Irish corned beef: a culinary history', *Journal of Culinary Science & Technology*, March 2011, p. 11.

[5] Jean Heffer, *Le Port de New York et le Commerce extérieur américain, 1860–1900* [The Port of New York and American Foreign Trade, 1860–1900], Publications de la Sorbonne, Paris, 1991, p. 39; Division of Foreign Markets, *Meat in Foreign Markets*, USDA, Washington, 1905.

[6] Division of Foreign Markets, *Meat in Foreign Markets*, op. cit., p. 23.

[7] Joseph Anderson, *Capitalist Pigs: Pigs, Pork, and Power in America*, West Virginia University Press, Morgantown, 2019, p. 121.

[8] Heffer, *Le Port de New York et le Commerce extérieur américain*, op. cit., pp. 65–8, 184–211.

[9] Anderson, *Capitalist Pigs: Pigs, Pork, and Power in America*, op. cit., pp. 113, 120.

[10] See Alessandro Stanziani, *Histoire de la qualité alimentaire* [History of Food Quality], Seuil, Paris, 2005, pp. 194–5.

[11] Holmes, *Meat Supply and Surplus with Consideration of Consumption and Exports*, op. cit., p. 4.

[12] Anderson, *Capitalist Pigs: Pigs, Pork, and Power in America*, op. cit., pp. 128–9.

13 Statement of Robert T. Lunham (5 June 1899), in Committee on Manufactures, *Adulteration of Food Products*, Senate, 56th Congress, 1st session, Report 516, 1900, p. 240.

14 *Illinois Farmer*, 1 October 1863, quoted by Anderson, *Capitalist Pigs: Pigs, Pork, and Power in America*, op. cit., p. 127.

15 Máirtín Mac Con Iomaire, 'The pig in Irish cuisine past and present', art. cit.

16 Abraham Denny and Edward Maynard Denny, 'Improvements in the manufacture of bacon', Letters Patent no. 2194, 1862.

17 Abraham Denny and Edward Maynard Denny, 'Improvements in the manufacture of bacon and in the apparatus connected therewith', Letters Patent no. 361, 1864.

18 Douglas, *Manual of the Pork Trade*, op. cit., p. xvi.

19 Douglas, *Douglas's Receipt Book*, op. cit., p. 86. See also Douglas, *Manual of the Pork Trade*, op. cit., pp. 54–5.

20 Douglas, *Manual of the Pork Trade*, op. cit., pp. 87–8.

21 Ibid., pp. 57–8.

22 Ibid., pp. 68, 71–2.

23 In 1893, Loudon Douglas began his *Manual of the Pork Trade* by noting that 'the consumer's tastes and ideas have undergone a complete revolution during the last twenty years' (Douglas, *Manual of the Pork Trade*, op. cit., p. XIII). The influence of American food products on the UK was not limited to meat products. See Ted Collins, 'The North American influence on food manufacturing in Britain, 1880–1939', in Segers, Bieleman and Buyst (eds), *Exploring the Food Chain*, op. cit.

24 'Les salaisons et la charcuterie anglo-danoise' [Anglo-Danish cured and processed meats], *La Conserve Alimentaire* [Preserved Food], 21, September 1910, p. 302.

25 Stanziani, *Histoire de la qualité alimentaire*, op. cit., pp. 226, 228.

26 Gobin, *Précis pratique*, op. cit., p. 214.

27 Auguste Valessert, *Traité pratique de l'élevage du porc et de charcuterie* [Practical Treatise on the Raising of Pork and Meat Curing], Garnier, 1891, p. 199.

28 Heffer, *Le Port de New York et le Commerce extérieur américain*, op. cit., p. 271.

29 Jérôme Bourdieu et al., 'Crise sanitaire et stabilisation du marché de la viande en France, XVIIIe–XXe siècles' [Health crisis and the stabilization of the meat market in France, 18th–20th centuries], *Revue d'histoire moderne et contemporaine* [Review of Modern and Contemporary History], 51, 3, 2004, p. 130; on trichinosis, see Stanziani, *Histoire de la qualité alimentaire*, op. cit.

30 Gary Libecap, 'The rise of the Chicago packers and the origins of the meat inspection and antitrust', *Economic Enquiry*, 30, 1992; Bourrier, *Le Porc et les produits de la charcuterie*, op. cit., pp. 383–90.

31 Alfred Picard, *Rapport général, Exposition universelle de 1889 à Paris* [General Report, 1889 Universal Exposition in Paris], vol. VIII, Imprimerie Nationale, Paris, 1892, p. 77.

32 See Daniel Fung et al., 'Meat safety', in Yiu Hui et al. (ed.), *Meat Science and Applications*, Marcel Dekker, New York, 2001, p. 176.

33 See Stanziani, *Histoire de la qualité alimentaire*, op. cit., p. 232.

34 'Borax and boric acid', *The National Provisioner*, 28 March 1908.

35 'Communications diverses concernant les denrées alimentaires' [Various communications concerning foodstuffs], *Revue internationale des falsifications* [International Review of Falsifications], 11, 6, November–December 1898, p. 200.

36 *La Boucherie* [The Butcher], 9, April 1895, p. 3.

37 'Give us the truth!', *The National Provisioner*, 2 June 1906, p. 21.

38 'Salaisons américaines' [American cured meats], in Bourrier, *Le Porc et les Produits de la charcuterie*, op. cit., pp. 383–93.

39 Picard, *Rapport Général*, op. cit., p. 386.

40 Holmes, *Meat Supply*, op. cit., p. 8.

41 'Introduction en fraude de lards américains en Allemagne' [Fraudulent import of American bacon into Germany], letter from the French Consulate (Danzig) to the Foreign Ministry, 5 March 1890, French National Archives, F/12/6837.

42 Ibid.

43 'Revue du marché, cours et mouvement de la navigation' [Review of the market, course and movement of shipping], 20 March 1890 ('Viandes salées d'origine Américaine' [Cured meats of American origin]), French National Archives, F/12/6837.

44 *Journal de Commerce de Belfast* [Belfast Commercial Journal], 11 March 1890 (as quoted by the French Consulate), French National Archives, F/12/6837.

45 'Revue du marché, cours et mouvement de la navigation', art. cit.

46 'American Groceries in France', *Consular Reports. Commerce, Manufactures, Etc.*, vol. XLIII, House of Representatives, 53rd Congress, 2nd session, Washington, 1893, p. 331.

47 Ibid.

48 Léon Arnou, *Manuel de l'épicier, produits alimentaires et conserves, denrées coloniales, boissons et spiritueux, etc.* [Grocer's Manual, Fresh and Preserved Food Products, Colonial Foodstuffs, Drinks and Spirits, etc.], Baillière, Paris, 1904, p. 117.

49 Ibid.

50 Douglas, *Manual of the Pork Trade*, op. cit., p. 90.

51 'Les salaisons et la charcuterie anglo-danoise', art. cit., p. 302.

52 Tenna Jensen, 'The nutritional transformation of Danish pork', in Derek Oddy and Alain Drouard, *The Food Industries of Europe in the Nineteenth and Twentieth Centuries*, Ashgate, Farnham, 2013, p. 134.

53 Jean Mauclère, *Sous le ciel pâle de Lithuanie* [Under the Pale Sky of Lithuania], Plon, Paris, 1926, chapter XXI: 'Au royaume de la viande' [In the kingdom of meat].

54 Koolmees, 'From stable to table: The development of the meat industry in the Netherlands, 1850–1990', art. cit.

55 See Curt Wagner, *Konserven und Konservenindustrie in Deutschland* [Tinned Food and the Tinned Food Industry in Germany], Gustav Fischer, Jena, 1907, p. 73.

56 Ibid., p. 80.

57 Ibid., pp. 79, 84–5, 88.

58 Ibid., p. 85.

59 Anderson, *Capitalist Pigs*, op. cit., p. 129.

60 'Suffolk hams' are not widely available today. Cured in beer and sugar, the meat is red-brown. 'Bradenham hams' used to be made in Buckinghamshire. They had a coal-black skin and dark red meat.

61 'Success in sausage making', *The National Provisioner*, 18 April 1908, p. 16.

62 Ganz (ed.), *Directory and Hand-Book of the Meat and Provision Trades*, op. cit., p. 197.

63 Ibid., p. 365.

64 'Souvenirs de Monsieur Austruy' [Memories of Mr Austruy] (1942), in Michel Rachline, *La saga Olida, un art de vivre à la française* [The Olida Saga: The Art of Living French Style], Olida & Albin Michel Communications, Paris, 1991, p. 21.

65 'Souvenirs de Monsieur Girot' [Memories of Mr Girot] (1946), in Rachline, *La saga Olida*, op. cit, p. 12.

66 Division of Foreign Markets, *Meat in Foreign Markets*, op. cit., pp. 22, 24, 35.

67 Federal Trade Commission, *Report on the Meat-Packing Industry*, Part V, *Profits of the Packers*, Washington, 1920, p. 38.

68 J. La Cérière, 'Recettes américaines pour la préparation du porc' [American recipes for the preparation of pork], *L'Alimentation moderne et les industries annexes* [Modern Food Production and Allied Trades], April 1924, p. 41–4.

69 William Richardson, 'Lecture VII – Science in the packing industry', in Institute of American Meat Packers, *The Packing Industry*, op. cit., p. 274.

70 Oscar Mayer, 'Lecture V – Pork operations', in Institute of American Meat Packers, *The Packing Industry*, op. cit., p. 193.

71 Federal Trade Commission, *Report on the Meat-Packing Industry*, op. cit., p. 13.

72 Denis-Placide Bouriat, *Bulletin de la Société d'encouragement* [Bulletin of the Society of Encouragement], 1854, quoted by Chevallier and Chevallier Jr, 'Recherches chronologiques', art. cit.

73 R. Horowitz, *Putting Meat on the American Table*, op. cit., p. 69.

74 Mayer, 'Lecture V – Pork operations', art. cit., p. 198.

75 Georges Chaudieu, *De la gigue d'ours au hamburger, ou la curieuse histoire de la viande* [From the Haunch of Bear to the Hamburger, or the Curious History of Meat], La Corpo, Chennevières, 1980, p. 94.

Chapter 5: A Chemical Success Story

1 See Charlie House, 'Pleasant Street. House on the street', *The Milwaukee Journal*, 29 November 1965.

2 Cushman, 'The packing plant and its equipment', art. cit., p. 114.

3 'Nebraska City's New Industry', *The Omaha Daily Bee* (Omaha, Nebraska), 17 July 1886, p. 1.

4 Marcia Poole, *The Yards, a Way of Life: A Story of the Sioux City Stockyards*, Lewis and Clark Interpretative Center, Sioux City, 2006.

5 Ganz (ed.), *Directory and Hand-Book of the Meat and Provision Trades*, op. cit. p. 360.

6 Manuscript note from Albert Heller, 1 March 1922, reproduced in Jay Pridmore, *Well Seasoned: A Centennial of Heller*, Heller Inc., Bedford Park, Ill., 1993.

7 Label *Rosaline Konservirungs Salz* [Rosaline Preserving Salt] (n.d.), Special Collections, Michigan State University Libraries.

8 Committee on Manufactures, *Adulteration of Food Products*, op. cit., pp. 114, 180.

9 *Past Times: Newsletter of Antique Advertising Association of America*, 10, 4, 2001.

10 Letter extracted in Pridmore, *Well Seasoned: A Centennial of Heller*, op. cit., p. 17.

11 Ibid.

12 Label *Rosaline Konservirungs Salz*, document cited.

13 Ibid.

14 Advertisement for 'Rosaline Berliner Konservirungs-Salze', in *The Scientific Meat Industry*, B. Heller & Co., June 1904, vol. 1, no. 1.

15 Advertisement for 'Rosaline Berliner Konservirungs-Salze', in *Secrets of Meat Curing*, B. Heller & Co., Chicago, 1904, p. 86.

16 See Pridmore, *Well Seasoned: A Centennial of Heller*, op. cit., p. 39.

17 See the label *Red Konservirungs Salz* reproduced in ibid., p. 6.

18 James Duff, *The Manufacture of Sausage*, op. cit, p. xxiv.

19 Label partly reproduced in Pridmore, *Well Seasoned: A Centennial of Heller*, op. cit., p. 26.

20 *The Scientific Meat Industry*, Heller, Chicago, June 1904, vol. 1, no. 1.

21 B. Heller & Co., advertisement, 'Freezine, the only scientific invention for keeping milk and cream sweet without ice' (n.d.), Special Collections, Michigan State University Libraries.

22 *Freezine* (n.d.), B. Heller & Co. Collection, Special Collections Research Center, University of Chicago Library.

23 Ibid.

24 *The Healthfulness of Freezine*, B. Heller & Co, Chicago, 1899, p. 3.

25 Massachusetts State Board (1899), quoted in *Adulteration of Food, Report*, op. cit., p. 21.

26 *Hoard's Dairyman*, 14 April 1899, quoted in 'Preservatives in Dairy Products', *Farmer's Voice*, 1899.

27 On the impact of formaldehyde, see Deborah Blum, *The Poison Squad: One Chemist's Single-Minded Crusade for Food Safety at the Turn of the Twentieth Century*, Penguin Press, New York, 2018 (chapter 3).

28 *The Healthfulness of Freezine*, op. cit.

29 Ibid., pp. 9–10.

30 Ibid.

31 *Adulteration of Food and Drugs*, House of Representatives, 52nd Congress, report 914, 1892, p. 2.

32 Ibid.

33 *Adulterated Articles of Food, Drink, and Drugs*, House of Representatives, 50th Congress, report 3341, 1888, p. 2.

34 H.C. Adams, *Dairy and Food Commissioner of Wisconsin Biennial Report (1901–02)*, quoted by Wisconsin Dairy and Food Commission, *Wisconsin Dairy and Food and Weights and Measures Department*, no. 14, Madison, 1914, p. 9.

35 Advertisement for 'Preservaline', reproduced in *The Chemical Composition of Food Preservatives*, Bulletin 67, Experiment station, Iowa State College, 1902, p. 263.

36 Extract from *The Chemical Composition of Food Preservatives*, Bulletin 67, Experiment station, Iowa State College, 1902, reproduced in *Farm Field and Fireside*, 26, 22, May 1903, p. 684.

37 'Chemical Food Preservatives', *Wallaces' Farmer* (Des Moines, Iowa), 27 February 1903, p. 309.

38 Ibid.

39 Quoted by Edward Keuchel, 'Chemicals and meat: The embalmed beef scandal of the Spanish-American War', *Bulletin of the History of Medicine*, 1974, 48, 2, p. 251.

40 See ibid. and Young, *Pure Food: Securing the Federal Food and Drugs Act of 1906*, op. cit., pp. 135–9.

41 *Food Furnished by Subsistence Department to Troops during Spanish-American War*, Record of a court of enquiry, Senate, 56th Congress, Washington, 1900. Testimony of Capt. J.G. Galbraith, pp. 978 and 981–2.

42 Ibid., testimony of Capt. William Stanton, p. 984.

43 Ibid., testimony of Capt. L.C. Allen, p. 989.

44 Ibid. p. 988.

45 Ibid., testimony of Sgt. Charles Boon.

46 Keuchel, 'Chemicals and Meat', art. cit., p. 263.

47 Duff, *The Manufacture of Sausage*, op. cit.

48 Ibid., p. 56.

49 Ibid., p. ii ('must' is underlined and in italics in the original).

50 Ibid., p. 27.

51 Ibid., p. 56.

52 Committee on Manufactures, *Adulteration of Food Products*, op. cit., p. 149

53 *The Farmer's Voice*, 13 May 1899.

54 Committee on Manufactures, *Adulteration of Food Products*, op. cit., p. 181.

55 Ibid., p. 184. On the process for obtaining formaldehyde, see p. 196.

56 Ibid., p. 172.

57 Ibid., pp. 172, 175.
58 Ibid., p. 179.
59 Ibid.
60 'Champion of Oleo', *Rock Island Argus*, 11 May 1899, p. 1.
61 'Food Adulteration', *Los Angeles Times*, 5 March 1900.
62 'Preservaline cases on trial. Experts say preparation is a poison', *Los Angeles Herald*, 2 September 1903, p. 8.
63 Ibid.; 'Freezem's Troubles', *The National Provisioner*, 3 October 1903, p. 3.
64 *Scientific Meat Industry*, B. Heller & Co., June 1904, p. 6.
65 *Los Angeles Herald*, 1 September 1905.
66 *Los Angeles Herald*, 18 October 1905.
67 *Los Angeles Herald*, 21 October 1905.
68 Reproduced in *Scientific Meat Industry*, B. Heller & Co., July 1905, p. 8.
69 *Secrets of Meat Curing*, B. Heller & Co., Chicago, 1922, p. 257.
70 *Secrets of Meat Curing*, B. Heller & Co., Chicago, 1908, p. 281.
71 *Secrets of Meat Curing*, B. Heller & Co., Chicago, 1911, p. 280.
72 *Secrets of Meat Curing*, B. Heller & Co., Chicago, 1908, p. 280.
73 *A New Way to Make Bologna and Frankfurters*, B. Heller & Co. (n.d.), Special Collections, Michigan State University Libraries.

Chapter 6: Pure Food Versus Nitro-Meat

1 Richard Fischer, 'Food adulterations', in *Proceedings of Fifth Annual Meeting of Wisconsin Buttermaker Association*, Fond du Lac, 1906, pp. 158–9.
2 Harvey Wiley, *Chemical Examination of Canned Food*, Circular 5, US Department of Agriculture, Division of Chemistry, Washington, 1899, p. 4.
3 Harvey Wiley, 'Food adulteration in its relation to the public health', paper presented at the meeting of the American Public Health Association in Minneapolis, 31 October and 1, 2, 3 November 1899 (reprinted from vol. XXV of *Transactions of the American Public Health Association*, 1900, p. 9).
4 To grasp the explosion of research on nitrate at the end of the nineteenth century, consult the twenty-page bibliography in Daniel Leech, *The Pharmacological Action and Therapeutic Uses of the Nitrites and Allied Compounds*, Sherratt & Hughes, Manchester, 1902, pp. 165–87, and especially the summary by Albert Mathews, 'The pharmacology of nitrates and nitrites', in Harry Grindley and Harold Mitchell (eds), *Studies in Nutrition: An Investigation on the Influence of Saltpeter on the Nutrition and Health of Man with Reference to its Occurrence in Cured Meats*, vol. 1, University of Illinois, Urbana-Champaign, 1918, pp. 518–19.
5 See Peter Koolmees, 'From stable to table: The development of the meat industry in the Netherlands, 1850–1990', art. cit., p. 131.
6 Karl Lehmann, *Methods of Practical Hygiene*, vol. II, Kegan Paul, Trench and Trübner, London, 1893, p. 44.

7   Charles Ainsworth Mitchell, *Flesh Foods, with Methods for their Chemical, Microscopical, and Bacteriological Examination*, Charles Griffin and Company, London, 1900, p. 108.

8   Robert von Ostertag, *Manual of Meat Inspection*, William Jenkins (New York) and Baillière, Tindall and Cox (London), 1904, p. 806.

9   United States Industrial Commission, *Adulteration of Food Products, Review and Topical Digest of the Evidence Taken Before the Senate Committee on Manufactures Between March, 1899, and February, 1900*, 56th Congress, 2nd session, Washington, 1901.

10 Testimony of Harvey Wiley to the Committee on Interstate and Foreign Commerce, *Hearings on the Pure Food Bills*, House of Representatives, Washington, 1906, p. 272.

11 Ibid.

12 Harvey Wiley, *The History of a Crime Against the Food Law*, Harvey Wiley M.D. Publisher, Washington, 1929, p. 59.

13 'Now a drink squad', *Washington Herald*, 30 October 1907, p. 2.

14 Among the many methodological flaws that have been noted, Roger Horowitz stressed that the composition of the groups of guinea pigs was unrepresentative, since they were made up entirely of young white men. (Horowitz, *Putting Meat on the American Table*, op. cit., p. 59.) Other authors have pointed out that there were no reliable control groups and that some of the guinea pigs had successively participated in experiments on different additives (see, for example, Clayton Coppin and Jack High, *The Politics of Purity*, University of Michigan Press, Ann Arbor, 1999).

15 Hearing of H. Wiley in Committee on Interstate and Foreign Commerce, *Hearings on the Pure Food Bills*, op. cit., p. 272

16 The story of the standards of food purity is covered in detail in Oscar Anderson, *The Health of a Nation: Harvey W. Wiley and the Fight for Pure Food*, University of Chicago Press, Chicago, 1958. See in particular chapter 7, mainly devoted to the elaboration of the standards of food purity by the Association of Official Agricultural Chemists (AOAC).

17 'Standards of Purity for Food Products', Circular 10 (20 November 1903) and Circular 13 (20 December 1904), United States Department of Agriculture, Washington.

18 'For Purity of Canned Goods. Discussed by Agents Before Food Commission', *The Boston Post*, 24 November 1905; 'Talks on Liquor. Hearing by Mr Wilson's Chemists this Morning', *The Boston Herald*, 24 November 1905.

19 'Oppose Heyburn Bill', *The Washington Post*, 29 January 1906, p. 4.

20 'Retail Section', *The National Provisioner*, 14 September 1907, p. 41.

21 'All Whisky White and Peas Yellow', *The Washington Times*, 23 February 1906, p. 4. This article shows that the manufacturers could still continue to use saltpetre in their products if they indicated it clearly on the packaging.

22 Adolphe Smith, 'Chicago: unhealthy work in the stockyards', *The Lancet*, 28 January 1905, p. 258.

23 Adolphe Smith, 'Chicago: the dark and insanitary premises used for the slaughtering of cattle and hogs', *The Lancet*, 14 January 1905, p. 122.

24 Smith, 'Chicago: unhealthy work in the stockyards', art. cit., p. 260.

25 'Packer's defense by Armour man, he says pump treatment of ham is to "preserve" the hams', *The Minneapolis Journal*, 29 May 1906, p. 3; 'In defense of packers: one of Armour men makes statement', *The Evening Statesman*, 3 June 1906, p. 6.

26 'Packer's Defense by Armour man', art. cit.

27 Ibid.

28 'Proof of packer's use of chemicals', *Evening World*, 13 June 1906, p. 3.

29 E.R. Randolph, 'False Ideas about Food Adulterations', *The National Provisioner*, 17 February 1906, p. 14.

30 *Congressional Record – Senate*, 16 January 1906, p. 1135.

31 'Meat industry and Heyburn food bill', *The National Provisioner*, 10 February 1906, p. 34.

32 'That "poison squad" again', *The National Provisioner*, 13 January 1906, p. 44.

33 *The National Provisioner*, 17 February 1906, p. 14.

34 Ibid., p. 35.

35 *The National Provisioner*, 28 April 1906, p. 15.

36 Letter from L.J. Callanan, 'Wants Wiley brought to book', *The National Provisioner*, 24 February 1906.

37 Committee on Interstate and Foreign Commerce, Report no. 2118, House of Representatives, 59th Congress, 1st session, 7 March 1906, p. 7.

38 Congressman Adams, in Committee on Agriculture on the so-called 'Beveridge Amendment' to the Agricultural Appropriation Bill (H.R. 18537) as passed by the Senate, May 25, 1906, *Hearings, to which are added various documents bearing upon the 'Beveridge Amendment'*, House of Representatives, 59th Congress, 1st session, 1906, pp. 91–2.

39 Statement by Thomas Wilson in Committee on Agriculture on the so-called 'Beveridge Amendment' to the Agricultural Appropriation Bill (H.R. 18537) as passed by the Senate, May 25, 1906, *Hearings*, op. cit., p. 92.

40 Ibid.

41 Ibid., p. 61.

42 'District meat slaughtered by best methods', *The Washington Times*, 24 June 1906, p. 1.

43 'Packers line up for final fight', *Chicago Daily Tribune*, 11 June 1906; 'All sides busy on meat bill', *The Washington Post*, 12 June 1906.

44 'Packers line up for final fight', art. cit., p. 2.

45 Harvey Wiley, 'The Pure Food battle. Looking backward and forward' (December 1915), in Harvey Wiley and Anne Lewis Pierce, *1001 Tests of Foods, Beverages and Toilet Accessories, Good and Otherwise*, Hearst International Library Co., New York, 1916, pp. xi–xii.

46 On the main food bill, see 'Agrees on food bill', *The Sun*, 6 March 1906, p. 7. On the bill specific to meat products, see 'Meat bill ready', *The New York Times*, 14 June 1906, p. 1. The historian Oscar Anderson has

described (virtually hour by hour) the legislative battle over the 'standards of food purity'. See Oscar Anderson, *The Health of a Nation: Harvey W. Wiley and the Fight for Pure Food*, op. cit., pp. 197–8.

47 In the *National Provisioner* of 7 July 1906 (p. 14), the meatpackers were explicit that the text adopted in the Meat Inspection Bill was the version modified by James Wadsworth, which was favourable to them.

48 'Vindicated', *The National Provisioner*, 7 July 1906, p. 21.

49 'Points in the pure food law', *The National Provisioner*, 7 July 1906, p. 16.

50 'No More Food Standards', *The National Provisioner*, 21 July 1906, p. 37.

51 'Dr Wiley's sacrifice squad to test effect of saltpeter', *The Washington Times*, 21 October 1906, p. 14.

52 'Standards of purity for food products', USDA, Office of the Secretary (Circular 19), 26 June 1906.

53 'Dr Wiley's sacrifice squad to test effect of saltpeter', art. cit.

54 The text of the regulation reads as follows: 'Dyes, chemicals and preservatives. Regulation 39 (a) No meat or meat food product for interstate commerce, or for foreign commerce except as hereinafter provided, shall contain any substance which lessens its wholesomeness, nor any drug, chemical, or dye (unless specifically provided for by a Federal statute), or preservative, other than common salt, sugar, wood smoke, vinegar, pure spices, and, pending further inquiry, saltpeter.' (Regulations Governing the Meat Inspection of the United States Department of Agriculture, Order no. 137 of 25 July 1906, Bureau of Animal Industry.)

55 *The Prairie Farmer*, 78, 41, 11 October 1906, p. 14 (our emphasis).

56 'Dr Wiley's sacrifice squad to test effect of saltpeter', art. cit.

57 Ibid.

58 Ibid.

59 Ibid. See also '"Poison Squad" Problem', *The Rock Island Argus*, 21 September 1906, p. 9. The experiment wouldn't actually begin until January 1907. (See Dr Wiley, statement (21 January 1907), in *Hearings Before the Committee on Expenditures in the Department of Agriculture*, House of Representatives, 59th Congress, 2nd session, p. 465.)

60 'Dr Wiley's sacrifice squad to test effect of saltpeter', art. cit.

61 'Eat saltpeter', *The Courier-Journal* (Louisville), 25 March 1907, p. 3.

62 'Poison Squad Quarters', *Oswego Daily Palladium*, 13 October 1906.

63 '"Poison Squad" problem', art. cit.

64 See, for example, 'Flaw in a food bill', *The National Provisioner*, 3 February 1906, p. 21.

65 'Wiley on Preservatives', *The National Provisioner*, 12 January 1907, p. 14.

66 'Saltpeter is not prohibited', *The National Provisioner*, 19 January 1907, p. 13.

67 Harvey Wiley, 'Influence of preservatives and other substances added to foods upon health and metabolism', *Proceedings of the American Philosophical Society*, 47, 189, 1908, p. 308.

68 Ibid.

69 'Eat Saltpeter', *The Courier-Journal* (Louisville), art. cit. See also 'The latest', *The Courier-Journal* (Louisville), 25 March 1907, p. 1.

70 'To stop saltpeter sandwiches', *The Rock Island Argus*, 22 April 1907, p. 1.

71 Wiley, 'Influence of preservatives and other substances added to foods upon health and metabolism', art. cit., pp. 324–5.

72 Wiley, 'The Pure Food battle', art. cit., p. xiii.

73 Harvey Wiley, *Foods and Their Adulteration*, Blakiston's Son & Co., Philadelphia, 1907, pp. 50, 53. See also pp. 55–6.

74 Ibid., p. 56.

75 Wiley, *The History of a Crime Against the Food Law*, op. cit., pp. 62–3. Wiley's text says specifically: 'The seventh part treated of the use of saltpeter, particularly in meats. Owing to the well known results of the depressing effects of saltpeter on the gonads, and for other reasons, the Bureau refused to approve the use of this coloring agent in cured meats.'

76 Letter from Henry Granville Sharpe, Commissary General (War Department) to the Secretary of War, 21 February 1908. Endorsed and forwarded to the Secretary of Agriculture 24 February 1908, National Archives and Record Administration, RG16 191 E17 box 9.

77 'Saltpeter squad next', *The Sun*, 1 April 1907, p. 5.

78 Ibid.

79 Harry Grindley and Thomas Burrill, 'Facts about Chicago Plants', *The National Provisioner*, 9 June 1906, pp. 17, 30.

80 Ibid., p. 17.

81 See 'Statement made by Chicago Packers', *The National Provisioner*, 9 June 1906, p. 18.

82 'Human Test For Food Preservers', *The New York Times*, 31 March 1907, p. 16.

83 Ibid.

84 See 'Report of the Executive Committee', *The National Provisioner*, 12 October 1907, pp. 57–8.

85 'Will eat saltpeter', *The Washington Bee*, 27 April 1907, p. 3.

86 'Saltpeter experiments are now under way', *The National Provisioner*, 21 September 1907, p. 15; see also 'Saltpeter tests going on', *The National Provisioner*, 11 January 1908 (p. 16) and 'Scientific tests successful', *The National Provisioner*, 1 August 1908, p. 16.

87 'Saltpeter experiments are now under way', art. cit., p. 15.

88 See 'Cold storage tests', *The National Provisioner*, 18 April 1908, p. 19. On 6 December 1906, the report submitted to the Board of Trustees of the University of Illinois indicates that the American Meat Packers Association promised funds of 'fifteen thousand dollars or more if necessary' for the programme on saltpetre. (See University of Illinois, *Proceedings of the Board of Trustees*, 1906, p. 29.) In 1910, the official cost had doubled. (See Michel Ryan, hearing before Senate Committee on Cost of Living, reproduced in *The National Provisioner*, 9 April 1910, p. 18.)

89 Michel Ryan, hearing before Senate Committee on Cost of Living, art. cit., p. 18.

90 Letter from Harvey Wiley to H.M. Huston, 8 April 1907, National Archives and Record Administration, RG88 FDA – Bureau of Chemistry miscellaneous records, box 2.

91 Ibid.

92 'Blacklist out; food men halt manufacturers', *The Washington Times*, 13 July 1907, pp. 1, 9.

93 Commissioners of the District of Columbia, *Annual Report, Vol. III, Report of the Health Officer*, Washington, Government Printing Office, 1907, pp. 67–8.

94 'Now a drink squad', *The Washington Herald*, 30 October 1907, p. 2.

95 See Committee on Expenditures in the Department of Agriculture, *Hearings*, Government Printing Office, Washington, 1911, p. 868.

96 'Saltpeter sausage violation of the law?', *The Washington Times*, 6 December 1907, p. 8.

97 See 'Germans verified Wiley poison tests', *The New York Times*, 19 August 1911, p. 4; 'Wiley shows Rusby charges are false', *New York Journal of Commerce*, 17 August 1911.

98 Letter from Henry Granville Sharpe, Commissary General (War Department) to the Secretary of War, 21 February 1908, letter cited.

99 Regulations Governing the Meat Inspection of the United States Department of Agriculture, Bureau of Animal Industry, Order no. 150, March 1908.

100 'Features of the revised regulations', *The National Provisioner*, 28 March 1908, p. 16.

101 'New federal meat regulations', *The National Provisioner*, 28 March 1908, p. 1.

102 Harry Grindley and Ward MacNeal (eds), *Studies in Nutrition: An Investigation of the Influence of Saltpeter on the Nutrition and Health of Man with Reference to Its Occurence in Cured Meats*, vol. 2, University of Illinois, Urbana-Champaign, 1929. See also Richard Swiderski, *Poison Eaters: Snakes, Opium, Arsenic, and the Lethal Show*, Universal Publishers, Boca Raton, 2010, p. 246.

103 Harvey Wiley, *The History of a Crime Against the Food Law*, op. cit, pp. 57, 62.

Chapter 7: Wonder Product of the Twentieth Century: Sodium Nitrite

1 Bureau of Animal Industry, 'Notice regarding meat inspection – sodium nitrite for curing meat', *Service and Regulatory Announcements*, no. 223, November 1925, USDA, Washington.

2 'Utilisation du sel nitrité pour la préparation des viandes et des denrées à base de viande' [Use of nitrited curing salt for the preparation of meats and meat-based foodstuffs], decree of 8 December 1964, *Journal officiel de la République Française* [Official Journal of the French Republic], 5 January 1965, p. 119.

3 Arthur Gamgee, 'Mémoire sur l'action des nitrites sur le sang' [Memoir on the action of nitrites on the blood], *Compte rendu des séances de l'Académie des Sciences* [Report on Sessions of the Academy of Sciences], session of 2 March 1869.

4 Antoine Rabuteau, *Éléments de toxicologie et de médecine légale appliquée à l'empoisonnement* [Elements of Toxicology and Legal Medecine Applied to Poisoning], Lauwereyns, Paris, 1873, pp. 196–8.

5 Henri Huchard, 'Propriétés physiologiques et thérapeutiques de la trinitrine (note sur l'emploi du nitrite de sodium)' [Physiological and therapeutic properties of trinitrin (note on the use of sodium nitrite)], *Bulletin général de thérapeutique médicale et chirurgicale* [General Bulletin of Medical and Surgical Therapy], 104, 1883, p. 347. The first page of the article is reproduced in facsimile in Liliane Pariente, *Angine de poitrine et trinitrine* [Angina and Trinitrin], L. Pariente, Paris, 1980, p. 82.

6 See Henry Ralfe, 'Seventeen cases of epilepsy treated by sodium-nitrite', *British Medical Journal*, 2 December 1882, p. 1095; Matthew Hay, 'Nitrite of sodium in angina pectoris', *The Practitioner*, 30, 17, 1883, p. 194; see also *Journal of Nervous and Mental Disease*, 10, April 1883, p. 353.

7 Hay, 'Nitrite of sodium in angina pectoris', art. cit., p. 194. Matthew Hay claims he was the first to have the idea of using sodium nitrite in place of amyl nitrite, which was commonly used in cardiology at the time. (See J. Reeves, 'Brunton's use of amyl nitrite in angina pectoris: an historic root of nitric oxide research', *News in Physiological Science*, 10, June 1995.)

8 See Sulimen Al Omar et al., 'Therapeutic effects of inorganic nitrate and nitrite in cardiovascular and metabolic diseases', *Journal of Internal Medicine*, 279, 2016, pp. 317–18; 'Du nitrite de sodium dans le traitement de l'épilepsie' [On sodium nitrite in the treatment of epilepsy], *Bulletin général de thérapeutique médicale et chirurgicale*, 104, 1883, p. 285; 'Sodium nitrite in epilepsy', *Chicago Medical Review*, 6, 1882, p. 296.

9 Sydney Ringer and William Murrell, 'Nitrite of sodium as a toxic agent', *The Lancet*, 3 November 1883, pp. 766–7.

10 Report by the Royal Society of Medicine and Surgery, London, 'Du nitrite de sodium dans le traitement de l'épilepsie' [On sodium nitrite in the treatment of epilepsy], *Bulletin général de thérapeutique médicale et chirurgicale*, 104, 1883, p. 92.

11 Ringer and Murrell, 'Nitrite of sodium as a toxic agent', art. cit.; and W. Murrell, 'Nitrite of sodium as a toxic agent', *The Lancet*, 17 November 1883, p. 880.

12 Ringer and Murrell, 'Nitrite of sodium as a toxic agent', art. cit., p. 767.

13 Ibid.

14 Ibid.

15 Ibid.

16 Ibid.

17 Murrell, 'Nitrite of sodium as a toxic agent', art. cit.

18 'Puissance toxique du nitrite de sodium' [Toxic power of sodium nitrite], *Bulletin général de thérapeutique médicale et chirurgicale*, 106, 1884,

pp. 40–1; note on a letter from T. Law in *The Lancet*, 17 November 1883, p. 880; A. Baines, 'On nitrite of sodium in the treatment of epilepsy and as a toxic agent', *The Lancet*, 1 December 1883.

19   See the letters from W. Murrell, T. Law and M. Hay in the article 'Nitrite of sodium', *British Medical Journal*, November 1883, p. 997. On the polemic around the usage of therapeutic nitrite, see also Edith Smith and Francis Dudley Hart, 'William Murrell, physician and practical therapist', *British Medical Journal*, 11 September 1971.

20   Erich Harnack, 'Die Vergiftung durch saltpetrigsaure Alkalien und ihr Verhältnis zur Ammoniakvergiftung' [Poisoning by nitrous acid alkalis and their relationship to ammonia poisoning], *Archives Internationales de Pharmacodynamique*, 13, 1903, p. 185.

21   Merck, *Nitrite und Nitroverbindungen* [Nitrite and Nitro-Compounds], Merck Chemische Fabrik, Darmstadt, 1929, p. 14. The experiments on animals were carried out by G. Armstrong Atkinson at the University of Edinburgh. See G. Armstrong Atkinson, 'The pharmacology of the nitrites and nitro-glycerine', *Journal of Anatomy and Physiology*, January 1888, 22, part 2.

22   Merck, *Nitrite und Nitroverbindungen*, op. cit., p. 13; Erich Hesse, 'Entgiftung der Nitrite' [Nitrite Poisoning], *Naunyn-Schmiedeberg's Archives of Pharmacology*, 126, 3–4, 1927, p. 210.

23   Bait-Rite pellets, see www.connovation.co.nz

24   Jessica Morrison, 'Counterattacking the wild pig invasion with bacon preservative sodium nitrite', *Chemical and Engineering News*, vol. 92, no. 41, 2014, p. 23.

25   Jessica Workum, Laurens Bisschops and Maarten van den Berg, 'Auto-intoxicatie met "zelfmoordpoeder"' [Auto-intoxication with "suicide powder"], *Nederlands Tijdschrift voor Geneeskunde* [Dutch Journal of Medicine], 2019, 163, D3369.

26   Ibid.

27   Truffert and Cheftel, 'À propos de l'emploi de nitrite de sodium', art. cit. Truffert and Cheftel quote Indian findings which establish the lethal dose at 2g. According to other estimates, the lethal dose falls between 33mg/kg (children and older people) and 250mg/kg (adults). See L. Schuddeboom, *Nitrates et nitrites dans les denrées alimentaires* [Nitrates and Nitrites in Foodstuffs], Council of Europe, 1993, p. 103. While M. Ranken set the fatal dose at 1g (Michael Ranken, *Handbook of Meat Product Technology*, Blackwell, London, 2000, p. 54), other authors have placed it at 4g (G. Rentsch et al., 'Über Vergiftung mit reinem Natriumnitrit sowie einen orientierenden Nitritnachweis am Krankenbett' [On poisoning with pure sodium nitrite and an indicative nitrite detection at the hospital bedside], *Archiv für Toxicologie* [Archive of Toxicology], 17, 1958, p. 18) and others still (probably because the product wasn't pure) at 17.5g (Louis Blanchard, Armand Névot and Jean Pantaléon, 'Considérations sur l'emploi d'un nitrite alcalin dans la salaison des viandes' [Considerations on the use of an alkaline nitrite in the curing of

meats], *Revue de pathologie générale et comparée* [Review of General and Comparative Pathology], 655, 1954, p. 339).

28 See Minori Nishiguchi et al., 'An autopsy case of fatal methemoglobinemia due to ingestion of sodium nitrite', *Journal of Forensic Research*, 6, 1, 2015; for a list of poisonings: Sally Bradberry et al., 'Methemoglobinemia caused by the accidental contamination of drinking water with sodium nitrite', *Clinical Toxicology*, 32, 2, 1994.

29 'Govt halts distribution of Dutch salts', *Exit International newsletter*, January–April 2018, p. 8; Maud Effting and Anneke Stoftelen, 'Was de zelfdoding van hun 19-jarige dochter een oprechte doodswens of werd ze op het idee gebracht?' [Was the suicide of their 19-year-old daughter a sincere death wish or was she put up the idea?], *De Volkskrant*, 16 March 2018; Erik Luiten, 'Ximena uit Uden kon voor 1,55 euro zelfmoordpoeder kopen, vader reageert op stop verkoop aan particulieren' [Ximena from Uden could buy suicide powder for 1.55 euro, father responds to stop sales to individuals], *Brabants Dagblad*, 16 March 2018.

30 Workum, Bisschops and van den Berg, 'Auto-intoxicatie met "zelfmoord-poeder"', art. cit.

31 'Kindergarten teacher in China arrested for allegedly poisoning pupils' porridge, sending 23 to hospital', *South China Morning Post*, 2 April 2019.

32 'A malicious hand at pillow side', China Radio International, 30 August 2010.

33 'Woman accused of poisoning neighbor', CNN, 27 March 2006.

34 See, for example, Jean Hurel, *À propos de cinq cas d'intoxication par le nitrite de soude* [Concerning Five Cases of Intoxication by Sodium Nitrite], medical thesis, Paris, 1945.

35 Dr Andrieu et al., 'Intoxication collective par le nitrite de sodium' [Collective intoxication by sodium nitrite], *Bulletin de l'Académie de médecine* [Bulletin of the Academy of Medicine], 127, 11–12, 1943.

36 Bradberry et al., 'Methemoglobinemia caused by the accidental contamination of drinking water with sodium nitrite', art. cit.

37 G. Barton, 'A fatal case of sodium nitrite poisoning', *The Lancet*, 23 January 1954, pp. 190–1.

38 Anna Kaplan et al., 'Methaemoglobinaemia due to accidental sodium nitrite poisoning. Report of 10 cases', *South African Medical Journal*, 77, 17, March 1990.

39 Siddula Gautami et al., 'Accidental Acute Fatal Sodium Nitrite Poisoning', *Clinical Toxicology*, 33, 2, 1995.

40 Morris Greenberg, William Birnkrant and Joseph Schiftner, 'Outbreak of sodium nitrite poisoning', *American Journal of Public Health*, 35, 1945.

41 Berton Roueché, *Eleven Blue Men, and Other Narratives of Medical Detection*, Little, Brown and Company, Boston, 1954, p. 95.

42 Ibid., p. 88.

43 Ibid., p. 96.

44 W. Ten Brink et al., 'Nitrite poisoning caused by food contaminated with cooling fluid', *Clinical Toxicology*, 19, 1982.

45 Thomas Chan, 'Food-borne nitrates and nitrites as a cause of methemo-globinemia', *Southeast Asian Journal of Tropical Medicine*, 27, 1, March 1996.

46 G. Askew et al. 'Boilerbaisse: an outbreak of methemoglobinemia in New Jersey in 1992', *Pediatrics*, 94, 1994.

47 Bradberry et al., 'Methemoglobinemia caused by the accidental contami-nation of drinking water with sodium nitrite', art. cit.

48 W. Gowans, 'Fatal methaemoglobinaemia in a dental nurse. A case of sodium nitrite poisoning', *British Journal of General Practice*, 40, November 1990.

49 Ibid.

50 John Haldane et al., 'The action as poisons of nitrites and other physio-logically related substances', *Journal of Physiology*, 21, 2–3, 1897, p. 161. Concerning the toxicity of nitrite, Haldane and his co-authors pondered how this worked: 'As regards the poisonous effects produced by nitrites the question still remains open whether nitrites cause death by rendering the blood corpuscles incapable of carrying sufficient oxygen to support life, or whether death occurs from a direct action of the nitrites on the nervous system and other tissues.'

51 John Haldane, 'The red colour of salted meat', *Journal of Hygiene*, 1, 1, 1901.

52 E. Polenske, 'Über den Verlust, welchen das Rindfleisch an Nährwerth durch das Pökeln erleidet, sowie über die Veränderungen Salpeterhaltiger Pökellaken' [On the loss of nutritional value sustained by beef through curing, as well as the changes in saltpeter-containing brine], *Arbeiten aus dem Kaiserlichen Gesundheitsamte* [Works of the Imperial Health Department], 7, 1891. See also Polenske, 'Chemische Untersuchung ver-schiedener, im Händel vorkommender Konservirungsmittel für Fleisch und Fleischwaaren', art. cit.

53 Karl Kisskalt, 'Beiträge zur Kenntnis der Ursachen des Rothwerdens des Fleisches beim Kochen, nebst einigen Versuchen über die Wirkung der schwefligen Saüre auf die Fleischfarbe' [Contributions to knowledge of the causes of the reddening of meat during cooking, together with experi-ments on the effect of sulphuric acid on the colour of the meat], *Archiv für Hygiene* [Archive for Hygiene], 35, 1899.

54 Haldane, 'The red colour of salted meat', art. cit., p. 121.

55 Ibid., pp. 121–2.

56 See Robert Moulton and Winford Lewis, *Meat Through the Microscope, Applications of Chemistry and the Biological Sciences to Some Problems of the Meat Packing Industry*, Institute of Meat Packing and The University of Chicago, Chicago, 1940, pp. 206, 208.

57 S. Orlow, 'La coloration des saucisses et des jambons' [The coloration of sausages and hams], *Revue internationale des falsifications* [International Review of Falsifications], 2, 1903, p. 36.

58 Ibid.

59 Ibid.

60 Ibid., p. 38.

61 Ibid., p. 37

62 *Le Mois scientifique et industriel* [The Scientific and Industrial Month], 2 October 1903.

63 'The red color in salted meats chemically explained', *The National Provisioner*, 25 April 1903.

64 Hoagland, 'The action of saltpeter upon the color of meat', art. cit.

65 See Osman Jones, 'Nitrite in cured meats', *Analyst*, 684, 1933, p. 141.

66 Friedrich Glage, *Die Konservierung der roten Fleischfarbe. Eine einfache Methode zur Erzeugung hochroter Fleisch- und Wurstfarbe* [The preservation of red meat colour. A simple method for producing a deep red colour in meat and sausages], R. Schoetz, Berlin, 1909, p. 5.

67 Quoted in Koller, *Salz, Rauch und Fleisch*, op. cit., p. 177.

68 Glage, *Die Konservierung der roten Fleischfarbe*, op. cit., p. 5.

69 Ibid., p. 25.

70 See Ralph Hoagland, 'Coloring matter of raw and cooked salted meats', *Journal of Agricultural Research*, 3, 3, 1914. The experiments of F. Glage are also mentioned in the essential work justifying the use of nitrite: Winford Lewis et al., 'Use of sodium nitrite in curing meats', *Industrial and Engineering Chemistry*, December 1925, p. 1243.

71 The invention of the Haber-Bosch process has been the subject of several books. For an overview, see, for example, Thomas Hager, *The Alchemy of Air*, Three Rivers Press, New York, 2008.

72 Koller, *Salz, Rauch und Fleisch*, op. cit.

73 George Doran, 'Art of curing meats', US Patent 1212614 (January 1917); French version: 'Procédé pour la conservation des viandes', Patent INPI No. 486077 (July 1917).

74 See Pridmore, *Well Seasoned: A Centennial of Heller*, op. cit.

75 Jan Budig, 'Prague ham: the past and the present', *Maso International Journal of Food Science and Technology*, 1, 2012; the history of Praganda was explored by Eben Van Tonder, bacon producer and author of numerous articles on the history of nitrite curing: see E. Van Tonder, 'The Master Butcher from Prague, Ladislav Nachmüllner', Cape Town, 2014–2016.

76 Robert von Ostertag, *Handbuch der Fleischbeschau* [Handbook of Meat Inspection], vol. 2, Enke, Stuttgart, 1913, p. 720. Also quoted in W. Lewis et al., 'Use of sodium nitrite in curing meats', art. cit., p. 1243.

77 Georg Lebbin, patent FR 512608 'Procédé de salage des viandes' [Procedure for salting meats] (France); patent 73375 'Pökelsalz' [Curing salt] (Switzerland); patent FI7908 'Förfarande för saltning av köttvaror' [Procedure for the salting of meat products] (Finland).

78 Lewis et al., 'Use of sodium nitrite in curing meats', art. cit., p. 1243.

79 Dietrich Milles, 'History of toxicology', in Hans Marquardt et al. (eds), *Toxicology*, Academic Press, 1999, p. 21. See also Koller, *Salz, Rauch und Fleisch*, op. cit., p. 190.

80 'Bundesratsverordnungen betr. gesundheitsschädlicher und täuschender Zusätze zu Fleisch und dessen Zubereitungen, und betr. Ergänzung der Ausführungsbestimmungen D. zum Schlachtvieh- und Fleischbeschaugesetze' [Federal Council Ordinances on Harmful and Deceptive Additives in Meat and Its Preparations, and on the Supplemental Provisions D. to the Livestock Slaughter and Meat Inspection Act], 14 December 1916, in *Reichsgesetzblatt* [Imperial Law Gazette], p. 1359, *Zentralblatt für das Deutsche Reich* [Central Gazette for the German Empire], p. 532, et *Verörf. d. kais. Gesundheitsamts* [Publications of the Imperial Health Department], 1917, p. 28.

81 Leopold Pollak, 'Über vergleichende Pökelversuche von Fleisch unter Zusatz von Salpeter und Natriumnitrit zur Lake' [On comparative curing of meat with the addition of saltpetre and sodium nitrite to the brine], *Zeitschrift fur Angewandte Chemie* [Journal of Applied Chemistry], 39, 1922, p. 229.

82 Ibid.

83 See Dean Griffith, *The Griffith Story*, Griffith Press, Alsip, 2006, p. 15.

84 Friedrich Auerbach and Gustav Riess, 'Das Verhalten von Salpeter und Natriumnitrit bei der Pökelung von Fleisch' [The behaviour of saltpetre and sodium nitrite in the curing of meat], *Zeitschrift für Angewandte Chemie*, 35, 19, 1922.

85 E.C. Squire, 'Danish Bacon Displaces American Product on British Markets', USDA, January 1922.

86 'American exporters of meat operate under big handicap', *Meat & Live Stock Digest*, 2, 12, July 1922, p. 1.

87 Ibid.

88 Ibid.

89 Robert Kerr, Clarence Marsh, Walter Schroeder and Edward Boyer (USDA), 'The use of sodium nitrite in the curing of meat', *Journal of Agricultural Research*, 33, 6, 1926, p. 543.

90 On the nitrite curing experiments conducted in 1923–1925, see Pegg and Shahidi, *Nitrite Curing of Meat*, op. cit., pp. 12–13, and Binkerd and Kolari, 'The history and use of nitrate and nitrite in the curing of meat', art. cit., p. 657; see also Horowitz, *Putting Meat on the American Table*, op. cit., p. 60.

91 Lewis et al., 'Use of sodium nitrite in curing meats', art. cit., p. 1243.

92 Kerr et al. 'The use of sodium nitrite in the curing of meat', op. cit., p. 543.

93 Ibid., p. 547.

94 Ibid., p. 550.

95 Ibid.

96 Moulton et Lewis, *Meat through the Microscope*, op. cit., p. 224.

97 Kerr et al., 'The use of sodium nitrite in the curing of meat', op. cit., p. 544, and Moulton and Lewis, *Meat through the Microscope*, op. cit. The initial results showed that the content of 'residual nitrite' in nitrite-cured hams was greater than that for nitrate-cured hams, but when it came to bacon levels were lower ...

98 Bureau of Animal Industry, 'Amendment 4 to B.A.I. Order 211 – Revised', USDA, 19 October 1925; Bureau of Animal Industry, 'Notice regarding meat inspection – sodium nitrite for curing meat', art. cit.

99 'Packing industry's progress is told at Chicago meeting', *Meat & Live Stock Digest*, 6, 5, December 1925, p. 2.

100 See *Report of the Chief of the Bureau of Animal Industry*, USDA, 1924, p. 20, and 'Scientists find better method for curing meats', *USDA Official Record*, 6, 8, 23 February 1927, p. 3.

101 Bureau of Animal Industry, 'Notice regarding meat inspection – sodium nitrite for curing meat', art. cit.

102 In 1972, an American congressional committee confirmed that the authorized norms for dosages of sodium nitrite had been established according to the levels of nitrite necessary for coloration. (See: Committee on Government Operations, *Regulation of Food Additives – Nitrites and Nitrates, Nineteenth Report*, US GPO, Washington, 1972, p. 23.)

103 Bureau of Animal Industry, 'Notice regarding meat inspection – sodium nitrite for curing meat', art. cit.

104 Ibid.

105 Mazÿck Ravenel et al., 'Nitrites permitted in meats', *American Journal of Public Health*, February 1926, p. 157.

106 Ibid.

107 Kerr et al., 'The use of sodium nitrite in the curing of meat', op. cit, p. 550; Lewis et al., 'Use of sodium nitrite in curing meats', art. cit., pp. 1243–5.

108 See British Food Manufacturers' Research Association, 'The Curing of Meat', *Food Research Reports*, No. 6, op. cit., pp. 11–12.

109 Ibid.

110 O. Kapeller, 'Über den Verkehr mit Nitritpökelsalzen' [On the traffic of nitrited curing salts], *Zeitschrift für Fleisch- und Milchhyg.* [Journal of Meat and Milk Hygiene], 41, 205, 1931; Koller, *Salz, Rauch und Fleisch*, op. cit., p. 186.

111 Koller, *Salz, Rauch und Fleisch*, op. cit., p. 181.

112 Ibid., p. 186.

113 The article was about the heart problems that the worker developed as a consequence of this: Otto Schulz, 'Herzmuskelschädigung durch Nitriteinwirkung bei der Herstellung von Pökelsalz aus Natriumnitrit und Kochsalz' [Heart muscle damage due to nitrite action in the production of curing salt from sodium nitrite and table salt], *Sammlung von Vergiftungsfällen* [Collection of Poisoning Cases], 6, 1, 1935.

114 Rentsch et al., 'Über Vergiftung mit reinem Natriumnitrit', art. cit., p. 17.

115 Koller, *Salz, Rauch und Fleisch*, op. cit., p. 191.

116 'Nitritgesetz', 18 June 1934 (R.G.B.Z. 1934 I 5/3); 'Begründung des Nitritgesetzes' [Establishment of the nitrite law], *Reichsanzeiger* [Reich Gazette], 144 of 23 June 1934. Since 1934, the rules in force in Germany have hardly changed at all, despite certain adjustments to do

with European harmonization. See Walter Vösgen, 'Gift für die schöne dauerhafte Rotfärbung' [Poison for lovely permanent red colouring], *Fleischwirtschaft* [Meat Business], 2, 2010, pp. 51–2.

117 In technical terms, this dose constitutes 100 ppm (parts-per-million). See Honikel, 'The use and control of nitrate and nitrite for the processing of meat products', art. cit., p. 70; O. Büch, 'Massenvergiftung durch Natriumnitrit' [Mass poisoning by sodium nitrite], *Archives of Toxicology*, 14, 2, 1952.

Chapter 8: Nitrite in the United Kingdom and in France

1 First schedule, part I of Public Health (Preservatives, etc., in Food) Regulations, 1925. On the lead-up to the law of 1925, see French and Phillips, *Cheated not poisoned?*, op. cit. (Chapter 5: 'Assessing food additives: regulating chemical preservatives, 1888–1938'). The question of sulphur dioxide in sausages was the subject of a major commentary in 'Preservatives and colouring matters in food: The Departmental Committee's Report', *British Medical Journal*, November 1924, p. 829.

2 Louis Truffert and Henri Cheftel, *Rapport sur l'emploi des nitrites alcalins dans la salaison des viandes* [Report on the Use of Alkaline Nitrites in the Curing of Meat], December 1955; Minute of the Ministry of Health, 20 April 1929, MAF 101/326.

3 Letter from R.W. Greef and Co. Ltd. to the Ministry of Health, March 1926, MAF 101/326.

4 Notes of 1 April 1926, MAF 101/326.

5 Ministry of Health, memorandum addressed to Mr Neville, 15 February 1938 (referring to the report by Dr Gordon Monier-Williams dated 31 October 1929), MAF 101/327.

6 Ibid.

7 Letter from the City of Liverpool (City Analyst's Department), 17 October 1929, MAF 101/326.

8 Minutes of the Ministry of Health, 3 May 1929, MAF 101/326.

9 Minutes of the Ministry of Health, 4 November 1929, MAF 101/326.

10 Minutes of the Ministry of Health, 25 January 1929, MAF 101/326.

11 British Food Manufacturers' Research Association, 'The Curing of Meat', *Food Research Reports*, art. cit., p. 14. The experiments conducted in the Association's laboratory showed that the nitro-additive 'does not appear to impart flavour to meat'.

12 Ibid., p. 27

13 British Food Manufacturers' Research Association, letter to the Ministry of Health, 7 October 1931. MAF 101/327.

14 British Food Manufacturers' Research Association, 'The curing of meat', art. cit., p. 11.

15 Minutes of meeting on 27 September 1938, MAF 101/327. NB: the file bears the label MAF (Ministry of Agriculture and Fisheries) but the minutes are from the Ministry of Health.

16 The history of the Billingham factory is described in detail in Anthony Travis, *Nitrogen Capture: The Growth of an International Industry (1900–1940)*, Springer, Cham, 2018.

17 T. McQuiston, 'Fatal poisoning by sodium nitrite', *The Lancet*, 14 November 1936, p. 1153.

18 'Three deaths from sodium nitrite', *The Times*, 19 August 1936, p. 7.

19 Ibid.

20 A. Scholes, 'Poisoning by sodium nitrite', *The Analyst*, 1936, vol. 61, p. 685.

21 'Three deaths from sodium nitrite', art. cit.

22 Ibid.

23 Arnold Tankard, 'Sodium nitrite poisoning', *The Analyst*, 62, 1937, p. 735.

24 12 April 1926, MAF 101/326.

25 Letter from the Ministry of Food to Imperial Chemical Industries, 20 January 1940, MAF 72/165, no. 55.

26 Ibid.

27 Minute of 23 July 1935, MAF 101/327.

28 Ibid.

29 Ibid.

30 'Bacon, ham and other cured meat and pickled meats. Use of nitrites and nitrates. Representations by USA and Food Manufacturers Federation', MH 56/176.

31 'Public Health (Imported Food) Regulations, 1937'.

32 Bureau of Animal Industry (B.A.I.) Circular 2052 dated 29 September 1937, quoted in MAF 72/165, no. 12.

33 As confirmed by a British official: 'The U.S.A. accepted this position but in December 1937, asked for it to be altered' ('Bacon, ham and other cured meat and pickled meats', doc. cit.); see also MAF 101/327 (15 February 1938 and 19 July 1938).

34 MAF 101/327 (25 July 1938).

35 Letter from the Foreign Office to the Ministry of Health, 19 January 1938.

36 Letter from the Ministry of Health to the Secretary of State, 2 August 1938, MAF 101/327.

37 MH 56/176, minutes of 28 April 1939.

38 MH 56/176, minutes of 2 June 1939.

39 'Notes on an informal discussion held on June 16th on the subject of the possibility of importing American bacon at short notice', MAF 72/165, no. 1.

40 Ibid.

41 Letter from the Food (Defence Plans) Department to the Ministry of Health, 22 June 1939, MAF 72/165, no. 5, and letter from the Ministry of Health to the Food (Defence Plans) Department, 28 June 1939, MAF 72/165, no. 6.

42 Ibid.

43 Letter from the Food (Defence Plans) Department, 30 June 1939, MAF 72/165, no. 10.

44 'Public Health (Preservatives etc., in Food) Amendment Regulations, 1939' (Circular 1892 of 25 October 1939).

45 See the minutes for March 1940, MH 56/176.

46 Derek Oddy, *From Plain Fare to Fusion Food: British Diet from the 1890s to the 1990s*, The Boydell Press, Woodbridge, 2003, p. 160.

47 See, for example, the letter from the British Food Manufacturers' Research Association of 10 January 1940, MAF 72/165, no. 49.

48 See, for example, the letter from ICI, 'Bacon production – sodium and potassium nitrites' of 20 January 1940, MAF 72/165, no. 55.

49 Ministry of Food, Bacon and Ham Branch, 'Authority to use added nitrite in the production of bacon', MAF 72/165, no. 75A and no. 43–no. 49.

50 See the reasons set out in the letter from the British Food Manufacturers' Research Association addressed to the Ministry of Food, 10 January 1940, MAF 72/165.

51 Statutory Rules and Orders 1944 no. 164. The initial text establishing the system of licences for use of nitrite was the Bacon (Licensing of Producers) no. 2 Order, 1939 of 16 December 1939 (clause no. 10).

52 Quoted from a letter from the Bacon and Ham Division of 15 September 1949, MAF 72/165.

53 Ministry of Food, Bacon and Ham Branch, 'Authority to use added nitrite in the production of bacon', document cited. 'Poison' is capitalized in the letter of 15 September 1949, MAF 72/165.

54 See Louis Tanon, 'F. Bordas', *Annales d'hygiène publique, industrielle et sociale* [Annals of Public, Industrial and Social Hygiene], 14, October 1936.

55 Frédéric Bordas, 'L'unification des méthodes d'analyse des produits alimentaires' [Unification of methods of analysis of food products], *Annales des falsifications* [Annals of Falsifications], 58, August 1913, quoted by Stanziani, *Histoire de la qualité alimentaire*, op. cit., p. 71.

56 See, for example, Frédéric Bordas, 'Enrobage des produits de la charcuterie' [Coating of cured meat products], *Annales d'hygiène publique, industrielle et sociale*, 1933, pp. 282–5.

57 F. Bordas quoted by Brouardel, *Cours de médecine légale*, op. cit., p. 241.

58 Frédéric Bordas, 'Les nitrites dans les saumures' [Nitrites in brine], *Annales d'hygiène publique, industrielle et sociale*, 13, February 1935.

59 Ibid., p. 60. Bordas also recommended on numerous occasions that saltpetre curing should be banned. See, for example, *Annales des falsifications*, 312, December 1934, p. 579.

60 Koller, *Salz, Rauch und Fleisch*, op. cit., p. 181. See also Brooks et al., 'The function of nitrate, nitrite and bacteria in the curing of bacon and hams', art. cit.

61 See, for example, Damazy Tilgner, *L'Industrie moderne de la conserve* [The Modern Canning Industry], translated by Henri Cheftel, Enault, Paris, 1933.

62 L. Truffert, in report on the session of 12 January 1953 of the Conseil supérieur d'hygiène publique de France (CSHPF) [French High Council of

Public Hygiene] (CAEF – Centre des Archives Economiques et Financières [Centre for Archives of Economics and Finance], B64521).

63 Truffert and Cheftel, 'À propos de l'emploi de nitrite de sodium', art. cit., p. 12.

64 H. Cheftel, in 'Commission d'examen des sels conservateurs employés en charcuterie', 17 November 1953, pp. 5–6 (CAEF, B64521).

65 H. Cheftel, opinion in report on the session of 12 January 1953 of the Conseil supérieur d'hygiène publique de France, p. 5 (CAEF, B64521).

66 Ibid.

67 Charles Jaeger, *Manuel Pratique du Charcutier Moderne* [Practical Manual for the Modern Pork Butcher], Paris, 1954, p. 89.

68 Ibid.

69 Commission d'examen des sels conservateurs en charcuterie [Commission on Use of Preservative Salts in Meat-Processing], 'Observation sur les conclusions du rapport de MM. L. Truffert & H. Cheftel concernant l'emploi des nitrites et des phosphates dans certaines conserves de viande' [Observation on the conclusions of the report by Messrs L. Truffert and H. Cheftel concerning the use of nitrites and phosphates in certain preserved meats], November 1953.

70 Truffert and Cheftel, 'À propos de l'emploi de nitrite de sodium', art. cit.

71 Blanchard, Névot and Pantaléon, 'Considérations sur l'emploi d'un nitrite alcalin dans la salaison des viandes', art. cit., p. 338.

72 Armand Névot, 'Discussion', *Revue de pathologie générale et comparée* [Review of General and Comparative Pathology], no. 655, 1954, p. 343.

73 Henry et al., 'Opportunité de l'addition de nitrates alcalins ou de nitrites alcalins ou des deux sels simultanément aux salaisons de viandes', art. cit., p. 1292.

74 Conseil supérieur d'hygiène publique de France, report on the session of 12 January 1953, p. 5 (CAEF, B64521).

75 Letter from the minister of public health to the minister of agriculture, 16 March 1953 (French National Archives, 880495/7P533).

76 Letter from the Ministry of Agriculture to the Ministry of Public Health, 7 January 1957 (French National Archives, 880495/7P533).

77 Ibid.

78 See F. Spanzaro, 'Le Marché commun et le marché de la viande' [The Common Market and the meat market], *Revue de la conserve*, May–June 1961; F. Spanzaro, 'Les projets d'organisation du marché du porc au sein de la Communauté économique européenne' [Plans to organize the pork market within the European Economic Community], *Revue de la conserve*, July 1961.

79 Letter from the minister of agriculture to the minister of public health, 31 October 1962 (French National Archives, 880495/9DGS2424).

80 Léon Gruart, 'Nourritures terrestres et chimiques. Querelles sur le jambon: nitrate ou nitrite?' [Fruits of the earth and chemical food. Quarrels over ham: nitrate or nitrite?], *Le Figaro*, 17 May 1963.

81 Letter from the Fédération nationale de l'industrie de la salaison, de la charcuterie en gros et des conserves de viandes [National Federation of the Meat-Curing, Wholesale Cured Meat and Preserved Meat Industry] to the Ministry of Agriculture, 4 June 1963 (French National Archives, 830510/11-3Cons11) (our emphasis).

82 See, for example, 'Malgré plusieurs condamnations, la charcuterie alsacienne aux phosphates et aux nitrites continue à empoisonner les consommateurs' [Despite several convictions, meat processed using phosphates and nitrites in Alsace continues to poison consumers], *Libération*, 25 October 1957.

83 J. Orgeron et al., 'Methemoglobinemia from eating meat with high nitrite content', *Public Health Report*, 72, 3, 1957.

84 James Cribbett, 'Division of field operations – nitrite poisoning from wieners', February 1956, reproduced in Subcommittee of the Committee on Government Operations, *Regulation of Food Additives and Medicated Animal Feeds*, US GPO, Washington, 1971, p. 220.

85 Rentsch et al., 'Über Vergiftung mit reinem Natriumnitrit', art. cit., p. 17.

86 M. Freedman, 'Seven cases of poisoning by sodium nitrite', *South African Medical Journal*, 1, 36, 1962.

87 W. Paulus and F.L. Schleyer, 'Eine wiederholte Massenvergiftung mit Natriumnitrit' [A repeated mass poisoning with sodium nitrite], *International Journal of Legal Medicine*, 39, 1–2, 1948.

88 O. Büch, 'Massenvergiftung durch Natriumnitrit' [Mass poisoning by sodium nitrite], art. cit.

89 Karl Braunsdorf, 'Über einige Vergiftungsfälle mit Natriumnitrit (Natrium nitrosum)' [On a number of cases of poisoning with sodium nitrite], *Sammlung von Vergiftungsfällen*, *Archiv für Toxicologie* [Collection of Cases of Poisoning, Archive of Toxicology], 13, 1944, p. 216.

90 'Gift in der Wurst', *Die Zeit*, 30 January 1958.

91 'Angeklagte Drogisten kannten das Nitritgesetz nicht', *Pharmazeutische Zeitung* [Pharmaceutical Journal], 28, 1958, p. 730.

92 'Nitrit-Skandal immer grösser, Gift zentnerweise verwendet, Verhaftungen am laufenden Band', *Neues Deutschland*, 14 March 1958.

93 'Natriumnitrit: Maria hilf in der Wurst' [Sodium nitrite: Maria help in the sausage], *Der Spiegel*, 5 February 1958, pp. 28–9.

94 'Des bouchers allemands vendaient des saucisses au nitrite de sodium' [German butchers sold sausages made with sodium nitrite], *France soir*, 21 February 1958.

95 Letter to the Fraud Prevention Office, 21 February 1958 (French National Archives, 830510/11-16).

96 As witnessed by a letter addressed on 4 June 1963 to the Ministry of Agriculture by the Fédération nationale de l'industrie de la salaison, de la charcuterie en gros et des conserves de viandes (French National Archives, 880495/P590).

97 Blanchard et al., 'Considérations sur l'emploi d'un nitrite alcalin dans la salaison des viandes', art. cit., p. 340.

98 Ibid., p. 341.

99 Report on the session of Conseil supérieur d'hygiène publique de France, 30 April 1963.

100 'Sel nitrité' [Nitrited curing salt], letter from the minister of agriculture to the minister of public health, 9 June 1964 (French National Archives, 880495/P590).

101 'Modification du tableau C (section I) des substances vénéneuses' [Modification of table C (section 1) of poisonous substances], order of 15 September 1964, *Journal officiel de la République Française*, 3 October 1964, p. 8924; 'Conditions de délivrance et d'étiquetage des nitrites métalliques et du sel nitrité' [Conditions of delivery and labelling of metal nitrites and nitrited curing salt], order of 15 September 1964, *Journal officiel de la République Française*, 3 October 1964, pp. 8923–4; 'Utilisation du sel nitrité pour la préparation des viandes et des denrées à base de viande' [Use of nitrited curing salt for the preparation of meats and meat-based foodstuffs], order of 8 December 1964, *Journal officiel de la République Française*, 5 January 1965, p. 119.

102 Académie nationale de médecine, *Rapport. Au sujet du sel nitrité* [Report. On the subject of nitrited curing salt], 23 June 1964 (French National Archives, 880495/P590).

103 Ibid.

104 Ibid.

105 Ibid.

106 Ibid.

Chapter 9: The Nitroso Surprise

1 Doran patents, 'Art of curing meats' and 'Procédé pour la conservation des viandes', art. cit.

2 B. Heller & Co., 'Partial list of B. Heller & Co. products for butchers and packers', 1927.

3 Brochure *Prague Salt. The Safe, Fast Cure*, The Griffith Laboratories, Chicago/Toronto, 1940.

4 'DQ curing salt' available at www.butcher-packer.com

5 Letter from T. Macara, director of research, British Food Manufacturers' Research Association, 26 September 1939, MH 56/176.

6 Levi Paddock (assigned to Swift & Co.), 'Meat curing method', US Patent 1951436, March 1934.

7 Horowitz, *Putting Meat on the American Table*, op. cit., p. 66.

8 B. Heller & Co., 'Partial list of B. Heller & Co. products for butchers and packers', art. cit.

9 Vernon Ruttan, *Technological Progress in the Meatpacking Industry, 1919–1947*, Marketing Research Report no. 59, USDA, 1954, p. 3.

10 Pallu, *La Charcuterie en France. Tome III*, op. cit., p. 191.

11 Horowitz, *Putting Meat on the American Table*, op. cit., p. 66.

12 Bruce Kraig, *Hot Dog: a Global History*, Reaktion Books, London, 2009, p. 53.

13 Paul Aldrich (ed.), *The Packers's Encyclopedia*, vol. 3: *Sausage and Meat Specialties: A Practical Operating Handbook for the Sausage and Meat Specialty Manufacturer*, National Provisioner, Chicago, 1938, pp. 47, 100.

14 Bruce Kraig, *Hot Dog: a Global History*, op. cit., p. 53.

15 Peter Magee and John Barnes, 'The production of malignant primary hepatic tumours in the rat by feeding dimethylnitrosamine', *British Journal of Cancer*, 10, 1956.

16 The chronology of the discovery (in 1974) of nitrosamines in cigarette smoke is outlined at the start of P. Finster, *Literature Study, N-Nitrosamines in Tobacco Products*, British American Tobacco, Hamburg, 1986.

17 N. Böhler, 'En ondartet leversykdom hos mink og rev' [A malignant liver disease in mink and foxes], *Norsk Pelsdyrblad* [Norwegian Journal of Fur Animals], 34, 1960.

18 Nils Koppang, 'An outbreak of toxic liver injury in ruminants', *Nordisk Veterinaermedicin* [Nordic Veterinary Medicine], 16, 1964; Ferenc Ender et al., 'Isolation and identification of a hepatotoxic factor in herring meal produced from sodium nitrite preserved herring', *Naturwissenschaften* [Natural Sciences], 51, 1964.

19 Raymond Bonnett et al., 'Reactions of nitrous acid and nitric oxide with porphyrins and haems. Nitrosylhaems as nitrosating agents', *Journal of the Chemical Society, Chemical Communications*, 21, 1975.

20 Bastide et al., 'A central role for heme iron in colon carcinogenesis associated with red meat intake', art. cit.; Bastide et al., 'Heme iron from meat and risk of colorectal cancer: a meta-analysis and a review of the mechanisms involved', art. cit.

21 Committee on Medical Aspects of Food Policy, Pharmacology Sub-Committee, 'Minutes of the 15th Meeting Held at the BFMIRA, Randalls Road, Leatherhead, on 18 October 1968', MAF 260/559 part 1 (intervention by F.J.C. Roe).

22 William Lijinsky, 'Possible formation of N-nitroso compounds from amines and nitrites', in B.J. Tinbergen and B. Krol (eds), *Proceedings of the Second International Symposium on Nitrite in Meat Products*, Pudoc – Centre for Agricultural Publishing and Documentation, Wageningen, 1977, p. 276.

23 D. Heath and P. Magee, 'Toxic properties of dialkylnitrosamines and some related compounds', *British Journal of Industrial Medicine*, 19, 1962, p. 278.

24 Committee on Medical Aspects of Food Policy, Pharmacology Sub-Committee, 'Minutes of the 15th Meeting Held at the BFMIRA, Randalls Road, Leatherhead, on 18 October 1968', doc. cit. (intervention by Peter Magee).

25 Ibid. (intervention by John Barnes).

26 See Nathalie Jas, 'Adapting to "reality": the emergence of an international expertise on food additives and contaminants in the 1950s and early 1960s', in Soraya Boudia and Nathalie Jas (eds), *Toxicants, Health and Regulation since 1945*, Pickering & Chatto, London, 2013.

27 René Truhaut, 'Survey of the hazards of the chemical age', *Pure and Applied Chemistry*, 21, 3, 1970.

28 Ibid.

29 Ibid.

30 Ibid.

31 Interview with René Truhaut in 'Dangereuses nitrosamines' [Dangerous nitrosamines], *Médecine Mondiale* [World Medicine], 58, 1970, p. 91.

32 Leon Golberg, 'Chemical and biological implications of human and animal exposure to toxic substances in food', *Pure and Applied Chemistry*, 21, 3, 1970.

33 Ibid.

34 Ibid.

35 See J. Glover, J.F. Pennock, G.A.J. Pitt and T.W. Goodwin, 'Richard Alan Morton, 22 September 1899 – 21 January 1977', *Biographical Memoirs of the Fellows of the Royal Society*, Vol. 24, November 1978.

36 Letter from Richard Morton to Cledwyn Hughes (Minister of Agriculture, Fisheries and Food), 2 December 1968, MAF 260/559, part 1.

37 Richard Morton, 'Food Additives and Contaminants Committee, Possible Report on Nitrites and Nitrosamines', July 1968, MAF 260/559, part 1.

38 Ibid.

39 Ibid.

40 L.G. Hanson, 'Note for Mr Propper for the Informal Meeting with Lord Trenchard, et al. Nitrosamines in Food', 22 January 1970, MAF 260/559 part 2. Another version of this document can be found in MAF 260/630.

41 Minutes of BIBRA meeting, February 1969, in MAF 260/559 part 1, 58/3.

42 R.F. Giles, confidential minutes, 11 February 1971, F.S. 3731, MAF 260/631.

43 Letter from the Food Manufacturers' Federation, 6 January 1970, MAF 260/630.

44 L.G. Hanson, 'Nitrates, Nitrites and Nitrosamines', 23 October 1968, F.S. 3314, MAF 260/559 part 1.

45 'Nitrites, nitrosamines, and cancer', *The Lancet*, 18 May 1968, pp. 1071–2.

46 Ibid.

47 Ibid.

48 H. Truffert in report on the session of 12 January 1953 of the Conseil supérieur d'hygiène publique de France.

49 M. Ravenel et al., 'Nitrites permitted in meats', art. cit.

50 Hermann Druckrey et al., 'Panel discussion on regulations governing dyes and other additives in foods', *Acta – Unio Internationalis Contra Cancrum*, 13, 2, 1957, p. 145.

51 Hermann Druckrey, 'Report of Symposium on Potential Cancer Hazards from Chemical Additives and Contaminants to Foodstuffs', *Acta – Unio Internationalis Contra Cancrum*, 13, 2, 1957, p. 66.

52 Dr Fabre in report on the session of 12 January 1953 of the Conseil supérieur d'hygiène publique de France.

53 Dr Dreyfus in ibid.

54 Ibid.

55 M. Henry et al., 'Opportunité de l'addition de nitrates alcalins ou de nitrites alcalins ou des deux sels simultanément aux salaisons de viandes', art. cit., p. 1296.

56 Ibid.

Chapter 10: Botulism, the 'Blood Sausage Poison'

1 Tom Addiscott, 'Is it nitrate that threatens life or the scare about nitrate?', *Journal of the Science of Food and Agriculture*, 86, 2006, p. 2005.

2 Gladwin et al., 'Meeting report: The emerging biology of the nitrite anion', art. cit., p. 313.

3 Otto-Joachim Grüsser, 'Der "Wurstkerner" Justinus Kerners Beitrag zur Erforschung des Botulismus' [The 'sausage' Kerner, Justinus Kerner's contribution to the study of botulism], in Heinz Schott (ed.), *Justinus Kerner Jubiläumsband zum 200. Geburtstag* [Volume on the 200th Anniversary of Justinus Kerner's Birth], Stadt Weinsberg, Weinsberg, 1991; Frank Erbguth, 'Historical notes on botulism', *Movement Disorder*, 19, 2004.

4 Julius Schlossberger, 'Das Gift verdorbener Würste mit Berücksichtigung seiner Analogen in andern thierischen Nahrungsmitteln' [The poison of spoiled sausages with consideration of analogues in other animal-based foods], *Archiv für physiologische Heilkunde* [Archive for Physiological Medicine], Supplement 1852. Also extracted in *Medicinisches Correspondenz-Blatt des Würtembergischen Ärztlichen Vereins* [Medical Correspondence Bulletin of the Würtemberg League of Doctors], 23, 19, 1853; Robert von Ostertag, *Handbook of Meat Inspection*, 3rd edn, Jenkins, New York, 1912, p. 759.

5 Dr Von Faber, 'Ueber Wurstvergiftungen' [On sausage poisonings], *Medicinisches Correspondenz-Blatt des Würtembergischen Ärztlichen Vereins*, 24, 33, 1854, pp. 249–50.

6 Ibid.

7 Michael Peck and Sandra Stringer, 'The safety of pasteurised in-pack chilled meat products with respect to the foodborne botulism hazard', *Meat Science*, 70, 2005.

8 See, for example, Dr Bosch, 'Ueber Wurstvergiftungen, besonders deren Behandlung' [On sausage poisonings and particularly their treatment], *Medicinisches Correspondenz-Blatt des Würtembergischen Ärztlichen Vereins*, 23, 37, 1853; Von Faber, 'Ueber Wurstvergiftungen', art. cit.; Dr Berg, 'Ueber Wurstvergiftungen' [On sausage poisonings], *Medicinisches Correspondenz-Blatt des Würtembergischen Ärztlichen Vereins*, 25, 41 and 42, 1855.

9 Justinus Kerner, *Neue Beobachtungen über die in Würtemberg so häufig vorfallenden tödlichen Vergiftungen durch den Genuss geräucherter Würste* [New Observations about the Fatal Poisonings Caused

by the Consumption of Smoked Sausages, Which Are So Common in Würtemberg], Ofiander, Tübingen, 1820, p. 2.

10 J. Kerner quoted by Frank Erbguth, 'Justinus Kerner und das Wurstgift, medizin-historische Aspekte' [Justinus Kerner and the sausage poison, medical-historical aspects], in Frank Erbguth and Markus Naumann (eds), *Botulinumtoxin. Visionen und Realität* [Botulism Toxin: Visions and Reality], Wissenschaftsverlag Wellingsbüttel, Hamburg, 2003, p. 29.

11 See, for example, remarks by Dr Wolshofer, *Medicinisches Correspondenz-Blatt des Würtembergischen Ärztlichen Vereins*, 25, 20, 1855, p. 159. See also Erbguth, 'Historical notes', art. cit., p. 3.

12 J. Schlossberger and H. Müller, quoted by Émile Van Ermengem, 'Contribution à l'étude des intoxications alimentaires. Recherches sur des accidents à caractère botulinique provoqués par du jambon' [Contribution to the study of food intoxications: research on botulinum-related accidents caused by ham], *Archives de pharmacodynamie* [Archives of Pharmacodynamics], 3, 1897, p. 255.

13 Justinus Kerner, *Das Fettgift oder die Fettsäure und ihre Wirkungen auf den thierischen Organismus* [Fat Poison or Fatty Acid and its Effect on the Animal Organism], J.G. Cotta, 1822. The law on sausages by Emperor Leon VI which J. Kerner discusses is described in Jean Théodoridés, *Des miasmes aux virus. Histoire des maladies infectieuses* [From Miasma to Virus: History of Infectious Diseases], L. Pariente, 1991, p. 153. The ban of 1258 is quoted in Kurt Nagel et al., *L'Art et la Viande* [Art and Meat], Erti, Paris, 1984, p. 26.

14 Gladwin et al., 'Meeting report: The emerging biology of the nitrite anion', art. cit., p. 313.

15 Kerner, *Das Fettgift oder die Fettsäure und ihre Wirkungen auf den thierischen Organismus*, op. cit.

16 See ibid., pp. 361–6.

17 Grüsser, 'Der "Wurstkerner"', art. cit., p. 250.

18 See 'Bekanntmachung des Medicinal-Collegiums' [Announcement of the College of Medicine], *Medicinisches Correspondenz-Blatt des Würtembergischen Ärztlichen Vereins*, 22, 48, 1852.

19 Report by Dr Von Faber, 'Ueber Wurstvergiftungen', art. cit., p. 260.

20 Jacob Gailer, *Nouveau Orbis Pictus à l'usage de la jeunesse* [New Orbis Pictus for Use by Youth], Loeflund, Stuttgart, 1832 (illustration 122 and p. 221, 'Le boucher' [The butcher]).

21 Von Kurzböck, *Spectacle de la nature*, art. cit. (pages on 'Les saucisses' [sausages]).

22 David Ryckaert, *Ein Schlachter bietet einer Frau ein Glas Bier an* [A Slaughterer Offers a Woman a Glass of Beer], Städel Museum, Frankfurt.

23 Engraving by J. Louis reproduced in K. Nagel et al., *L'Art et la Viande*, op. cit., p. 17.

24 See, for example, Belleforest (1571) 'Comment conserver la viande' [How to preserve meat], in Ferrières, *Histoires de cuisines et trésors des fourneaux*, op. cit.

25 See, for example, 'Allegories' for December (thirteenth, sixteenth and seventeenth centuries), reproduced in Nagel et al., *L'Art et la Viande*, op. cit., pp. 67, 69.

26 Ernest Dickson, *Botulism, a Clinical and Experimental Study*, Rockefeller Institute for Medical Research, New York, 1918, p. 11; Van Ermengem, 'Contribution à l'étude des intoxications alimentaires', art. cit., pp. 254–7.

27 *Schwäbischer Merkur* report in *Medicinisches Correspondenz-Blatt des Würtembergischen Ärztlichen Vereins*, 24, 16, 1854, p. 128.

28 'Chronik' [Chronicle], *Medicinisches Correspondenz-Blatt des Würtembergischen Ärztlichen Vereins*, 24, 30, 1854, pp. 239–40.

29 Report by Dr Müller, *Medicinisches Correspondenz-Blatt des Würtembergischen Ärztlichen Vereins*, 24, 30, 1854, p. 239.

30 Report by Dr Schüz, 'Wurstvergiftungen' [Sausage poisonings], *Medicinisches Correspondenz-Blatt des Würtembergischen Ärztlichen Vereins*, 24, 21, 1855, p. 163.

31 Report by Dr Von Faber, 'Ueber Wurstvergiftungen', art. cit., p. 261.

32 J. Schlossberger, 'Das Gift verdorbener Würste', art. cit., p. 5.

33 Ibid.

34 Ibid. See also the series 'Überwachung und Kontrolle von Nahrungs- und Genussmitteln' [Oversight and control of foodstuffs and beverages] in the archives of the Medizinalkollegium [Medical College] of Tübingen stored in the Landesarchiv Baden-Württemberg.

35 Report by Dr Schroter, 'Ein Tödlich abgelaufener Wurstvergiftungsfall' [A deadly case of sausage poisoning], *Medicinisches Correspondenz-Blatt des Würtembergischen Ärztlichen Vereins*, 30, 29, 1860.

36 Ibid.

37 Ibid.

38 Schlossberger, 'Das Gift verdorbener Würste', art. cit.

39 Grüsser, 'Der "Wurstkerner"', art. cit., p. 233.

40 Report by Dr Röser, 'Vergiftung durch Leberwürste' [Poisoning by liver sausages], *Medicinisches Correspondenz-Blatt des Würtembergischen Ärztlichen Vereins*, 12, 1, 1842, p. 1.

41 Archives of the Medizinalkollegium of Tübingen, Landesarchiv Baden-Württemberg. See also Johann Autenrieth, in Grüsser, 'Der "Wurstkerner"', art. cit., p. 233.

42 Édouard Van den Corput, *Du poison qui se développe dans les viandes et dans les boudins fumés* [On Poison That Develops in Meats and Smoked Blood Sausages], Tircher, Brussels, 1855, p. 6.

43 Ibid.

44 Ibid.

45 Van Ermengem, 'Contribution à l'étude des intoxications alimentaires', art. cit., p. 225.

46 Ibid., p. 216.

47 Pieter Devriese, 'On the discovery of Clostridium Botulinum', *Sartoniana. Journal of the Sarton chair of the History of Sciences at Universiteit Ghent*, 13, 2000, p. 133.

48 Van Ermengem, 'Contribution à l'étude des intoxications alimentaires', art. cit., p. 226.

49 Ibid., pp. 222–3.

50 Ibid., p. 582.

51 Ibid., p. 522.

52 Ibid., pp. 237, 342.

53 E. Dickson, *Botulism, a Clinical and Experimental Study*, op. cit., p. 5. In France, 'we see the word for the first time in hygiene publications in the early 1900s [...] In France, where botulism was recorded only after 1875, there were only twenty-one cases, of which only three were fatal, between 1875 and 1924.' (M. Ferrières, *Histoire des peurs ...*, op. cit., p. 414.) In the UK, botulism was virtually unknown until the first case identified, which occurred in 1922. See Brandon Horowitz, 'The ripe olive scare and hotel Loch Maree tragedy: Botulism under glass in the 1920s', *Clinical Toxicology*, 49, 2011.

Chapter 11: The Botulism Pretext

1 www.hanegal.dk/viden-om/nitrit

2 Ibid.

3 www.mcleanmeats.com/nitrates-nitrites-and-your-health/

4 See, for example, Claire Walbecque, 'La Fict dit non à la taxe sur les charcuteries' [Fict says no to a tax on processed meats], *PorcMag*, 23 October 2019.

5 See Emmanuelle Espié, *Caractéristiques épidémiologiques du botulisme humain en France de 2001 à 2003* [Epidemiological Characteristics of Human Botulism in France 2001–2003], Institut de veille sanitaire [Institute of Sanitary Surveillance], 2003. See also Sylvie Haeghebaert et al., 'Caractéristiques épidémiologiques du botulisme humain en France, 1991–2000', *Bulletin épidémiologique hebdomadaire* [Weekly Epidemiological Bulletin], 14/2002, p. 6.

6 See, for example, René André, *Contribution à l'étude du botulisme* [Contribution to the Study of Botulism], De Bussac, Clermont-Ferrand, 1945.

7 André-Louis Vittoz, *Contribution à l'étude des frontières du botulisme* [Contribution to the Study of the Limits of Botulism], Foulon, Paris, 1944, p. 17.

8 René Legroux et al., 'Statistique du botulisme de l'Occupation 1940–1944' [Statistics of botulism during the Occupation 1940–1944], *Bulletin de l'Académie de médecine*, 129, November 1945.

9 Ibid.

10 René Legroux and Colette Jéramec, 'Le botulisme et les jambons salés' [Botulism and salted hams], *Bulletin de l'Académie de médecine*, 128, March 1944.

11 Ibid.

12 Ibid.

13 René Legroux and Colette Jéramec, 'L'infection botulique du porc' [Botulism infection in pigs], *Bulletin de l'Académie de médecine*, 128, June 1944.

14 Louise-Aimée Dewé, *Le Botulisme. À propos de cinq cas observés à l'Hôtel-Dieu* [Botulism: Concerning Five Cases Observed at the Hôtel-Dieu], Foulon, Paris, 1943, pp. 34, 78. See also François Émile-Zola, *Formes mortelles du botulisme* [Fatal Forms of Botulism], Le François, Paris, 1946, pp. 11, 82; André, *Contribution à l'étude du botulisme*, op. cit., pp. 33–7; Vittoz, *Contribution à l'étude des frontières du botulisme*, op. cit., p. 17.

15 See, for example, Donald Emmeluth, *Botulism*, 2nd edn, Chelsea House, New York, 2010, pp. 83, 86; Friedrich-Karl Lücke and Terry Roberts, 'Control in meat and meat products', in Andreas Hauschild and Karen Doods, *Clostridium Botulinum, Ecology and Control in Foods*, Marcel Dekker, New York, 1993, pp. 182, 192.

16 Michel-Robert Popoff, 'Botulinum neurotoxins: more and more diverse and fascinating toxic proteins', *Journal of Infectious Diseases*, 209, 2, 2014.

17 Dickson, *Botulism, A Clinical and Experimental Study*, op. cit., pp. 17, 21, 23ff; Christelle Mazuet et al., 'Le botulisme humain en France, 2007–2009' [Human botulism in France, 2007–2009], *Bulletin épidémiologique hebdomadaire* [Weekly Epidemiology Bulletin], 6, February 2011, p. 53; Friedrich-Karl Lücke and Peter Zangerl, 'Food safety challenges associated with traditional foods in German speaking regions', *Food Control*, No. 43, 2014; Kashmira Date et al., 'Three outbreaks of foodborne botulism caused by unsafe home canning of vegetables', *Journal of Food Protection*, 74, 12, 2011.

18 Jim McLauchlin et al., 'Food-borne botulism in the United Kingdom', *Journal of Public Health*, 28, 4, August 2006.

19 See Incident Management Team, *An Outbreak of Food-Borne Botulism in Scotland, November 2011*, Health Protection Scotland, Glasgow, 2013.

20 M. O'Mahony et al., 'An outbreak of foodborne botulism associated with contaminated hazelnut yoghurt', *Epidemiology & Infection*, 1990, 104, 3.

21 Jiu-Cong Zhang et al., 'Botulism, where are we now ?', *Clinical Toxicology*, 48, 2010, p. 872; Peck and Stringer, 'The safety of pasteurized in-pack chilled meat products with respect to the foodborne botulism hazard', art. cit., p. 465.

22 Date et al., 'Three outbreaks of foodborne botulism caused by unsafe home canning of vegetables', art. cit.

23 David Paterson et al., 'Severe botulism after eating home-preserved asparagus', *Medical Journal of Australia*, 157, 4, 1992.

24 Pierre Abgueguen et al., 'Nine cases of foodborne botulism type B in France and literature review', *European Journal of Clinical Microbiology and Infectious Diseases*, 22, 2003.

25 Jacques Tourret, *Quelques aspects actuels du botulisme: à propos de 16 observations* [Some Current Aspects of Botulism: Concerning

16 Observations], medical thesis, Clermont-Ferrand, 1973. The deaths from botulism following consumption of industrially processed asparagus are described in observations 11 and 12, pp. 55-7. Earlier that same year, in another town, there were other cases of botulism linked to preserved asparagus, this time home-produced (see ibid., observations 1-3, pp. 43-6).

26 Gary Barker et al., 'Probabilistic representation of the exposure of consumers to Clostridium botulinum neurotoxin in a minimally processed potato product', *International Journal of Food Microbiology*, 100, 1-3, 2005.

27 Valérie Delbos et al., 'Botulisme alimentaire, aspects épidémiologiques' [Food botulism, epidemiological aspects], *La Presse médicale* [Medical Press], 34, 2005, p. 459.

28 Barker et al., 'Probabilistic representation of the exposure of consumers to Clostridium botulinum neurotoxin ...', art. cit.

29 Mandy Seaman et al., 'Botulism caused by consumption of commercially produced potato soups stored improperly', *Morbidity and Mortality Weekly Report*, 60, 26, 2011, p. 890.

30 Frederick Angulo et al., 'A large outbreak of botulism: the hazardous baked potato', *Journal of Infectious Diseases*, 178, 1998.

31 Charles Armstrong et al., 'Botulism from eating canned ripe olives', *Public Health Reports*, 34, 51, 1919.

32 Horowitz, 'The ripe olive scare and hotel Loch Maree tragedy: Botulism under glass in the 1920s', art. cit.

33 Dwight Sisco, 'An outbreak of botulism', *JAMA*, February 1920. See James Young, 'Botulism and the ripe olive scare of 1919-1920', *Bulletin of the History of Medicine*, 50, 3, 1976.

34 'Improperly prepared olives can cause botulism poisoning', *Lodi News Sentinel*, 22 November 1978, p. 17.

35 Luca Padua et al., 'Neurophysiological assessment in the diagnosis of botulism: usefulness of single-fiber EMG', *Muscle and Nerve*, October 1999.

36 Amy Cawthorne et al., 'Botulism and preserved green olives', *Emerging Infectious Diseases*, 11, 5, 2005.

37 See *Euro Surveillance*, 15, 14, 2010; 16, 49, 2011.

38 See Tilgner, *L'Industrie moderne de la conserve*, op. cit., pp. 69-70.

39 For example, in March 2007, withdrawal of 'Dal Raccolto' and 'Cibo specialties' by the FDA; March 2011, alert on olives from Greece; July 2012, withdrawal by the UK Food Standards Agency of 'I Divini di Chicco Francesco' olives; June 2013, withdrawal of tinned 'Délices d'olives noires' by the regional health authority for Provence-Alpes-Côte d'Azur, etc.

40 Emmeluth, *Botulism*, op. cit., pp. 21-2.

41 Jeff Rush and Jean Kinsey, *Castleberry's: 2007 Botulism Recall, a Case Study*, University of Minnesota, 2008; Julie Schmit, 'Management problems cited in botulism case', *USA Today*, 29 June 2008.

42 Niels Skovgaard, 'Microbiological aspects and technological need: techno-logical needs for nitrates and nitrites', *Food Additives and Contaminants*, 9, 5, 1992.

43 Peter Dürre, 'From Pandora's box to Cornucopia: Clostridia – a his-torical perspective', in Hubert Bahl and Peter Dürre (eds), *Clostridia: Biotechnology and Medical Applications*, Wiley, New York, 2001, pp. 6–7.

44 Hamid Tavakoli et al. 'A survey of traditional Iranian food products for contamination with toxigenic Clostridium botulinum', *Journal of Infection and Public Health*, 2, 2009, p. 94.

45 Van Ermengem, 'Contribution à l'étude des intoxications alimentaires', art. cit., p. 346.

46 See Stéphane Horel, *Lobbytomie, comment les lobbies empoisonnent nos vies et la démocratie* [Lobbytomy: how lobbies poison our lives and democracy], La Découverte, Paris, 2018.

47 See Corporate Europe Observatory, *Exposed: Conflicts of Interest Among EFSA's Experts on Food Additives*, CEO, Brussels, 2011; 2015–2018 data are given in the report by Corporate Europe Observatory, *Recruitment Errors: The European Food Safety Authority (EFSA) Will Probably Fail, Again, to Become Independent from the Food Industry. June 17 Update.* CEO, Brussels, 2017.

48 EFSA Panel on Food Additives and Nutrient Sources Added to Food (ANS), 'Re-evaluation of sodium nitrate (E 251) and potassium nitrate (E 252) as food additives', and 'Re-evaluation of potassium nitrite (E 249) and sodium nitrite (E 250) as food additives', *EFSA Journal*, 2017, 15(6).

49 Commission Decision (EU) 2015/826 of 22 May 2015 concerning national provisions notified by Denmark on the addition of nitrite to certain meat products, *Official Journal of the European Union*, 58, 28 May 2015, p. L 130/15 (our emphasis).

50 Ibid. (our emphasis). See also remarks by Denmark in EFSA Panel on Food Additives and Nutrient Sources Added to Food, 'Statement on nitrites in meat products', *EFSA Journal*, 8, 5, 2010, p. 5.

51 EFSA Panel on Food Additives and Nutrient Sources Added to Food (ANS), 'Re-evaluation of potassium nitrite (E 249) and sodium nitrite (E 250) as food additives', art. cit., p. 106 ('Conclusions').

52 Ibid.

53 'Existing safe levels for nitrites and nitrates intentionally added to meat and other foods are sufficiently protective for consumers, the European Food Safety Authority (EFSA) has concluded', www.farminguk.com, 4 July 2017.

54 EFSA, 'Opinion of the Scientific Panel on Biological Hazards on the request from the Commission related to the effects of nitrites/nitrates on the microbiological safety of meat products', art. cit., p. 1 (our emphasis).

55 Commission regulation 1129/2011, 11 November 2011, *Official Journal of the European Union*, L 295/1.

56 EFSA Panel on Food Additives and Nutrient Sources Added to Food, 'Statement on nitrites in meat products', art. cit., p. 8.

57 EFSA, 'Opinion of the Scientific Panel on Biological Hazards on the request from the Commission related to the effects of nitrites/nitrates on the microbiological safety of meat products', art. cit., pp. 11, 24.

58 Roberto Chizzolini et al., 'Biochemical and microbiological events of Parma ham production technology', *Microbiologia SEM*, 9 N, 1993, pp. 26–34. For an exact description of the banning of nitro-additives, see also Giovanni Parolari, 'Review: achievements, needs and perspectives in dry-cured ham technology: the example of Parma ham', *Food Science and Technology International*, 2, 2, 1996, pp. 70–1.

59 Giovanni Parolari et al., 'Extraction properties and absorption spectra of dry cured hams made with and without nitrate', *Meat Science*, 64, 2003, p. 483.

Chapter 12: A Case Study: Sodium Nitrite and Smoked Fish

1 *British Nutrition Foundation Information Bulletin*, 3, March 1969, p. 14.

2 In the nineteenth century, investigations of cases of botulism showed that deaths due to toxic sausages resembled poisonings linked to the consumption of raw or undercooked fish which had been kept for months without being correctly salted. In Germany, smoked herrings regularly used to cause botulism poisoning. In each case it was due to botched salting and smoking. (See the list provided by Dickson, *Botulism, a Clinical and Experimental Study*, op. cit., p. 6.) Innumerable cases have been recorded in Russia, mostly connected to sturgeons, pike and smoked salmon. The fish in question had often been preserved using an especially rudimentary method: according to Émile Van Ermengem (1897), fish were 'rubbed with salt and buried in the ground packed in ice, inside wooden boxes'. See Van Ermengem, 'Contribution à l'étude des intoxications alimentaires', art. cit., p. 258.

3 Bruce Tompkin, 'Nitrite', in P. Michael Davidson et al. (eds), *Antimicrobials in foods*, 3rd edn, Taylor & Francis, Boca Raton, 2005, pp. 210–12.

4 FAO.org, online version of D.C. Cann, *Botulism and Fishery Products*, Ministry of Agriculture, Fisheries and Food/Torry Research Station (revised), n.d. The sampling is described by D.C. Cann et al., 'Incidence of Clostridium botulinum type E in fish products in the United Kingdom', *Nature*, 211, 1966.

5 'Smoked fish products recalled over botulism concerns', BBC.com, 14 April 2016; Aidan Fortune, 'Salmon recalled over clostridium botulinum fears', foodmanufacture.co.uk, 20 April 2018, etc.

6 Dirk Dressler, 'Botulismus durch Räucherlachsverzehr' [Botulism through consumption of smoked salmon], *Nervenarzt* [Neurologist], 6, 2005.

7 Delbos et al., 'Botulisme alimentaire', art. cit., p. 456.

8 Vittoz, *Contribution à l'étude des frontières du botulisme*, op. cit., pp. 53–7.

9 McLauchlin et al., 'Food-borne botulism in the United Kingdom', art. cit.

10 Peck and Stringer, 'The safety of pasteurized in-pack chilled meat products with respect to the foodborne botulism hazard', art. cit. (table p. 465).

11 E. Jeffery Rhodehamel et al., 'Incidence and heat resistance of Clostridium botulinum type E spores in menhaden surimi', *Journal of Food Science*, 56, 6, 1991, pp. 1562–3.

12 Susan Gilbert et al., *Risk Profile: Clostridium Botulinum in Ready-to-eat Smoked Seafood in Sealed Packaging*, Institute of Environmental Science & Research, Christchurch, 2006, pp. 28, 33.

13 Michael Peck, 'Clostridium botulinum', in Vijay Juneja and John Sofos, *Pathogens and Toxins in Foods: Challenges and Interventions*, ASM Press, Washington, 2010, p. 37.

14 www.botulismblog.com

15 Jørgen Lerfall and Marianne Østerlie, 'Use of sodium nitrite in salt-curing of Atlantic salmon (Salmo salar L.): impact on product quality', *Food Chemistry*, February 2011.

16 See, for example, Service de la Répression des Fraudes, 'Notes sur l'emploi du salpêtre pour la conservation du poisson' [Notes on the use of salt-petre for the preservation of fish], 8 December 1948 (CAEF – Centre des Archives Economiques et Financières [Centre for Archives of Economics and Finance], B64525).

17 See Hans Huss et al., 'Control of biological hazards in cold smoked salmon production', *Food Control*, 6, 6, 1995, pp. 335–40; Gilbert et al., *Risk Profile: Clostridium Botulinum in Ready-to-eat Smoked Seafood in Sealed Packaging*, op. cit.

18 Adria, 'Etude de conservation du saumon fumé prétranché, emballé sous vide et réfrigéré et tests de quelques conservateurs' [Study of the preservation of pre-sliced smoked salmon, which is vacuum-packed and refrigerated, and tests on various preservative agents], Adria, Quimper, 1979.

19 Tompkin, 'Nitrite', art. cit., pp. 210–11.

20 See, for example, the case of producers in Greenland: Nauja Moller, 'Grønlandsk firma bag nitrit-skandale' [Greenland company behind the nitrite scandal], KNR-radio Toqqaannartoq, Maajip 06-at 2010.

21 See Anthony Rowley (ed.), *Les Français à table. Atlas historique de la gastronomie française* [The French at the Table: Historical Atlas of French Gastronomy], Hachette, Paris, 1997, pp. 32–3.

22 Stevenson, *The Preservation of Fishery Products*, op. cit., pp. 557–63.

23 Lerfall and Østerlie, 'Use of sodium nitrite in salt-curing of Atlantic salmon', art. cit.

24 Memorandum of FDA meeting published in Subcommittee of the Committee on Government Operations, *Regulation of Food Additives and Medicated Animal Feeds*, op. cit., p. 393.

25 Committee on Government Operations, *Regulation of Food Additives – Nitrites and Nitrates, Nineteenth Report*, op. cit., pp. 13–14. The firm opposition of the FDA is described in Subcommittee of the Committee on Government Operations, *Regulation of Food Additives and Medicated Animal Feeds*, op. cit., pp. 210–12.

26 Hearing of Commissioner Georges Larrick at the Subcommittee of the Committee on Interstate and Foreign Commerce, House of Representatives, *Food Additives*, US GPO, Washington, 1958, p. 449.

27 Ibid., p. 450.

28 Ibid.

29 Letter from the FDA to Norman Armstrong Ltd., 9 September 1959, National Archives and Record Administration, RG88 FDA General Subject files 1924–1978, box 23.

30 Letter from the FDA to Otten's Meat Curing Supplies, 23 December 1957, National Archives and Record Administration, RG88 FDA General Subject files 1924–1978, box 23.

31 Letter from the FDA to USDA, 23 December 1957, National Archives and Record Administration, RG88 FDA General Subject files 1924–1978, box 23.

32 See Subcommittee of the Committee on Government Operations, *Regulation of Food Additives and Medicated Animal Feeds*, op. cit., p. 215.

33 Letter from John Schnably (Bureau of Enforcement) to the General Counsel, FDA, 12 May 1959, National Archives and Record Administration, RG88 FDA General Subject files 1924–1978, box 31.

34 Ibid.

35 Ibid.

36 Hearing of Commissioner Charles Edwards at the Subcommittee of the Committee on Government Operations, *Regulation of Food Additives and Medicated Animal Feeds*, op. cit., p. 169.

37 Ibid. In the same volume see also the hearing of the FDA counsel William Goodrich (p. 256).

38 Ibid., p. 169; Code of Federal Regulations: regulations 21 CFR 121.1063 and 121.1064.

39 Johan Sakshaug et al., 'Dimethylnitrosamine; its hepatotoxic effect in sheep and its occurrence in toxic batches of herring meal', *Nature*, 206, 4990, June 1965.

40 K.G. Weckel and Susan Chien, 'Use of sodium nitrite in smoked Great Lakes chub', *Research Report 51*, University of Wisconsin, September 1969, p. 1; Mel Eklund, 'Control in fishery products', in Hauschild and Doods, *Clostridium Botulinum*, op. cit.

41 H. Tarr, 'The action of nitrites on bacteria', *Journal of the Fisheries Research Board of Canada*, 5, 1941; H. Tarr, 'The action of nitrites on bacteria: further experiments', *Journal of the Fisheries Research Board of Canada*, 6, 1942.

42 Code of Federal Regulations: Regulation 21 CFR 121.1230 (sodium nitrite in processing smoked chubs to aid in inhibiting the outgrowth of toxin formation from Clostridium botulinum type E); see the hearing of Dr Wodicka (FDA) in Subcommittee on Public Health and Environment, *FDA Oversight – Food Inspection*, US GPO, Washington, 1972, pp. 9, 21.

43 David Perlman, 'Suit seeks US data on food additives', *San Francisco Chronicle*, 8 February 1972; 'FDA is criticized on food additives', *Washington Post*, 25 April 1972.

44 Dale Hattis, 'The FDA and nitrate – a case study of violations of the food, drug and cosmetic act with respect to a particular food additive', extensively reproduced in Select Committee on Nutrition and Human Needs, *Nutrition and human needs. Part 4C-Food additives*, US GPO, Washington, 1972, p. 1715.

45 Subcommittee of the Committee on Government Operations, *Regulation of Food Additives and Medicated Animal Feeds*, op. cit., pp. 557–8 ; Michael Jacobson, *Don't Bring Back the Bacon: How Sodium Nitrite Can Affect Your Health*, CSPI, Washington, 1973, pp. 31–2.

46 Roy Morton (National Fisheries Institute) quoted by M. Jacobson, *Don't Bring Back the Bacon*, op. cit., pp. 32–3. See also Subcommittee of the Committee on Government Operations, *Regulation of Food Additives and Medicated Animal Feeds*, op. cit., p. 560.

47 Hattis, 'The FDA and nitrate', art. cit., p. 1716.

48 Ibid.

49 Dr Delphis Goldberg, in Subcommittee of the Committee on Government Operations, *Regulation of Food Additives and Medicated Animal Feeds*, op. cit., p. 561.

50 Ibid.

51 Morton Mintz, 'FDA says it erred on nitrite hazard', *Washington Post*. Facsimile reproduced in Jacobson, *Don't Bring Back the Bacon*, op. cit.

52 Committee on Government Operations, *Regulation of Food Additives – Nitrites and Nitrates*, op. cit., pp. 16–17, 19.

53 Ibid.

54 Elinor Ravesi, 'Nitrite additives – harmful or necessary?', *Marine Fisheries Review*, 38, 4, 1976, p. 28.

55 USA *vs* Nova Scotia Food Products Corp., 568 F. 2d 240 (2d Cir. 1977).

Chapter 13: Warnings and Denials

1 'Do we make carcinogens from nitrites we eat?', *Medical World News*, 10 April 1970, pp. 18–19.

2 'Meat color additive linked to cancer', *Washington Post*, 17 March 1971.

3 'Health warning on meat cosmetic', *Washington Daily News*, 16 March 1971.

4 'Cured meats yield cancer causatives', *Washington Evening Star*, 6 February 1972.

5 Leo Freedman quoted by Judy Blitman, 'Food and health experts warn against bringing home the bacon', *New York Times*, 8 August 1973, p. 42.

6 See Marvin Hayenga et al., *The U.S. Pork Sector: Changing Structure and Organization*, Iowa State University Press, Ames, 1985, p. 23.

7 Ibid.

8 On the history of the transformation of the pig species to suit the processes of industrial meat curing, see Jocelyne Porcher, *Cochons d'or: L'industrie*

porcine en question [Golden Pigs: The Pork Industry Examined], Quae, Versailles, 2012.

9 Max Brunck, 'Pork demand in 1980 – impact of economic and social changes', and Earl Butz, 'An optimist looks at the swine industry', in Robert Schneidau and Lawrence Duewer (eds), *Vertical Coordination in the Pork Industry*, Avi Publishing, Westport, 1972, p. 5.

10 Earl Butz quoted by CAST (Council for Agricultural Science and Technology), 'What is this thing called food', Report 61, 15 September 1976, p. 3.

11 John Romans, 'Factual look at bacon scare', *Farmers Weekly Review*, 20 November 1975, p. 3.

12 Gardner, 'Sowbelly blues', art. cit., p. 142 (emphasis in the original press release).

13 See the correspondence reproduced in Committee on Small Business, *Food Additives: Competitive, Regulatory, and Safety Problems*, U.S. GPO, Washington, 1977, pp. 864–865.

14 See *Congressional Record – Senate*, 19 December 1975 (pp. 1–2), 28 January 1976 (pp. 1–2). It was only in 1977 that the panel was finally opened up to independent experts and scientists representing consumer associations.

15 Michael Jacobson, 'Statement to USDA's expert panel on nitrites and nitrosamines', 10 December 1975, *Congressional Record – Senate*, 19 December 1975.

16 Nancy Ross, 'Disclosure of nitrate-nitrite ordered for cured meats', *Washington Post*, 23 August 1972.

17 Marian Burros, 'Meat official reports carcinogens in bacon', *Washington Post*, 11 October 1975.

18 Ibid.

19 Nrisinha Sen et al., 'Nitrosamines in cured meat products', *IARC Scientific Publications*, 14, 1976, pp. 333–42.

20 IARC, *Environmental N–nitroso Compounds Analysis and Formation*, IARC, Lyon, 1975.

21 IARC, *Some N–nitroso Compounds*, IARC monographs, vol. 17, IARC, Lyon, 1978.

22 Assistant Agriculture Secretary Carol Foreman quoted in *Food Chemical News*, 16 January 1978.

23 Donald Kennedy, 'Statement', 27 July 1978, FDA, p. 5.

24 Ibid.

25 Woodrow Aunan and Olaf Kolari (AMI Foundation), 'Functions of nitrite in cured meats', 33rd Annual Meeting IFT Symposium: Nitrate, Nitrite and Nitrosamines, Miami, June 1973, p. 3.

26 Ibid., pp. 4ff.

27 Anita Johnson interviewed by Joseph Winski, 'It's cost versus health risk as nitrite debate heats up', *Chicago Tribune*, 26 November 1978, p. 3.

28 Dennis Buege and Nitrite-Free Processed Meats Committee of the American Meat Science Association, 'Final report: Nitrite-free processed meats', *Reciprocal Meat Conference Proceedings*, 33, 1980.

29 Aaron Wasserman and Walter Kimoto, 'Consumer evaluation of the flavor of bacon cured with and without sodium nitrite', in B.J. Tinbergen and B. Krol (eds), *Proceedings of the Second International Symposium on Nitrite in Meat Products*, op. cit., p. 73.

30 Ibid., p. 74.

31 John Birdsall (AMI) quoted by Mary McKenzie, 'Nitrite additive in meat', in Subcommittee on Agricultural Research and General Legislation, *Food Safety and Quality – Nitrites*, op. cit., p. 162.

32 Richard Lechowich (CAST) quoted by Keith Wilkins, 'Nitrite and cured meat: a red hot issue', *Virginia Tech*, 1978.

33 R. Lechowich in Subcommittee on Agricultural Research and General Legislation, *Food Safety and Quality – Nitrites*, op. cit., pp. 57–8, 63.

34 Stephen Krut (American Association of Meat Processors), in Subcommittee on Dairy and Poultry, *Nitrite Restrictions on Poultry*, US GPO, Washington, 1978, p. 170.

35 Ibid.

36 Quoted in *Congressional Record – Senate*, 30 April 1981, p. 4276.

37 'U.S. study indicates botulism inhibition by nitrite', *Food Manufacture*, July 1972, p. 8.

38 Ibid.

39 See, for example, Albert Leach, *Food Inspection and Analysis*, 4th edn, Wiley, New York, 1941, p. 222.

40 C.E. Calm, *Sulphurous Acid and Sulphites as Food Preservatives*, Hygeian Chemical and Research Laboratory, Chicago, 1904, p. 18.

41 Ibid.

42 Ibid.

43 *The Scientific Meat Industry*, Heller, Chicago, 1, 1, June 1904, p. 11.

44 See in particular the arguments developed by Robert Eccles, *Food Preservatives, Their Advantages and Proper Use, the Practical Versus the Theoretical Side of the Pure Food Problem*, Van Nostrand, New York, 1905. Eccles' views formed the cornerstone of the meatpackers' case: see R. Eccles, 'The preservative situation', *The National Provisioner*, October 1908.

45 'Credit where it is due', *The National Provisioner*, 12 March 1904, p. 27.

46 On Robert Eccles, see Harvey Wiley, *The History of a Crime Against the Food Law*, op. cit., pp. 14–15, and H. Wiley, 'Memorandum regarding Dr Eccles's statement', in Committee on Interstate and Foreign Commerce, *Hearings on the Pure Food Bills*, op. cit., pp. 355–6.

47 Robert Eccles, 'The consumer interest in food preservatives', *The National Provisioner*, 12 October 1907, p. 76.

48 Eccles, *Food Preservatives, Their Advantages and Proper Use*, op. cit., p. 153.

49 Ibid., p. 197.

50 Ibid., p. 75.

51 Ibid.

52 Ibid., p. 35.

53 Harvey Wiley, 'A debate on the preservative question', *The Bulletin of Pharmacy*, December 1905.

54 C. Howard (New Hampshire Board of Health), 'Meat "improvers"', *Health*, 1936.

55 Mildred Maddocks, *The Pure Food Cook Book*, Hearst, New York, 1914, p. 189.

56 *56th Report on Food Products*, Connecticut Experiment Station, New Haven, 1953, pp. 31–2.

57 William Longgood, *The Poisons in Your Food*, Simon & Schuster, New York, 1960, pp. 162–3.

58 M. Jacobson, *Don't Bring Back the Bacon*, op. cit., p. 8 (emphasis in the original).

59 William Lijinsky, 'Nitrites in food', *Science*, Vol. 182, 21 December 1973.

60 'Need for nitrites in meat to prevent botulism questionable', in Committee on Government Operations, *Regulation of Food Additives – Nitrites and Nitrates*, op. cit., p. 11.

61 Congressman L.H. Fountain's Intergovernmental Relations Subcommittee, quoted by Gardner, 'Sowbelly blues', art. cit., p. 142.

62 *Congressional Record, Extensions of Remarks*, 2 March 1972, p. 6749.

63 Peter Schuk and Harrison Wellford, 'Botulism and nitrites', *Science*, 180, 29 June 1973, p. 1322.

64 'Household worries', *Newsweek*, 31 October 1977, p. 109.

65 'Potassium sorbate proposed as answer to nitrite reduction in cured meats', *Food Processing*, June 1976; Liz Forrestal, 'Nitrosamines and nitrite: nightmare for regulators', *Chemical Week*, October 1978.

66 A.M. Al-Shuibi and Bassem Al-Abdullah, 'Substitution of nitrite by sorbate and the effect on properties of mortadella', *Meat Science*, 62, 2002.

67 'The urgent drive for nitrite substitute', *Business Week*, 11 September 1978.

68 See 'Potassium sorbate proposed as answer to nitrite reduction in cured meats', art. cit.

69 Clyde W. Sweet, 'Additive composition for reduced particle size meats in the curing thereof', US Patent 3899600 assigned to Eastman Kodak Co., 12 August 1975.

70 R. Lechowich in Subcommittee on Agricultural Research and General Legislation, *Food Safety and Quality – Nitrites*, op. cit., p. 53.

71 Consultative Committee of Directors of Food Research Organizations, 'Minutes of meeting', 13 December 1971, MAF 260/632, no. 54.

72 'Nitrates and nitrites', 12 February 1971, MAF 260/631.

73 Ian Adams, 'Group discussion on nitrosamines in food', FDA Washington, 21 July 1972, MAF 260/632, p. 90b.

74 Marian Burros, 'Controversial decision on bacon safety', *Washington Post*, 16 December 1976.

75 Gardner, 'Sowbelly blues', art. cit., p. 114.

76 Yet the antibacterial effects of nitro-additives were well known: in their filthy factories, the meatpackers used them as an economical method of combatting the putrefying bacteria *Bacillus putrefaciens* and *Bacillus foedans* (see Chapter 3). In a study on the effects of nitro-additives on bacteria capable of developing on rotten fish (see Chapter 10), a chemist discovered that nitrite also had a bacteriostatic effect on *Clostridium botulinum* (H. Tarr, 'The action of nitrites on bacteria', art. cit.) and recalled that on meat the additive was meant to produce a desirable colour in the finished product (H. Tarr, 'Bacteriostatic action of nitrates', *Nature*, 147, April 1941, p. 418).

77 Bureau of Animal Industry, 'Notice regarding meat inspection – sodium nitrite for curing meat', art. cit.

78 FDA and USDA, 'Memorandum of interview', 8 July 1948, National Archives and Record Administration, RG88 FDA General Subject files 1924–1978, box 23.

79 'Section 318.7 – Approval of substances for use in the preparation of meat food products', Bureau of Animal Industry, *Regulations Governing the Meat Inspection of the United States Department of Agriculture*, USDA, Washington, 1965, pp. 101–06.

80 MAF 260/559 part 1, no. 37.

81 Minute of 24 February 1938 (report addressed to the secretary), MAF 101/327.

82 Letter from the Ministry of Health to the Ministry of Food, 3 July 1942, MH 56/176.

83 Académie nationale de médecine, *Rapport. Au sujet du sel nitrité*, art. cit.

84 Letter from the Fédération nationale de l'industrie de la salaison, de la charcuterie en gros et des conserves de viandes to the Ministère de l'Agriculture, 4 June 1963, art. cit.

85 Confédération de la charcuterie de France et de l'Union Française, *Manuel de l'apprenti charcutier*, op. cit., p. 124.

86 Ibid., pp. 154, 157. The same text, in its list of food poisonings, also notes that botulism is non-existent in French charcuterie (p. 78). It makes no mention of any link between the use of nitro-additives and protection against botulism.

87 Ibid., p. 125.

88 See, for example, in 1971 in Pallu, *La Charcuterie en France. Tome I: Généralités, charcuterie crue*, op. cit., p. 145. In the same vein, 'Session d'étude sur les produits de la salaison et de la charcuterie' [Study session on cured meat products], report by Monsieur Niel (Massy Central Laboratory) following a seminar held at the Lyon Veterinary School, 9–13 October 1972 (French National Archives, 830510/8-3cons8).

89 According to a report by the US National Academy of Science, the first authors to mention a specific anti-botulinic role for sodium nitrite were P. Steinke and E. Michael Foster (University of Wisconsin at Madison).

Maclyn McCarty et al., *The Health Effects of Nitrate, Nitrite and N-nitroso Compounds*, vol. 2, National Academy Press, Washington, 1982, pp. 2–10.

90 On the 'Committee of Food Protection' and the accusations of conflicts of interest, see: Select Committee on Nutrition and Human Needs, *Nutrition and human needs. Part 4C – Food Additives*, op. cit., pp. 1565, 1679–81, 1726.

91 P. Steinke and E.M. Foster, 'Botulinum toxin formation in liver sausage', *Journal of Food Science*, January 1951, Vol. 16.

92 'Feeding at the company trough', *Congressional Record – House*, 24 August 1976, p. 8975.

93 Advertisement for Wm. J. Stange Co., 'Pick a color – any color!', *American Soft Drink Journal*, February 1957.

94 'Feeding at the company trough', *Congressional Record – House*, art. cit.

95 See Hayenga et al., *The U.S. Pork sector*, op. cit., pp. 48–50.

96 Ibid.

97 'Feeding at the company trough', *Congressional Record – House*, art. cit.

Chapter 14: The 'No Alternative' Lie

1 See, for example, Robert Frank Kelly, 'Processing meat products without nitrates or nitrites', *Food Product Development*, October 1974.

2 Richard Lyng quoted in 'The urgent drive for nitrite substitute', *Business Week*, art. cit.

3 Richard Lechowich quoted in V. McElheny, 'The controversy over nitrites in meats', *New York Times*, 15 March 1978.

4 Aaron Wasserman quoted in 'The urgent drive for nitrite substitute', *Business Week*, art. cit.

5 Dennis Buege and Nitrite-Free Processed Meats Committee of the American Meat Science Association, 'Final report: Nitrite-free processed meats', art. cit., p. 130.

6 'U.S. label rules eased on nitrite meats', *Chicago Tribune*, 21 August 1979.

7 Robert Lenahan and Clark Burbee, *Regulatory Impact Statement: Nitrate and Nitrite, National Economics Division Staff Report*, USDA, Washington, 1979, p. 2.

8 Robert Gunn and William Terranova, 'From the Center for Disease Control. Botulism in the US – 1977', *Reviews of Infectious Diseases*, 4, 1979, p. 723. See also Committee on Small Business, *Food Additives: Competitive, Regulatory, and Safety Problems*, op. cit., pp. 800–08.

9 Marian Burros, 'Nitrites, nitrates in meat products targeted by US', *Washington Post*, 1 September 1977.

10 Marian Burros, 'Nitrite-free bacon spices controversy', *Washington Post*, 14 March 1979, p. D8.

11 Kelly, 'Processing meat products without nitrates or nitrites', art. cit.

12 Schuk and Wellford, 'Botulism and nitrites', art. cit.

13 'Nitrates, nitrites: readers speak up', *The New York Times*, 11 February 1976, p. 52.

14 Buege and Nitrite-Free Processed Meats Committee of the American Meat Science Association, 'Final report: Nitrite-free processed meats', art. cit.

15 Blitman, 'Food and health experts warn against bringing home the bacon', art. cit., p. 42.

16 Marian Burros, 'The nitrite question: what can you eat?', *The New York Times*, 23 December 1981, p. C1.

17 Ibid.

18 Burros, 'Nitrite-free bacon spices controversy', art. cit.

19 Ibid.

20 Ibid.

21 Food Safety and Quality Service, *Background – Nitrite in Cured Meat*, US Department of Agriculture, July 1980, p. 6.

22 Britton Central School District, quoted in Subcommittee on dairy and poultry, *Nitrite Restrictions on Poultry*, op. cit., p. 13.

23 Paul Ingrassia, 'Meats without nitrites could be labeled conventionally in change mulled by U.S.', *Wall Street Journal*, 12 September 1978, p. 38.

24 Ibid.

25 Nicholas Von Hoffman, 'Nitrite ban aftereffects', *The Washington Post*, 7 September 1978.

26 'The urgent drive for nitrite substitute', *Business Week*, art. cit.

27 Ingrassia, 'Meats without nitrites could be labeled conventionally in change mulled by U.S.', art. cit.

28 'Ray Kennedy: the rebel meat-packer', *Esquire*, November 1976, p. 144. See also M. Burros, 'Controversial decision on bacon safety', art. cit.

29 Ibid.

30 Ingrassia, 'Meats without nitrites could be labeled conventionally in change mulled by U.S.', art. cit.

31 Hearing of Carol Tucker-Foreman (USDA), in Subcommittee on Agricultural Research and General Legislation, *Food Safety and Quality – Nitrites*, op. cit., p. 162.

32 Hearing of Bill Schultz in ibid., p. 85.

33 'Cites Carol Foreman's proposed nitrite-free meat products a health hazard', *Congressional Record — House*, 25 May 1978, p. 15464.

34 'National Pork Producers Council et al. *vs* Bergland et al.' (No 80-1229, judgement of 23 September 1980), *United States Attorneys Bulletin*, 28, 22, 24 October 1980, p. 764; Food Safety and Quality Service, *Background – Nitrite in Cured Meat*, art. cit, p. 4.

35 'USDA defends safety of nitrite-less meats in court suit', *Food Chemical News*, Vol. 21, No. 31, 15 October 1979.

36 Ibid.

37 Virginie Payette, 'Consumer Alert battling regulations', *The Lewiston Journal*, 10 November 1981, p. 3.

38 'No-nitrite meats', Consumer Alert, Inc., (n.d.), (capitals in the original), document znpc0137 available at www.industrydocuments.library.ucsf.edu

39 'Nitrates, Nitrites: Readers Speak Up', art. cit.

40 Ibid.

41 Ibid.

42 Daniel Gray and L.W. Summers, *Curing Meat on the Farm*, Alabama Polytechnic Institute, Auburn, 1912, pp. 202, 204.

43 George Washington Carver, *The Pickling and Curing of Meat in Hot Weather*, Tuskegee Institute, Tuskegee, 1912, p. 6.

44 Carl Dunker and Orville Hankins, *A Survey of Farm Meat-Curing Methods*, USDA, Washington, 1951, pp. 2, 3, 6.

45 Ibid., pp. 2, 6.

46 Clayton Yeutter (administrator, Department of Agriculture) in Subcommittee of the Committee on Government Operations, *Regulation of Food Additives and Medicated Animal Feeds*, op. cit., p. 388.

47 Gail Dack, 'Characteristics of botulism outbreaks in the United States', University of Chicago (n.d.), reproduced in Subcommittee of the Committee on Government Operations, *Regulation of Food Additives and Medicated Animal Feeds*, op. cit.

48 Subcommittee of the Committee on Government Operations, *Regulation of Food Additives and Medicated Animal Feeds*, op. cit., p. 383.

49 Paul Newberne quoted in Philip Hilts, 'The day bacon was declared poison', *Washington Post Magazine*, 26 April 1981.

50 Hearing of R. Lechowich in Subcommittee on Agricultural Research and General Legislation, *Food Safety and Quality – Nitrites*, op. cit., p. 53.

51 Hearing of William Lijinsky (National Cancer Institute) in Committee on Small Business, *Food Additives: Competitive, Regulatory and Safety Problems*, op. cit., p. 641.

52 Hearing of W. Lijinsky in Subcommittee on Agricultural Research and General Legislation, *Food Safety and Quality – Nitrites*, op. cit., p. 36

53 'Still a jungle', *New York Times*, 16 November 1967; 'Compromised meat', *New York Times*, 21 November 1967, reproduced in *Congressional Record*, 27 November 1967.

54 Committee on Agriculture, *Amend the Meat Inspection Act*, Washington, 1967.

55 Inspection report, 21 September 1962, *Congressional Record – Senate*, 31 October 1967.

56 See *Congressional Record – Senate*, 3 August 1967; *Congressional Record*, 27 November 1967.

57 Reports by Dr J. Klein and the inspector Ruben Baumgart reproduced in *Congressional Record – Senate*, 31 October 1967, pp. 30528–30.

58 See the intervention of Senator Mondale, *Congressional Record – Senate*, 27 November 1967.

59 See the report of R. Baumgart reproduced in *Congressional Record – Senate*, 31 October 1967.

60 In a Senate hearing in September 1978, the vice president of the National Pork Producers Council indicated that 75–80% of production was controlled by the major industrial meat-processing firms: see the hearing of

Marvin Garner in Subcommittee on Agricultural Research and General Legislation, *Food Safety and Quality – Nitrites*, op. cit., p. 69.

61 N. Von Hoffman, 'Nitrite ban aftereffects', art. cit.

62 Ibid.

63 Ross Hall, 'The great nitrite scandal', *En-Trophy Institute Hamilton*, Ontario, Vol. 2, No. 1, reprinted in *The Ecologist*, 9, 3, May–June 1979, p. 96.

64 M. Jacobson, 'Statement to USDA's expert panel', art. cit.

65 Richard Greenberg (Swift), in Expert Panel on Nitrites and Nitrosamines, *Final Report on Nitrites and Nitrosamines*, op. cit., p. 93.

66 Ibid.

67 Expert Panel on Nitrites and Nitrosamines, *Final Report on Nitrites and Nitrosamines*, op. cit., p. 39.

68 Ibid., p. 50.

69 M. Jacobson, 'Statement to USDA's expert panel', art. cit.

Chapter 15: Contesting the Health Argument

1 *Opinion of the Attorney General of the United States*, Vol. 43, opinion no. 19, 30 March 1979, p. 12 (available at www.industrydocuments. library.ucsf.edu, reference gzxj0060).

2 'Nitrates and nitrites in meat products. Statement of policy, request for data', *Federal Register*, Vol. 42, 18 October 1977.

3 AMI, 'Statement', *Extensions of Remarks – House*, 6 April 1978.

4 See the hearing of Marvin Garner, vice president of the National Pork Producers Council, in Subcommittee on Agricultural Research and General Legislation, *Food Safety and Quality – Nitrites*, op. cit., p. 67.

5 Hearing of S. Krut (American Association of Meat Processors), in ibid., p. 240.

6 Ibid.

7 Richard Lyng (AMI) interviewed by J. Winski, 'It's cost versus health risk ...', art. cit., p. 4; also in 'Nitrites in meat: is the danger clear?', *The Montreal Gazette*, 29 November 1978.

8 Congressman Kelly in Subcommittee on Dairy and Poultry, *Nitrite Restrictions on Poultry*, op. cit., p. 17.

9 Steven Tannenbaum, interviewed by WMT-stations (Iowa), 16 August 1978, reproduced in *Extensions of Remarks*, 23 September 1978, p. 31163.

10 Robert Cassens et al., 'The use of nitrite in meats', *BioScience*, 28, 10, October 1978, p. 633.

11 Richard Lechowich et al., *Nitrite in Meat Curing: Risks and Benefits*, Council for Agricultural Science and Technology, Ames, 1978.

12 McElheny, 'The controversy over nitrites in meats', art. cit.

13 Cassens et al., 'The use of nitrite in meats', art. cit., p. 633.

14 Hayenga et al., *The U.S. Pork sector*, op. cit., pp. 99–102. See also US Department of Agriculture, *An Analysis of a Ban on Nitrite Use in Curing Bacon*, USDA, Washington, March 1979.

15 Buege and Nitrite-Free Processed Meats Committee of the American Meat Science Association, 'Final report: Nitrite-free processed meats', art. cit., p. 123.

16 Ibid., p. 130.

17 McElheny, 'The controversy over nitrites in meats', art. cit.

18 Wilkins, 'Nitrite and cured meat: a red hot issue', art. cit.

19 Aaron Wasserman and Ivan Wolff, 'Nitrites and nitrosamines in our environment: an update', USDA, Philadelphia, 1979.

20 R. Lechowich (CAST) in Subcommittee on Agricultural Research and General Legislation, *Food Safety and Quality – Nitrites*, op. cit., pp. 57–8.

21 See, for example, the presentation of the CAST report by the Representative Wampler: 'Nitrite in meat curing: risks and benefits. Part I', *Congressional Record – House*, 7 March 1978.

22 See the hearing of Senator Luger (p. 56) and Marvin Garner, vice-president of the National Pork Producers Council (p. 68), in Subcommittee on Agricultural Research and General Legislation, *Food Safety and Quality – Nitrites*, op. cit.

23 Senator Leahy in Subcommittee on Agricultural Research and General Legislation, *Food Safety and Quality – Nitrites*, op. cit., pp. 161–2.

24 Hearing of Carol Tucker-Foreman (USDA) in Subcommittee on Agricultural Research and General Legislation, *Food Safety and Quality – Nitrites*, op. cit., p. 162.

25 S. Krut (American Association of Meat Processors) in Subcommittee on Dairy and Poultry, *Nitrite Restrictions on Poultry*, op. cit., p. 170.

26 Lechowich (CAST), in Subcommittee on Agricultural Research and General Legislation, *Food Safety and Quality – Nitrites*, op. cit., pp. 57–8, 62.

27 For example, Boston Brisket Co., Boston; on grey corned beef, see Linda Bock, 'Red or Gray', *Telegram* (Worcester, MA), 16 March 2009.

28 Krut (American Association of Meat Processors), in Subcommittee on Dairy and Poultry, *Nitrite Restrictions on Poultry*, op. cit., p. 169.

29 Ibid.

30 Theodore Labuza, 'Food additives. Novelty or necessity', in Committee on Agriculture, Nutrition and Forestry, *Food Safety: Where Are We?*, US GPO, Washington, 1979, p. 314.

31 Theodore Labuza, 'The risks and benefits of our food supply', in Committee on Agriculture, Nutrition and Forestry, *Food Safety: Where Are We?*, op. cit., p. 328.

32 Charles Nyberg (Hormel), 'Sola dosis facit venenum (Only the dose makes the poison)', reproduced in *Congressional Record – Extensions of Remarks*, 14 September 1979.

33 Ibid.

34 Lechowich (CAST), Subcommittee on Agricultural Research and General Legislation, *Food Safety and Quality – Nitrites*, op. cit., p. 58.

35 Robert Proctor, *Cancer Wars: How Politics Shapes What We Know and Don't Know About Cancer*, op. cit., p. 128.

36 See Peter Harnik, *Voodoo Science, Twisted Consumerism: The Golden Assurances of the ACSH*, CSPI, Washington, 1983. A few years later, in his book *Cancer Wars*, the historian Robert Proctor dedicated several pages to the information campaigns conducted by Whelan and her organization. Proctor, *Cancer Wars: How Politics Shapes What We Know and Don't Know About Cancer*, op. cit., pp. 89–92.

37 Wayne Pines (FDA) quoted by Nannie Collins, *People Magazine*, 27 August 1979, p. 84.

38 ACSH, brochure 'Does Nature Know Best?', 1985.

39 Interview with E. Whelan by Marjorie Rice, 'Stop banning things at the drop of a rat', *New York Newsworld*, 24 November 1979.

40 Ibid.

41 See, for example, E. Whelan, 'The shopping bag diet', *ACSH News & Views*, September–October 1985, pp. 14–15.

42 E. Whelan in 'Caution about precautions', *Youngstown Vindicator*, 29 August 1978.

43 Ibid.

44 On Whelan's campaigns in defence of other carcinogenic agents, see Proctor, *Cancer Wars: How Politics Shapes What We Know and Don't Know About Cancer*, op. cit., especially pp. 91, 128, 150. See also the remarks on the way in which the carcinogenic industries developed the idea of 'irrational fear', for example the 'fear of nuclear radiation' ('radiophobia').

45 'Bacon, hamburgers and cancer. Bacon: new rules for chemicals', *Washington Star*, 16 May 1978.

46 See the basic patent for this technique: Hollenbeck, 'Curing of meat', art. cit.; see also the observations of Pegg and Shahidi, *Nitrite Curing of Meat*, op. cit., p. 12.

47 Richard Lyng interviewed by J. Winski, 'It's cost versus health risk …', art. cit., p. 4.

48 Robert Bogda, 'U.S. bacon industry seems to have won nitrite battle with federal regulators', *The Wall Street Journal*, 21 June 1978.

49 Goody Solomon, 'Consumer groups want ban. USDA faces sodium nitrite suit', *Chicago Tribune*, 1 June 1978.

50 Betty Stevens (*National Provisioner*) quoted by Bogda, 'U.S. bacon industry seems to have won nitrite battle with federal regulators', art. cit.

51 Robert Madeira (American Association of Meat Processors) quoted in ibid.

52 Richard Lyng (American Meat Institute) quoted in ibid.

53 Ibid.

54 Paul Newberne, 'Nitrite promotes lymphoma incidence in rats', *Science*, 204, 8 June 1979; see 'Statement on nitrites', FDA/USDA, 11 August 1978.

55 Donald Kennedy, 'Statement', 27 July 1978, FDA.

56 Bogda, 'Most pork-belly futures prices increase, indicating ban on nitrite isn't seen soon', art. cit.

57 Rich Jaroslovsky, 'U.S. plans total ban on use of nitrites in foods but on schedule to assure safety', *Wall Street Journal*, 17 August 1978.

58 Motions: H.R.13899 (16 August 1978), H.R.14213 (2 October 1978), H.R.563 (15 January 1979), H.R.1818 (1 February 1979), H.R.1879 (2 May 1979), etc.

59 Letter from the British embassy (Washington, DC) to the Ministry of Agriculture (London), 2 November 1978, MAF 260/897.

60 Declaration of Congressman Martin, 'UPI unearths secret plan at Agriculture and FDA to ban nitrite', *Congressional Record – House*, 17 August 1978.

61 'Capitol comment', *Farmers Weekly Review*, 57, 15, 21 September 1978, p. 4. On the same theme, see, for example, S. Krut (American Association of Meat Processors), in Subcommittee on Dairy and Poultry, *Nitrite Restrictions on Poultry*, op. cit., p. 170.

62 'Government asks delay in elimination of nitrites', *New York Times*, 31 March 1979, p. 6.

63 Bob Bergland quoted in 'Congress asked to delay nitrites control but to approve phase-out of use in food', *Wall Street Journal*, 2 April 1979, p. 12.

64 Lynne Olson, 'Califano, Bergland ask delay on nitrites ban', *The Sun*, 31 March 1979, p. A16.

65 'Government asks delay in elimination of nitrites', art. cit.

66 Ibid.; Olson, 'Califano, Bergland ask delay on nitrites ban', art. cit.

67 Ellen Haas quoted by Olson, 'Califano, Bergland ask delay on nitrites ban', art. cit.

68 Ibid.

69 Ibid.

70 'Senators ask FDA, USDA to end "confusion" on nitrite announcements', *Food Chemical News*, 15 October 1979, p. 15.

71 General Accounting Office, *Does Nitrite Cause Cancer?*, Washington, January 1980. The report was produced at the express request of Senator Grassley of Iowa, as he would admit later (see *Congressional Record*, 30 April 1981).

72 Victor Cohn, 'US agencies reject banning nitrite in cured meats', *Washington Post*, 20 August 1980.

73 Jere Goyan (FDA) quoted in ibid.

74 'Nitrites get a nod – at least for now', *Newsweek*, 1 September 1980, p. 62; 'USDA and FDA issue joint statement on nitrite study', *USDA News*, 19 August 1980.

75 James Martin in Committee on Agriculture, *USDA/FDA Announcement on Nitrites and Related Issues*, US GPO, Washington, 1980, pp. 5–6.

76 Senator Grassley, *Congressional Record – Senate*, 14 December 1981, p. 31068 (our emphasis).

77 Paul Jacobs, 'No link between nitrite in meat, cancer, US claims', *Los Angeles Times*, 20 August 1980, p. 26.

78 'All clear for nitrites in food', *New Scientist*, 28 August 1980.

79 'Group seeking nitrite ban', *Sunday Journal*, 25 February 1979.

80 'Nitrites get a nod – at least for now', art. cit.

81 Ibid.

82 Donald Kennedy, 'Dr Kennedy replies', *The Wharton Magazine*, summer 1981, p. 55.

83 Committee on Agriculture, *USDA/FDA Announcement on Nitrites and Related Issues*, op. cit., pp. 14–15.

84 Karen De Witt, 'U.S. will not seek to ban nitrite from foods as a cause of cancer', *New York Times*, 20 August 1980.

85 Sydney Butler (USDA) quoted in Cohn, 'US agencies reject banning nitrite in cured meats', art. cit.

86 Sydney Butler (USDA) quoted in 'Nitrites get a nod – at least for now', art. cit.

87 'Nitrites get a nod – at least for now', art. cit.

88 J. Goyan (FDA) quoted in 'Nitrites get a nod – at least for now', art. cit.

89 R. Jeffrey Smith, 'Nitrites: FDA Beats a Surprising Retreat', *Science*, 209, September 1980, p. 1101.

90 Ibid.

91 Cohn, 'US agencies reject banning nitrite in cured meats', art. cit.; 'Study fails to link nitrites to cancer; US doesn't plan ban of preservative', *Wall Street Journal*, 20 August 1980.

Chapter 16: After the War

1 'Most farm officials named, but Reagan to appoint more', *Nevada Daily Mail*, 29 January 1981, p. 3.

2 Eric Schlosser, *Fast Food Nation, the Dark Side of the All-American Meal*, Perennial/HarperCollins, New York, 2001.

3 R.J. Smith, 'Nitrites: FDA Beats a Surprising Retreat', art. cit.

4 M. Burros, 'U.S. food regulation: tales from a twilight zone', *The New York Times*, 10 June 1987.

5 See, for example, David Treadwell, 'Up to 60% of cancer may be tied to diet, study says', *Los Angeles Times*, 17 June 1982.

6 See the threats made against the 1982 report by the National Academy of Science: Janet Neiman, 'Food groups cry foul over cancer link data', *Advertising Age*, 21 June 1982.

7 See what the Arizona health authorities said about this in Committee on Agriculture, Nutrition and Forestry, *Food Safety: Where Are We?*, op. cit., p. 414.

8 Nitrite Safety Council, 'A survey of nitrosamines in sausages and dry-cured meat products', *Food Technology*, July 1980, pp. 45–53.

9 Ralph Moss 'The nitrite fiasco – look what's back in our hot dogs', *The Nation*, 7 February 1981 (as quoted in Michael Sweeney, 'Whom can you trust? The nitrite controversy', in Aaron Wildavsky (dir.), *But is it true?*, Harvard University Press, Cambridge, MA, 1997.

10 Proctor, *Cancer Wars: How Politics Shapes What We Know and Don't Know About Cancer*, op. cit., pp. 75–100.

11 Ibid., p. 81 (our emphasis).

12 McCarty et al., *The Health Effects of Nitrate, Nitrite and N-nitroso Compounds*, op. cit.

13 In the press release, some sentences of the report were even rewritten to change the meaning. For example, on pages 1–4 of the report, the expression 'no greater hazard' was changed to 'no hazard' in the press release, changing the meaning of the sentence.

14 Ibid., Vol. 2, pp. 7.3–7.18.

15 Ibid., Vol. 2, pp. 1.4–1.5.

16 Jane Brody, 'A report on nitrites finds cured meats are relatively safe', *The New York Times*, 11 December 1981, pp. A1 and A32, also published 12–13 December 1981 in *International Herald Tribune* under the title 'Cured meat only a minor source of nitrite exposure, study finds'.

17 Senator Grassley, *Congressional Record – Senate*, 14 December 1981, p. 31068.

18 Sweeney, 'Whom can you trust? The nitrite controversy', art. cit., p. 245.

19 Burros, 'The nitrite question: what can you eat?', art. cit.

20 Brody, 'A report on nitrites finds cured meats are relatively safe', art. cit.

21 Burros, 'The nitrite question: what can you eat?', art. cit.; Marian Burros, 'Nitrite controversy rekindled by study', *Ocala Star-Banner* (*The New York Times*), 2 June 1982, p. 8E.

22 Burros, 'Nitrite controversy rekindled by study', art. cit.

23 Ibid.

24 Press release, 'Research council panel recommends limited reduction of nitrite in cured meats', 11 December 1981 (document qsfd0137 available at www.industrydocuments.library.ucsf.edu), p. 1.

25 Victor Cohn, 'Researchers urge food processors to cut nitrite, nitrate levels', *The Washington Post*, 11 December 1981.

26 Burros, 'The nitrite question: what can you eat?', art. cit.

27 Press release, 'Research council panel recommends limited reduction of nitrite in cured meats', art. cit., p. 3.

28 Ibid.

29 'Smokers risk more exposure to nitrosamine', *Tempe News* (Arizona), 9 June 1982

30 Ten years later, the tobacco lobby (the Tobacco Institute) would repeat virtually the same arguments to contest the dangers of nicotine addiction: the industry spent millions of dollars fighting the ban on smoking in offices, universities, restaurants and public places – on the basis that for non-smokers, the nitrosamines in the smoke represented only a small proportion of the total nitrosamines they were exposed to. In this curious game of relativism each lobby group used the carcinogenic agents of the others to distract attention from their own. See, for example, the many publications of A. Tricker (Philip Morris International): A. Tricker et al., 'Tobacco-specific and volatile n-nitrosamines in indoor air of smoker and nonsmoker homes' (document gzfw0084, available at www.industrydocuments.library.ucsf.edu); or A. Tricker, 'N-Nitroso compounds and man: sources of exposure, endogenous formation and occurrence in body fluids' (document hhwp0122); or H. Klus et al., 'Tobacco-specific and volatile

N-nitrosamines in environmental tobacco smoke of offices', *Indoor and Built Environment* 1992, 1, etc.

31 V. McElheny, 'The controversy over nitrites in meats', art. cit.

32 Bryce Nelson, 'Study minimizes danger of cancer from meat nitrites', *The Los Angeles Times*, 11 December 1981.

33 Brody, 'A report on nitrites finds cured meats are relatively safe', art. cit.

34 Hearing of Donald Kennedy (Commissioner of FDA) and of Richard Cooper (Chief Counsel), in Subcommittee on Agricultural Research and General Legislation, *Food Safety and Quality – Nitrites*, op. cit, part III, p. 161.

35 National Research Council – Assembly of Life Sciences – Committee on Diet, Nutrition, and Cancer, *Diet, Nutrition, and Cancer*, National Academy Press, Washington, 1982. p. 1.

36 Ibid., p. 256.

37 Ibid., p. 1.

38 Alvin Lazen, 'Diet, cancer and the American Meat Institute', *The Washington Post*, 10 July 1982.

39 Ibid.

40 Assistant Agriculture Secretary Carol Foreman quoted in *Food Chemical News*, 16 January 1978.

41 Steven Tannenbaum, interviewed by WMT-stations (Iowa), 16 August 1978, reproduced in *Extensions of Remarks*, 23 September 1978, art. cit.

42 Section 409(c)(3)(A) of the Federal Food, Drug and Cosmetic Act, 21 US C. Section 348(c)(3)(A), aka the 'Delaney Clause of the Food Additives Amendment of 1958'.

43 Congressman Clarence 'Bud' Brown, quoted in Elaine Williams and Robert Harkins, 'The nitrite controversy', Grocery Manufacturers of America, Inc., Washington, DC, November 1973 (revised February 1974), p. 20 (document lffv0228, available at www.industrydocuments.library. ucsf.edu).

44 Ibid.

45 See the hearing of Harry Mussman (USDA) in Committee on Small Business, *Food Additives: Competitive, Regulatory, and Safety Problems*, op. cit., pp. 717, 720–1.

46 'U.S. says banning of nitrites unneeded', *The Hartford Courant*, 20 August 1980, p. 8.

47 'Nitrosamines', 9 March 1971, MAF 260/631.

48 Hearing of Dr Virgil Wodicka (director, Bureau of Foods, FDA) in Subcommittee on Public Health and Environment, *FDA Oversight – Food Inspection*, op. cit., p. 10 (our emphasis).

49 For an insight into questions of the non-applicability of American anti-cancer measures ('Delaney Clause') to nitro-additives, see, for example: Expert Panel on Nitrites and Nitrosamines, *Final Report on Nitrites and Nitrosamines*, op. cit., especially pp. 25, 101, 104–5, etc.; also Committee on Government Operations, *Regulation of Food Additives – Nitrites and Nitrates*, op. cit.

50 Hall, 'The great nitrite scandal', art. cit., p. 94.

51 For an insight into the debates on the exclusion clause known as the 'grandfather clause', see, for example: Expert Panel on Nitrites and Nitrosamines, *Final Report on Nitrites and Nitrosamines*, op. cit., pp. 104–5; Committee on Government Operations, *Regulation of Food Additives – Nitrites and Nitrates*, op. cit.

52 See Thomas Neltner et al., 'Navigating the U.S. Food Additive Regulatory Program', *Comprehensive Reviews in Food Science and Food Safety*, 10, 2011, pp. 343, 349. The final decision to grant 'prior-sanctioned' status to nitro-additives has been published in the *Federal Register*, 14 January 1983, pp. 1702–5. See Code of Federal Regulations: regulations 21 CFR 181.33 and 181.34. (Subpart B – Specific Prior-Sanctioned Food Ingredients).

53 To understand the power of the judicial 'lock' that nitro-additives enjoyed thanks to 'prior-sanctioned' status, see, for example, the judgement of Judge Tamm, 631 F.2d 969, 203 U.S.App.D.C. 387, United States Court of Appeals, District of Columbia Circuit, 'Public Citizen *vs* Carol Tucker Foreman' (1980), available at law.ressource.org

54 Robert Proctor, *Cancer Wars: How Politics Shapes What We Know and Don't Know About Cancer*, op. cit., pp. 10, 104, 128.

55 American Meat Institute, *Processed Meats: Convenience, Nutrition, Taste – American Traditions and Iconic Foods*, 2013, available at www.meat-institute.org, p. 13 (accessed September 2019).

56 Ibid.

57 Jeffrey Sindelar and Andrew Milkowski, 'Rethinking nitrite safety concerns', AMI, July 2012.

58 American Meat Institute, 'AMI fact sheet: sodium nitrite, the facts', 2008, available at www.meatinstitute.org (accessed September 2019).

59 Ibid.

60 Betsy Booren – vice-president of scientific affairs of the North American Meat Institute (NAMI), the new name of the AMI – quoted in 'Processed meats rank alongside smoking as cancer cause – WHO', *The Guardian*, 26 October 2015. On Betsy Booren, see Horel, *Lobbytomie*, op. cit., pp. 20–3.

61 See, for example, 'Special report: how the World Health Organization's cancer agency confuses consumers', *Reuters*, 18 April 2016.

62 See the correspondence of nitro-meat lobbyist James Coughlin in Carey Gillam, 'Email of intrigue: IARC is killing us!', USRTK.org, 19 April 2018.

63 Randy Huffman (president of the AMI research foundation), quoted in 'Experts dispute meat–cancer link', porknetwork.com, 8 August 2008; also in 'Experts cast doubts on the meat and cancer hypothesis' on www.meatinstitute.org, 7 August 2008.

64 Pork Checkoff, *Sodium Nitrite: Essential to Food Safety*, Des Moines (Iowa), 2012.

65 'Nun erwiesen: positive Wirkungen des Nitritpökelsalzes dominieren' [Now proven: positive effects of nitrited curing salt dominate], www.aula.at,

accessed May 2011. The nitrited curing salt previously manufactured by Aula is currently distributed by Christl Gewürze Gmbh.

66 Richard Lane quoted by Toni Tarver, 'Are nitrates and nitrites misunderstood?', IFT.org, 1 January 2019.

67 Suleiman Matarneh quoted by Tarver, 'Are nitrates and nitrites misunderstood?', art. cit.

68 Ibid.

69 Pork Checkoff, *Sodium Nitrite: Essential to Food Safety*, art. cit.

70 Joint WHO/FAO expert consultation, *Diet, Nutrition and the Prevention of Chronic Disease*, op. cit., p. 101.

71 Andrew Milkowski et al., 'Nutritional epidemiology in the context of nitric oxide biology: A risk–benefit evaluation for dietary nitrite and nitrate', *Nitric Oxide*, 22, 2010, p. 117.

72 Ibid.

73 Pork Checkoff, *Sodium Nitrite: Essential to Food Safety*, art. cit.; see also Nathan Bryan quoted by Trent Loos and Sarah Muirhead, 'Cured meats beneficial', *Feedstuffs – The Weekly Newspaper for Agribusiness*, 81, 29, July 2009, p. 3.

74 See in Kennedy, 'Dr Kennedy replies', art. cit.

75 Richard Hickey and Richard Clelland, 'Hazardous food additives: nitrite and saliva?', *New England Journal of Medicine*, 4 May 1978.

76 Lazen, 'Diet, Cancer and the American Meat Institute', art. cit.

77 'Experts dispute meat–cancer link', art. cit.

78 'AMSA announces white paper on sodium nitrites in meat-processing', www.provisioneronline.com, 15 November 2011.

79 Ibid.

Chapter 17: Alice in Nitroland

1 Lewis Carroll, *Through the Looking-Glass*, quoted by Siddhartha Mukherjee, *The Emperor of All Maladies: A Biography of Cancer*, Scribner, New York, 2010.

2 See Håkan Björne, *The Nitrite Ion: Its Role in Vasoregulation and Host Defenses*, Karolinska Institutet, Stockholm, 2005, p. 20; see also Subcommittee of the Committee on Government Operations, *Regulation of Food Additives and Medicated Animal Feeds*, op. cit., p. 638.

3 Wiley, 'A debate on the preservative question', art. cit.

4 Otto Hehner, 'On the use of food preservatives', *The Analyst*, December 1890.

5 Pork Checkoff, *Sodium Nitrite: Essential to Food Safety*, art. cit.

6 AMI, 'AMI Fact sheet. Sodium nitrite, the facts', art. cit.

7 Ibid.

8 'Hot dog preservative could be new medication', 5 September 2005, www.nbcnews.com, accessed October 2019.

9 Ibid.

10 Ibid.

11 www.knowyournitrites.com, accessed January 2019.

12 Ibid.

13 'Nitrite in cured meat products', on the CMC website, www.cmc-cvc.com, accessed October 2014.

14 Ibid.

15 Ibid.

16 www.knowyournitrites.com, accessed January 2019.

17 AMI/NAMI, 'Media myth crusher – meat curing and sodium nitrite', February 2016.

18 Ibid.

19 www.chriskresser.com, 'The nitrate and nitrite myth: another reason not to fear bacon', 16 October 2013.

20 Ibid.

21 Ibid.

22 Toni Tarver, 'Are nitrates and nitrites misunderstood?', art. cit.

23 Ibid.

24 Anthony Butler, 'Nitrites and nitrates in the human diet: carcinogens or beneficial hypotensive agents?', *Journal of Ethnopharmacology*, 167, June 2015 (our emphasis).

25 Vivienne Lo et al., 'Potent substances – An introduction', *Journal of Ethnopharmacology*, 167, June 2015.

26 *La France agricole*, 3478, 15 March 2013, p. 41.

27 Ibid.

28 'Nitrates. De nombreux bénéfices avérés pour la santé (colloque médical)' [Nitrates: several proven health benefits (medical conference)], www.lafranceagricole.fr, accessed October 2019.

29 AMI, 'AMI Fact sheet. Sodium nitrite, the facts', art. cit., and AMI Foundation, 'AMI Foundation Fact sheet. Sodium nitrite, the facts', 2013.

30 Nathan Bryan quoted by Trent Loos, 'Cured meats found to hold health benefits', www.examiner.com (accessed October 2014).

31 Ibid.

32 Hong Jiang, Yaoping Tang and Nathan Bryan, 'Dose and stage specific effects of nitrite on colon cancer', *Free Radical Biology and Medicine*, 49, December 2010, p. S56; see also 'Nitrite may inhibit early stage colon cancer cell progression, study finds', available at www.meatinstitute.org

33 Nathan Bryan, 'Dietary nitrite and nitrate: from menace to marvel', AMSA Reciprocal Meat Conference in Lubbock (Texas), June 2010.

34 Deepa Parthasarathy and Nathan Bryan, 'Sodium nitrite: the "cure" for nitric oxide insufficiency', *Meat Science*, 92, 2012 (our emphasis).

35 Ibid.

36 *Agrodistribution*, 16, May 2011, p. 6.

37 AMI Foundation, *2011 Year in review*, AMI, Washington, 2012, p. 4; Betsy Booren (AMI Foundation), 'Meat and poultry industry: the intersection between safety and health', presentation to the Canadian Meat Council, 30 May 2013, pp. 55–68.

38 'Nitrite expert refutes review linking processed meats to health issues', *AMI Foundation News*, 13, 2, April 2011, p. 3; Nathan Bryan et al.,

'Ingested nitrate and nitrite and stomach cancer risk: an updated review', *Food and Chemical Toxicology*, 50, 2012.

39 B. Booren, 'Meat and poultry industry: the intersection between safety and health', art. cit., pp. 63, 68.

40 'Myth: nitrite in cured meat is linked to diseases like cancer', available at www.meatmythcrushers.com, accessed October 2019.

41 J. Ville and William Mestrezat, 'Origine des nitrites contenus dans la salive; leur formation par réduction microbienne des nitrates éliminés par ce liquide' [Origin of nitrites contained in saliva; their formation by microbial reduction of the nitrates eliminated by this liquid], *Compte rendu Société de Biologie* [Society of Biology Report], 63, 1907; see Mathews, 'The pharmacology of nitrates and nitrites', art. cit., pp. 518–19.

42 'Myth: Nitrite in cured meat is linked to diseases like cancer', video cited.

43 Ibid.

44 Ibid.

45 Meat myth crushers, 'Nitrite sources & benefits', available at www.youtube.com/watch?v=nfdgJvKhJ4I, accessed October 2019.

46 'Nitrite-cured meats, are they safe? Ask the meat-science guy', by Randy Huffman, available at www.youtube.com/watch?v=lXKSaspItv8, accessed October 2019.

47 Robert Proctor, *Golden Holocaust: Origins of the Cigarette Catastrophe and the Case for Abolition*, University of California Press, Berkeley, 2011, pp. 168–9, 422.

48 Quoted in ibid., p. 450.

49 Proctor, *Golden Holocaust: Origins of the Cigarette Catastrophe and the Case for Abolition*, op. cit., p. 303.

50 Mukherjee, *The Emperor of All Maladies*, op. cit.

51 Proctor, *Golden Holocaust: Origins of the Cigarette Catastrophe and the Case for Abolition*, op. cit., pp. 167, 279.

52 Ibid., pp. 262–3.

53 Ibid., p. 294.

54 Ibid., pp. 340, 490.

55 See, for example, the press campaign on the fake 'Hazleton' study (documents lrnf0107, jlmf0116, jzyf0107, kzyf0107, hlmf0116, etc. on www.industrydocuments.library.ucsf.edu).

56 Elisa Ong and Stanton Glantz, 'Tobacco industry efforts subverting IARC's second-hand smoke study', *The Lancet*, 355, 2000; Thomas Zeltner et al., *Tobacco Company Strategy to Undermine Tobacco Control Activities at the World Health Organization*, WHO, Geneva, 2000, pp. 193–227.

57 Norbert Hirschhorn, 'Shameful science: four decades of the German tobacco industry's hidden research on smoking and health', *Tobacco Control*, 9, 2000.

58 Ibid.

59 Jim Tozzi is a former civil servant who became a consultant. He was a formidable tactician for the tobacco lobby and some of his 'anti-science' manoeuvres have become legendary. See, for example, on his

role in the 'Data Quality Act', David Michaels and Celeste Monforton, 'Manufacturing uncertainty: contested science and the protection of the public's health and environment', *American Journal of Public Health*, 95, S1, 2005, p. S44.

60 See Chris Mooney, 'Paralysis by analysis: Jim Tozzi's regulation to end all regulation', *Washington Monthly*, May 2004.

61 Letter to Philip Morris International, 'On our work on the nitrosamines issue' (document qnky0098 on www.industrydocuments.library.ucsf.edu).

62 See, for example, his account in 'Reports on recent ETS and IAQ developments' (document jmph0143 on www.industrydocuments.library.ucsf. edu).

63 See 'Nitrosamines strategy paper' (document tnky0098) and 'Proposal to deal with the forthcoming release of research on tobacco and nitrosamines from cured meats conducted by Dr Susan Preston-Martin' (document yzjp0217 on www.industrydocuments.library.ucsf.edu).

64 'Proposal to deal with the forthcoming release ...', document cited, p. 2.

65 Ibid., p. 2.

66 Ibid., p. 3.

67 See, for example, John Peters et al., 'Processed meats and risk of childhood leukemia', *Cancer Causes and Control*, 5, 1994; S. Preston-Martin et al., 'Maternal consumption of cured meats and vitamins in relation to pediatric brain tumors', *Cancer Epidemiology, Biomarkers & Prevention*, 5, 1996.

68 Booren, 'Meat and poultry industry: the intersection between safety and health', art. cit., p. 57.

69 'Nitrosamine talking points' (document mlfv0175 on www.industry documents.library.ucsf.edu), p. 2.

70 Ibid., p. 4.

71 'Proposal to deal with the forthcoming release ...', document cited, p. 1.

72 Ibid., pp. 4–5.

73 Ibid., p. 6.

74 Mathias Girel, preface to R. Proctor, *Golden Holocaust. La conspiration des industriels du tabac* [The Tobacco Industry Conspiracy], Équateurs, Paris, 2014, p. 1.

75 David Michaels interviewed by Francie Diep, 'Why scientists defend dangerous industries', *The Chronicle of Higher Education*, 24 November 2019. See also Marion Nestle, *Unsavoury Truth. How Food Companies Skew the Science of What We Eat*, Basic Books, New York, 2018.

76 David Michaels, 'Science for sale', *Boston Review*, 28 January 2020.

77 See David Michaels, *Doubt Is Their Product: How Industry's Assault on Science Threatens Your Health*, Oxford University Press, New York, 2008, p. 46.

78 See David Klurfeld, 'Nitrite and nitrate in cancer', in Nathan Bryan and Joseph Loscalzo (eds), *Nitrite and Nitrate in Human Health and Disease*, Humana Press, New York, 2011. Mr Klurfeld has been a consultant for

the AMI. (See 'Oscar Mayer internal memorandum', document gthj0223 available at www.industrydocuments.library.ucsf.edu)

79 David Klurfeld and Stephen Kritchevsky, *Evaluation of Hot Dog Consumption and Childhood Cancer*, prepared for the American Meat Institute (n.d.), available at www.industrydocuments.library.ucsf.edu (document ago77d00).

80 Ibid., p. 10. Even today, Klurfeld protects the interests of the nitro-meat lobby and holds a key role in the US Department of Agriculture. See his interview with Sandrine Rigaud in 'Industrie agroalimentaire: business contre santé' [Agri-food industry: business versus health], *Cash investigation*, France 2, September 2016.

81 See Stéphane Foucart, *La Fabrique du mensonge* [Manufacturing Lies], Denoël, Paris, 2013.

Chapter 18: Bluff, Subterfuge, Hogwash: The Art of Nitro-Poker

1 WCRF, *Food, Nutrition, Physical Activity and the Prevention of Cancer*, op. cit. (Recommendation no. 5: 'Limit intake of red meat and avoid processed meat', p. XIX).

2 Michael Pariza (University of Wisconsin-Madison), quoted in 'Study discounts cancer-diet link', *Washington Post*, 16 March 1981.

3 M. Pariza, 'Smoked meats are safe, task-force conclude', 8 January 1998, available at www.news.wisc.edu

4 David Klurfeld, 'Nitrite and nitrate in cancer', art. cit., p. 263.

5 AMI/NAMI quoted in 'Processed meats rank alongside smoking as cancer cause – WHO', art. cit.

6 B. Booren quoted in Smyth, 'Processed meats blamed for thousands of cancer deaths a year', art. cit.

7 Andrew Milkowski, Joseph Sebranek and Jeffrey Sindelar, 'No, bacon is really not killing us or "the misguided persecution of bacon"', unpublished letter to *The Guardian*, 1 March 2018.

8 Amali Alahakoon et al., 'Alternatives to nitrite in processed meat: up to date', *Trends in Food Science & Technology*, 1, 2015.

9 Ibid., p. 46.

10 Jens Adler-Nissen et al., *Practical Use of Nitrite and Basis for Dosage in the Manufacture of Meat Products*, National Food Institute, Søborg, 2014.

11 Ibid., p. 8 on the 'necessity' of nitrite in frankfurters (publications by Gerald Hustad et al.): in fact, it was the work of teams from Oscar Mayer and its strategic allies.

12 Ibid., p. 7 on the history of nitrite-cured meats (publication by Evan Binkerd and Olaf Kolari). In fact, Binkerd was vice president of the Armour group and Kolari was an executive at Armour and the former head of information at the AMI Foundation.

13 Ibid., p. 9.

14 Food Chain Evaluation Consortium, *Study on the monitoring of the implementation ...*, op. cit.

15 Honikel, 'The use and control of nitrate and nitrite for the processing of meat products', art. cit.

16 R. Pegg and F. Shahidi, *Nitrite Curing of Meat*, op. cit., p. 1 (our emphasis).

17 See, for example, 'Manitoba inspectors seize farm's award-winning meats', CBC News, 30 August 2013; 'Pilot mound prosciutto makers start over', available at www.manitobacooperator.ca

18 EFSA, 'Opinion of the Scientific Panel on Biological Hazards on the request from the Commission related to the effects of nitrites/nitrates on the microbiological safety of meat products', art. cit., p. 1.

19 Centre technique de la salaison, de la charcuterie et des conserves de viandes, *Code des usages en charcuterie et conserves de viandes*, op. cit., section V, p. 7.

20 A number of manufacturers of nitrited curing salt have had publishing arms. For example, in France, the manuals produced and published by the directors of La Bovida (a food additive distributor) have long been a reference source in the professional schools.

21 Pallu, *La Charcuterie en France. Tome III*, op. cit., p. 113.

22 'Ackerlei-Metzgerei für Spitzenqualität ihrer Wurstwaren prämiert' [Ackerlei butcher given award for top-quality sausage products], Ackerlei (Bioland), 2017.

23 *Bio-Press-Fachmagazin für Naturprodukte* [Professional journal for natural products], 1 May 2012.

24 Demeter, *Cahier des charges transformation* [Processing Specifications], 2017, p. 34.

25 'Bio-Wurst natürlich hergestellt' [Organic sausage produced naturally], *Bio-Press-Fachmagazin für Naturprodukte*, 24 December 2014.

26 Demeter, 'Demeter Fleisch – Lebensqualität für alle' [Demeter meat – Quality of life for all], 2016.

27 Nature et Progrès, *Cahier des charges transformation alimentaire* [Food Processing Specifications], 2005, pp. 19, 32, 33, 38.

28 See Paul Nuki, 'Big stores sell "organic" food with additives', *The Sunday Times*, 4 October 1998.

29 Nature et Progrès, *La viande bio* [Organic Meat], Nature & Progrès Belgique asbl (n.d.).

30 Commission Regulation (EC) No 780/2006 of 24 May 2006 'amending Annex VI to Council Regulation (EEC) No 2092/91 on organic production of agricultural products and indications referring thereto on agricultural products and foodstuffs', *Official Journal of the European Union*, L137/10, 25 May 2006.

31 Friedrich-Karl Lücke, 'Nitrit und die Haltbarkeit und Sicherheit erhitzter Fleischerzeugnisse', art. cit.

32 Alexander Beck et al., *Pökelstoffe in Öko-Fleischwaren* [Curing Salt in Organic Meat Products], Forschungsinstitut für biologischen Landbau (FiBL) [Research Institute for Organic Agriculture], Frankfurt am Main, 2006, p. 11; and Friedrich-Karl Lücke, 'Einsatz von Nitrit und Nitrat in der ökologischen Fleischverarbeitung: Vor- und Nachteile' [Use of nitrite

and nitrate in organic meat-processing: advantages and disadvantages], *Fleischwirtschaft* [Meat Business], 83, 11, 2003.

33 Directorate-General for Agriculture and Rural Development, 'Conclusions from Group of Independent Experts on Food Additives and Processing Aids Permitted in Processing of Organic Food of Plant and Animal Origin', 5 July 2007.

34 Ibid.

35 Commission Regulation (EC) No 123/2008 of 12 February 2008 'amending and correcting Annex VI to Council Regulation (EEC) No 2092/91 on organic production of agricultural products and indications referring thereto on agricultural products and foodstuffs', *Official Journal of the European Union*, L38/3, 13 February 2008.

36 Ben Bouckley, 'DEFRA to Brussels: Ditch potentially ruinous nitrate ban', foodnavigator.com, 8 November 2010.

37 Ben Bouckley, 'Relief for organic meat industry as EC calls off nitrate ban', foodnavigator.com, 28 January 2011.

38 Liz Mulvey et al., *Alternatives to Nitrates and Nitrites in Organic Meat Products*, Campden BRI, Chipping Campden, 2010. See, for example, pp. 15, 57, 40–50 on the use of sorbate.

39 See, for example, Alexander Beck et al., *Herstellung von Öko-Fleisch- und Öko-Wurstwaren ohne oder mit reduziertem Einsatz von Pökelstoffen* [Manufacture of organic meat and sausages without or with reduced use of curing agents], Forschungsinstitut für biologischen Landbau (FiBL), Frankfurt am Main, 2008; Friedrich-Karl Lücke, 'Manufacture of meat products without added nitrite or nitrate – quality and safety aspects', *Proceedings of the 3rd Baltic Conference on Food Science and Technology*, Jelgava, 2008.

40 Mulvey et al., *Alternatives to Nitrates and Nitrites in Organic Meat Products*, op. cit.

41 Bouckley, 'Relief for organic meat industry as EC calls off nitrate ban', art. cit.

42 Rick Pendrous, 'Nitrate ban "could kill" UK's organic bacon', *Food Manufacture*, 11 August 2008.

43 Chloe Ryan, 'Life preservers: Focus on preservatives', *Food Manufacture*, 1 November 2010.

44 Satoriz, 'Charcuterie Rostain – entretien [interview]: Lionel Rostain', Satoriz.fr, May 2003.

45 Product catalogue for Rostain factory, Gap (Hautes-Alpes), 2014, p. 12.

46 Carol Foreman quoted in Goody, 'USDA faces sodium nitrite suit', art. cit.

47 Rogert Angelloti quoted in Goody, 'USDA faces sodium nitrite suit', art. cit.

48 Kim Severson, 'For Natural Dogs, a Growing Appetite', *The New York Times*, 5 July 2006.

49 Stephen McDonell (Applegate Farms) quoted in ibid.

50 Udo Gehring and Helmut Pohnl, Patent US 6689403, 'Mixture for reddening meat products', assigned to Karl Muller GmbH & Co., February 2004.

51 Ibid.

52 Lauren Hartman, 'Manufacturers seeking natural ways to extend foods' shelf life', *Food Processing*, 12 September 2016.

53 Ann Husgen et al., Patent US 20080305213A, 'Method and composition for preparing cured meat product', assigned to Kerry Group Services International, Ltd., December 2008.

54 Jess Halliday, 'New culture to speed up curing meat naturally', foodnavigator.com, 15 May 2009.

55 Ibid.

56 'Samix Parisien NAT – Saumure sans nitrites ajoutés [Brine without added nitrites] – clean label', Jaeger/Solina, 2017.

57 Donna Berry, 'To cure or not to cure?', meatpoultry.com, 13 October 2014.

58 Ibid.

59 'All-natural vegetable based ingredients creating innovative clean-label solutions for food products', kainoscapital.com (accessed October 2019).

60 Tae-Kyung Kim et al., 'Effect of fermented spinach as sources of pre-converted nitrite on color development of cured pork loin', *Korean Journal for Food Science of Animal Resources*, 37, 1, January 2017.

61 Ibid.

62 Dong-Min Shin et al., 'Effect of swiss chard (*Beta vulgaris var. cicla*) as nitrite replacement on color stability and shelf-life of cooked pork patties during refrigerated storage', *Korean Journal for Food Science of Animal Resources*, 37, 3, June 2017.

63 Ibid.

64 Jeffrey Sindelar, 'What's the deal with nitrates and nitrites used in meat products?', 2012, University of Wisconsin, Madison.

65 Gerhard Feiner, 'Nitrite free: Where does the truth end?', William Reed Business Media, 27 July 2007.

66 Ibid.

67 Nathan Bryan quoted in 'Bryan: Q & A on nitrites', *Feedstuffs – The Weekly Newspaper for Agribusiness*, 81, 29, July 2009, p. 3.

68 See Geoffrey Ras et al., 'Utilisation d'un ferment producteur de monoxyde d'azote comme alternative aux sels nitrités?' [Use of a culture producing nitric oxide as an alternative to nitrite curing salts?], 16èmes Journées Sciences du Muscle et Technologies des Viandes [16th Muscle Sciences and Meat Technology Days], *Viandes & Produits Carnés* [Meats and Meat Products], November 2016. Other scientists are investigating applications of plasma technology (applying energy to a gas, thereby initiating a breakdown of water molecules) in order to generate nitrite and other reactive nitrogen species. See Elena Sofia Inguglia et al., 'Searching for a cure', *TResearch*, Spring 2020, 15, 1.

69 See Deena Shanker and Lydia Mulvany, 'Hormel lawsuit reveals what "natural" meat really means', Bloomberg News, 10 April 2019; other legal proceedings have been attempted in 2016: see 'Putative class action targets

Hormel's "100% Natural", "No Preservatives" meat', *Shook, Hardy & Bacon Food & Beverage Litigation Update*, no. 620, 21 October 2016.

70 'Einsatz von unbedenklichen NPS-Alternativen verboten' [Use of harmless alternatives to nitrite curing salt prohibited], *BioPress-Fachmagazin für Naturprodukte*, 10 December 2015.

71 Herta v. Fleury-Michon company, Tribunal de commerce de Nanterre, account filing injunction, No. 2014R00839, 29 July 2014.

72 'Nitrites: les additifs dans la charcuterie scrutés par la répression des fraudes' [Nitrites: processed meat additives examined by the Fraud Prevention Service], *La Dépêche*, 29 May 2019.

73 'Use and removal of nitrite in meat products', www.fsai.ie (accessed May 2019).

74 Ibid.

75 Directorate-General for Health and Consumers, 'Summary record of the Standing Committee on the Food Chain and Animal Health held in Brussels on 14 December 2006', DG SANCO – D1(06)D/413447.

76 Directorate-General for Health and Consumers, 'Summary record of the Standing Committee on the Food Chain and Animal Health held in Brussels on 19 May 2010', DG SANCO – D1(2011)D/293964.

77 Directorate-General for Health and Food Safety, 'Opinion of the standing committee on plants, animals, food and feed held in Brussels on 17 September 2018'.

78 B. Heller & Co., *The Healthfulness of Freezine*, op. cit., p. 2 (our emphasis).

79 Committee on Manufactures, *Adulteration of food products*, op. cit., pp. 112, 184.

80 'Upton Sinclair on packing house meat', *Albuquerque Evening Citizen*, 4 June 1906, p. 4.

Conclusion

1 Keraudren, 'De la nourriture des équipages et de l'amélioration des salaisons dans la marine française', art. cit., p. 366.

2 Naomi Oreskes and Erik Conway, *Merchants of Doubt: How a Handful of Scientists Obscured the Truth on Issues from Tobacco Smoke to Global Warming*, Bloomsbury, New York, 2010, p. 33.

3 Horace (*Odes*, Book III, 2) quoted by David Kessler, *A Question of Intent: A Great American Battle with a Deadly Industry*, Public Affairs, New York, 2001.

4 Article 1245 of the French civil code, 'Responsibility in cases of defective products'. In French law, the producer can be held responsible for the fault even if the product was manufactured according to existing regulations (article 1245-9).

5 Interview with Patrice Drillet, president of the cooperative Cooperl Arc-Atlantique, *Porc Magazine*, 496, March 2015, p. 6.

6 'Mondialisation: les charcuteries françaises face à l'ardente obligation d'exporter' [Globalization: French processed meats and the pressing

need to export], *RIA La Revue de l'Industrie Agroalimentaire* [Review of Agrofood Industry], 6 November 2015.

7   See, for example, charcuterie-de-france.jp, or faguozhurouzhipin.com

8   Melissa Center et al., 'International trends in colorectal cancer incidence rates', *Cancer Epidemiology Biomarkers & Prevention*, 18, 6, 2009; Donald Parkin, 'International variation', *Oncogene*, 23, 2004.

9   David Stuckler et al., 'Manufacturing epidemics: the role of global producers in increased consumption of unhealthy commodities including processed foods, alcohol, and tobacco', *PLoS Medicine*, 9, 6, 2012.

10   Christopher Wild, 'IARC's global cancer research agenda to inform cancer policies', IARC, Lyon, May 2015.

11   Melina Arnold et al., 'Global patterns and trends in colorectal cancer incidence and mortality', *Gut*, 66, 4, 2016.

12   World Cancer Research Fund/American Institute for Cancer Research, *Continuous Update Project Expert Report 2018. Diet, Nutrition, Physical Activity and Colorectal Cancer*, WCRF/AICR, 2018.

INDEX